On Behalf of God

—— *A Christian Ethic for Biology* ——

Bruce R. Reichenbach and V. Elving Anderson

WILLIAM B. EERDMANS PUBLISHING COMPANY
GRAND RAPIDS, MICHIGAN
WITH
THE INSTITUTE FOR ADVANCED CHRISTIAN STUDIES

© 1995 Wm. B. Eerdmans Publishing Co.
255 Jefferson Ave. S.E., Grand Rapids, Michigan 49503
All rights reserved
Published in cooperation with
The Institute for Advanced Christian Studies

Printed in the United States of America

00 99 98 97 96 95   7 6 5 4 3 2 1

ISBN 0-8028-0727-5

Unless otherwise noted, Scripture quotations are from HOLY BIBLE: NEW INTERNATIONAL VERSION. Copyright © 1973, 1978, 1984 by the International Bible Society. Used by permission of Zondervan Bible Publishers.

What people do about their ecology depends on what they think about themselves in relation to things around them. Human ecology is deeply conditioned by beliefs about our nature and destiny — that is, by religion.

Lynn White, Jr.

Contents

Preface

THIS BOOK is part of a series — Studies in a Christian World View — that focuses on the relation of Christianity to the various academic disciplines — philosophy, psychology, economics, the arts, history, literature. The series addresses difficult questions that arise when we bring considerations derived from a specific religious worldview to bear on matters common to both Christians and non-Christians. Its intent is to assist readers in thinking through the ways in which a Christian might address important intellectual and practical issues of our day.

Our original assignment was to relate Christianity to science. Though this is a feasible task, it soon became obvious that the breadth of coverage required by this project would leave a host of specific problems that affect many lives untouched. Instead, we opted to narrow our focus to Christian ethics as it speaks to the science of biology.

Even this narrowing left our topic far too broad, for merely to survey the plethora of Christian ethical perspectives would create an enormous volume. Hence, we have chosen to adopt and develop a particular Christian ethic that we believe comports well with the biblical accounts and motifs. We then use this ethic to guide our consideration of a wide range of biological issues.

In Chapter One we sketch a science fiction story that incorporates many of the questions and problems we will address in the book. In Chapter Two we attempt to dispel the aura of superiority that frequently characterizes science as viewed by the layperson (and sometimes the scientist). We contrast the empiricist tradition of science that began with the seventeenth-century philosopher Francis Bacon, who stressed the objectivity and rationality of the scientific method, with the contemporary view of science developed by Thomas Kuhn, who believes that subjectivity and imaginative creativity play significant roles in science. Kuhn holds that, since science is not a value-free endeavor, values play an important role in the development, adoption, and implementation of specific scientific theories. This perspective legitimizes ethical reflection on science. We also borrow from Kuhn the notion of a paradigm that, when set in the context of a larger worldview, guides both scientific and ethical reflection.

In Chapter Three we develop the ethical paradigm that guides our subsequent discussion. We propose an ethic of stewardship built upon three commands found in the creation narratives of Genesis 1 and 2. First, there is the obligation to *fill* the land. Though initially this command primarily focused on the quantitative aspect of filling, we approach it largely from a qualitative perspective, arguing that it obligates us in our science to improve where possible both the environment and the human condition. The second obligation — to *rule over* the land as God's representative — involves gaining both knowledge and control over the earth and its inhabitants. According to the third obligation, the steward is to *care for* what is ruled over on behalf of the Lord. The last part of the chapter explores the tension between conservation and change and raises questions concerning the purpose of changing creation and whether there are limits to such change.

In Chapter Four we apply our ethical model to environmental concerns. We begin with the case study of the Brazilian rain forest, describing the current concern with extensive deforestation and its effects on the environment. We argue that the command to rule over or subdue the earth brings legitimacy to using the earth to benefit human beings. However, this subduing must extend beyond utilitarian-type considerations that treat the environment merely as an instrument for realizing human happiness and well-being. Stewards are to benefit not only the Lord but also the environment over which they are stewards, for the environment has a worth derived from God. This means that we ought to care for or tend the earth. We contrast a conservation with a preservation approach to ecology and argue that treating the environment as having derived worth implies an ethic of conservation. Finally, consideration of the command to fill introduces discussion of the importance of and justification for environmental diversity.

In Chapters Five and Six we apply our stewardship model of ethics to assisted reproduction. After exploring possible motives and methods for assisting reproduction, we consider what obligations the command to fill might create. In particular, we query whether persons have a right to have a child, what kind of right this might be, who would possess it, and who might be obligated by such a right. We then address the obligation to care for the childless. We respond to those who argue that it is immoral to share genetic materials because this genetic exclusivity is necessary to preserve the marital covenant. Donation of genetic material to others who are incapable of reproducing might further the realization that we are not completely self-sufficient, that there are ways in which members of a community can contribute to each other's quality and quantity of life. There are diverse ways to assist reproduction; we specifically address the morality of surrogacy, responding to the objections that it degrades the woman who bears the child by treating her as a means to some end and that it separates the decision to create children from the decision to raise them. Invoking the command to rule, we address questions arising from in vitro fertilization, including what to do with fertilized but unused blastocysts.

Chapter Seven begins with a discussion of the Human Genome Project and the possible uses of the information derived from it. We then turn to the question of how the project ought to be done. The command to fill, considered qualitatively, legitimizes scientific research that can lead to beneficial changes. Though this is uncontroversial with respect to somatic gene therapy, we argue that it can apply to germ line therapy as well.

The garden we are commanded to care for has both delights and snakes. One snake is the possible misuse of genetic information. In this regard we discuss who should have access to genetic information. Another snake concerns neglecting the human costs of genetic engineering. Here we consider the possibility that genetic testing might be used to discriminate against individuals with different genetic structures and the moral implications arising therefrom. Finally, the command to fill raises questions concerning scarce resources and the identification of those to whom we have obligations to improve. We inquire whether there are obligations to future generations and what they might be.

In Chapter Eight we turn from issues arising from diverse applications of science to questions concerning scientific knowledge. We note that both the Greek myth of Prometheus and the Genesis account of Adam and Eve express uneasiness with knowledge. This forces us to deal with the claim that knowledge is dangerous. The command to rule has implications for controlling the abuses of knowledge. However, since knowledge is personal, we reject censorship in favor of establishing controls to regulate the ability of persons to use this knowledge to carry out wrong beliefs and desires. We contend that Christianity has something to say about why we abuse knowledge and how that abuse might be ended, and that it has a prophetic message regarding the proper use and acquisition of scientific knowledge. As an application of this, we briefly consider the treatment of animals used in experimentation.

We proceed to consider possible justifications for conducting research. The command to fill can provide a Christian justification for scientific knowing, one that combines pragmatic concerns about the potential uses of this knowledge with the view that knowledge can satisfy the distinctively human desire to know.

The command to care for the earth and its inhabitants has implications for determining what obligations scientific researchers have to society in general and to particular sufferers. To what extent does the researcher have obligations to persons who are not part of the research project but who might be affected by it? Can the ill demand a cure, as some in the recent debate on AIDS advocate? We conclude with a reflection on the obligations that scientists have to their colleagues.

In Chapter Nine we go behind our ethical perspective to explore the very possibility of being moral. Research directed toward isolating the genetic factors that, through their influence on the brain, predispose persons to various behaviors indicates that not only physical but behavioral disorders can be traced, in some measure, to physiological deficiencies caused by genetic predispositions. This in turn raises the question whether and to what degree we can hold persons responsible for their behavior.

We address this concern by looking at the relations that hold between genes, brain, and behavior. Behavioral genetics, neuroscience, and sociobiology all move in the direction of reductionism, seeking to explain our behavior by biological factors mediated through the environment. This implies a determinism that tends to diminish belief in human responsibility. Though we do not offer a solution to the determinism-freedom and brain-mind dilemmas, we further the discussion by exploring various ways in which people have attempted to bridge biological and humanistic concerns, noting both their strengths and their weaknesses.

In Chapter Ten we turn our attention to human sexuality. First, we explore the biology of sexual determination and differentiation, concluding that sexuality is more than biological. We then use our stewardship paradigm to consider questions having to do with the proper understanding of human sexuality. The command to fill has both the obvious quantitative dimension and a less obvious qualitative aspect. This helps us not only to see sexuality as oriented toward procreation, but to make room for its bonding and pleasure functions as well. We apply the command to rule to the matter of sexual orientation; here we consider the roles complementarity, exclusivity, commitment, and love play in sexuality. In particular, we emphasize that sexuality must embody the rich dimensions of love that Christians find in distinguishing among *eros, philia,* and *agape* love. We argue that ruling over applies to ourselves as well as to others, that Christian morality centrally invokes the necessity of controlling and channeling our natural desires. Finally, from the perspective of caring for others, we explore matters having to do with attitudes toward those suffering from sexually transmitted diseases and with associated problems such as the testing for HIV antibodies and the allocation of scarce resources for AIDS research and education.

We have written the book both for ourselves and for others. Our goal is to help all of us carefully think through some of the important ethical issues that biology presents. Neither Christians nor non-Christians can hide from the moral issues; science and technology are changing so rapidly that to bury our minds in the sand is not only to make us into intellectual dinosaurs but also to remove us from speaking relevantly to issues that many soon will face personally.

Though the book consciously adopts a Christian ethical model, we invite others as well to join the dialogue, for much of what we say can be affirmed by both Christians and non-Christians. We make no pretense that this is the sole model of either ethics or Christian ethics, nor generally do we attempt to critique competing models. Rather, we assume that our model is consistent with Scripture, is reasonable and defensible within a pluralistic society, and can be used to provide significant insight into ethical responses to the problems posed. We want not only to pose the problems but also to encourage our readers to think with us about possible, reasoned responses. Our solutions are not meant to be dogmatic; rather, they are intended to stimulate further discussion. Carefully thought-through disagreement is welcome; debate is the beginning of personal growth and insight.

Thus we welcome you to join our dialogue — to understand and appreciate our stewardship model of Christian ethics, to investigate the issues with which biology confronts all of us, and to explore how we might conscientiously, creatively, and reflectively address the questions being raised. We challenge each to continue the discussion, not only about the topics considered below, but also about issues in biology extended beyond those we raise.

Finally, we wish to express our gratitude to a number of people who contributed to this book in a variety of ways. We owe a note of thanks to our wives, Sharon and Carol, who both encouraged us and endured our frequent departures to the computers in our basement offices. Sharon read and critiqued the entire manuscript, forcing clarification at many points. Robert and Rachel sacrificed uncounted hours with Dad so he could write, though their interruptions provided welcome breaks. We are grateful to Karen Mateer of the Augsburg College Library, who graciously helped locate and retrieve research materials. We also wish to thank IFACS for its sponsorship of and involvement in the project. Dr. Hessel Bouma's detailed critique of the manuscript for IFACS significantly helped to improve the final version. We also appreciate the work of our project editor, Jennifer Hoffman, who clarified our text in numerous places. Finally, we are grateful to the students in Bruce's Philosophy 410 course, Philosophy of Biology, who read the manuscript and submitted searching questions and comments that required revising the text.

CHAPTER ONE

Introduction

MILLIONS OF PEOPLE around the world sit with eyes glued to their wall-sized, stereo televisions. On their screens, neatly divided into four quadrants, 3-D pictures flash from locations in Russia, Japan, United Europe, and the United States. The televised scenes are practically identical: brightly colored, aerodynamic space planes sit poised to taxi down concrete runways, like gigantic birds waiting for the leader's signal to migrate. In two hours they will take off with a deafening roar, as each has done many times before in their frequent visits to the orbiting Earth Unity space station. Launched at fifteen-minute intervals, all four will rendezvous at Earth Unity, where their passengers and cargo will transfer to four interplanetary probe vehicles for the five-month journey to Mars.[1]

In the final hours before launching, the on-scene reporters scramble to find bits of factual information, little-known details, personal notes, and background stories to sustain the interest of the viewers, who otherwise would soon tire of the unchanging background of immobile vehicles and turn to the twenty-four-hour sports or soap-opera stations. The Russian reporter recounts that, though this was originally planned to be the first wave of a large-scale program to colonize Mars, today's launch probably will be the only attempt for several decades. Fifteen years of unrelenting

---

1. Conversations with Mark Engebretson, a colleague who has taught a simulation on the creation of a space station, James Burke's film series *After the Global Warming* (New York: Ambrose Video Publishing, Inc., n.d.), and a laboratory exercise in John W. Christianson, *Global Science: Energy, Resources and Environment* (Laboratory Manual, 3rd edition [Dubuque, IA: Kendal/Hunt Pub. Co., 1991], 1-3) provoked the scenario sketched in this chapter.

droughts and extensive ocean flooding of the coastal cities and agricultural deltas have forced drastic reductions in the international planetary colonization program. Governments have had to shift political and fiscal priorities as the global temperatures have increased to a point where the average temperature is now 5°C warmer than it was at the end of the twentieth century and warmer than it has been in the last forty million years.

Switching to visual flashbacks from news footage of the last fifty years, the reporter intones the history of the shift in governmental priorities. In the first forty years of the twenty-first century, the world governments could not agree on how to curtail emissions of carbon dioxide. Developed countries vacillated on implementing strict emission control standards as their economies flourished or stagnated. During prosperous times, they bowed to pressure from environmentalists to legislate higher standards that they then, by executive decree, relaxed during the periodic recessions. Thus, though now generally stabilized, carbon dioxide (CO_2) and nitrous oxide (NO) emissions from their industrial sectors doubled from what they were seventy-five years earlier. Emissions from transportation fared no better. Because the expense and relatively short range of the newly developed electric cars discouraged buyers, gasoline- and diesel-powered motor vehicles continued to spew large amounts of pollutants into the atmosphere.

In the developing world the rate of emissions soared dramatically. These countries offered to delay certain environmentally sensitive projects in exchange for the developed countries significantly reducing their own CO_2 and NO emissions, but the developed countries would not risk an economic recession. The developing countries requested a debt exchange, whereby the developed world would invest in the developing world's infrastructure and environment in exchange for the latter's preserving their own forests, endangered species, and ecology. But the offer generated little interest, attracting only a few northern European countries and foreign businesses such as pharmaceutical companies that could profit directly from projects that minimally disrupted the forests. Unable to obtain significant reduction of their burdensome foreign debt or investment in domestic clean technology, the developing countries launched programs to build and modernize their industrial capacities. Given the dated technology they could afford, pollution naturally accompanied the economic expansion.

To finance both their debt repayment and their development, countries with substantial forests, such as Indonesia, Cambodia, Zaire, Brazil,

Ecuador, Colombia, and Panama, harvested them for export to developed countries. Exponential population growth also increased the pressure on sensitive ecological regions, from the scrub grassland shared with animals in East Africa to the rain forests in Asia and South America. Wild animal populations declined from encroachment and poaching, bringing an end to much of the tourist industry. Timbered areas were cleared and opened to the increasing numbers of peasants who needed land for subsistence cultivation and to international agribusinesses for the development of large, export-oriented, monoculture plantations and ranches.

The television viewers watch as the camera pans to charts showing how the amount of carbon dioxide mushroomed from three hundred fifty parts per million (ppm) in 1990 to over six hundred ppm in eighty-five years. The charts record similar increases for other gases. Methane, the result in part of bacteriological action in rice paddies, doubled to roughly 3.5 ppm, as many developing countries struggled to maintain their green revolution by maximizing their paddy rice production. At the same time the burning of fossil fuels by factories, power plants, and vehicles increased the amount of nitrous oxide by 50 percent. This significantly contributed to the acidification and death of numerous lakes and rivers and destroyed large tracts of northern hemisphere forests.

The loss of the forests, the decline of ocean plankton, and the slowing of the ocean's circulation patterns severely limited the Earth's ability to absorb the increasing amounts of carbon dioxide and other emitted greenhouse gases. This led to the rise in average temperature as trapped CO_2 prevented the Earth's energy from escaping out into space. The effects were dramatic. What had been the main agricultural regions in the United States, Canada, Argentina, southern Africa, and western and eastern Europe gradually received significantly less rainfall each year as the temperature rose and wind and rain patterns shifted.

At first scientists found it difficult to determine whether the changes were long term or merely local or temporary variations. But by the fourth decade of the twenty-first century the trend became painfully obvious. The American plains, South African highlands, and eastern European steppe became desertified, subject to frequent grass fires and wind erosion. Irrigation helped for a decade or two, but the sources of water in the mountains dried up, the groundwater gradually was depleted, and salinization of the soil increased. For a while in the United States there was talk of floating water down from Alaska in huge nylon receptacles; a business firm actually made a few successful runs. But not only did this prove economically unfeasible, it at best temporarily satisfied the human water demands of the cities along the coast; it could not meet agriculture's insatiable thirst in the western United States. As parts of the northern latitudes received less rainfall, Alaska curbed its water export to the south.

In Africa the Sahara rapidly spread south, destroying ancient agricultural lands in a belt from western Africa to southern Sudan and Ethiopia. Famine and starvation stalked areas with some of the world's highest population growth. At the turn of the twenty-first century the sub-Saharan population had increased forty times faster than food production.[2] This trend spread well down into central Africa. Around the world, marginal farmers could not get the water, fertilizer, and herbicides necessary to survive the changes in the weather patterns. Increasingly they abandoned their unproductive farms and headed into the cities or tried to find more marginal land in wetter areas. Migrants to both encountered human hardship, the first from urban overcrowding and unemployment, the second from the devastating hurricanes and cyclones to which the wetter areas were prone.

The decline of farm production in the American and Canadian plains, South America, and central and eastern Europe brought grave food shortages. Many countries no longer could afford to purchase on the commodities market what little grain was available at four times the previous price, while government surplus to be shared with famine-stricken nations was scarce. This meant that world economies had to be redirected to cultivating areas which had not been previously farmed but which now had favorable climates and to creating a transportation network to get the limited food supplies to where they were needed. Constructing new transportation infrastructures absorbed immense amounts of capital.

The increase in CO_2 also caused ocean levels to rise. Warmer air over the poles, perhaps as much as 15°C warmer than before, melted the glaciers and sea ice. Viewers watch old news film showing how the seas flooded into coastal areas, covered the freshwater intakes for the major urban centers, seeped into and salinized underground aquifers used for drinking water, and drowned the remaining fertile river deltas. Major cities hastily constructed dikes to prevent ocean flooding, but smaller cities, unable to afford the expensive dike systems, began to look more like Venice than like their old selves. As a result, governments rapidly redirected money

2. Cheryl Simon Silver, *One Earth, One Future* (Washington, DC: National Academy Press, 1990), p. 78.

4

that had been targeted for space exploration to save the most highly populated areas and to resettle refugees from the land lost to sea rise and ocean erosion. Some countries suffered worse than others; the Maldives, for example, completely disappeared except as shoals on *National Geographic* maps of the Indian ocean.[3]

Following several lengthy commercials encouraging viewers to purchase the latest in fashion and style, the Japanese broadcaster introduces a segment on the moral dilemmas this shift in agriculture and population centers created. The debate on what to do with the tropical rain forests gradually subsided, not because consensus had been reached, but because precious little remained to be conserved. Each of the sides accused the other of irresponsible behavior: leaders of the developed countries blamed the developing countries for reckless treatment of the environment; leaders of the developing countries retorted that the failure of the developed countries to share their technology and forgive the enormous debts and exorbitant interest that consumed the developing world's economies had left them no other choice.

The reporter notes that all now agree that it would have been better had the developed countries not cut their forests. But disagreement rages on whether the harvesting could have been avoided, given population and economic pressures. Regardless, the debate has now shifted to questions of ecology in the northern parts of the northern hemisphere. Should the former tundra be turned into farmland? What ecological changes would such a shift produce? Should the resources of Siberia and Antarctica, two of the major untapped regions, be exploited, now that ice no longer covers parts of Antarctica year-round and large segments of Siberia are free from permafrost? Would this create serious and irreversible environmental damage? Hovering over all of these considerations are questions concerning how to feed the world population, whether governments should use food as an incentive to curb population growth, and whether individuals possess a right to reproduce.

With forty-five minutes remaining until launch time, the United Europe reporter focuses on the future of the colony on Mars. She is particularly interested in the ways in which humans will reproduce on this planet and what characteristics the colonists will genetically engineer in their descendants and transported species. One space capsule contains three thousand frozen animal embryos especially bred for durability, ability

3. Silver, p. 93.

to cope with gravity one-eighth that on Earth, cold temperatures, and low liquid requirements. It also holds one thousand human embryos carefully selected and tested for genes responsible for positive physical, social, and behavioral dispositions and free from various genetic diseases. Since the capsules could not carry mature animals to Mars and the astronauts themselves harbor only a narrow range of genes, these embryos will provide a bank for genetic diversity on the planet.

Eventually the colonists will implant some of the human embryos in the female astronauts, who will act as their surrogate mothers. However, since bearing children would make women less efficient in building and operating the space colony and could endanger their health, should complications develop, the space capsule also contains an Embryo Development Unit (EDU), designed to nurture both animal and human fetuses from embryo to birth. Several European universities had scheduled prelaunch tests of the EDU on humans, but massive protests prevented any human experimentation. When previously tested on mammals, including primates, the EDU produced severe deformations. This brought enraged protests from members of the animal rights community, who had successfully lobbied governments for years to curtail animal experimentation as being cruel to sentient beings. For them, further trials of the EDU only signaled a resurgence of animal abuse. Consistent with their past campaigns, they again took up the cause in mass demonstrations against the EDU. Later, secret government testing led to refinement of the EDU, resulting in a 75 percent success rate of normal animal births.

Once the colony is established and scientists can determine what traits are desirable, the geneticists will attempt to alter the embryos' genes to make living on Mars easier. They intend to breed animals to develop thermal properties in the skin to provide a greater tolerance to cold and to recycle toxic wastes. Even more important will be the colonists' attempts to breed resistance to solar ultraviolet and particle radiation, since Mars lacks a magnetic field and has only a thin atmosphere to protect it. The geneticists propose gene modification to enable more rapid repair of damage done to the organism by radiation, to retard or prevent progressive loss of red blood cells and bone mass, to adapt to cardiovascular changes brought on by the change in gravitational force, and to adjust to altered fluid needs and their resulting effects on electrolyte balance. As the colony develops and discovers its unique needs, the colonists will experiment with other changes, such as modifying genes to produce greater intelligence and sociability, important characteristics given the close quarters required by the size of the initial space colony housed under geodesic domes.

The American reporter concludes the pre-launch discussion with a piece on the question of who is being sent on the mission and what the members of the group will take with them. Since the group will be isolated for years, emphasis was placed not only on each member's specific skills but also on the members' mutual compatibility, the ratio of sexes, and the types of persons who would be needed to start a human colony on Mars. Debate raged over each of these items. The mission coordinators deemed age diversity particularly important, since the transition from the old to the new generation had to be phased to allow the development of new leadership. The mature, older generation would transmit the memories and wisdom needed to enable the colony to survive as the bearer of Earth's civilization. The coordinators also hotly argued over whether there should be strict sexual balance, finally agreeing on equality to placate political pressure groups. They found the issue of racial diversity more difficult to settle, since not all races or cultures could be represented. They finally decided that, to avoid any international political incidents, some participants would be selected at random from pools of qualified persons.

The mission coordinators also had difficulty selecting the abilities the astronauts were to have. For example, would only one colonist be a physician, or would all be trained in medicine? Since only twenty-four people were going on the mission, the colonists needed both multiplicity and redundancy of skills. What would be the balance between general skills and specialized training? The coordinators decided to provide long periods of training in diverse fields to each of the participants and their alternates. The more serious question was whether the mission should take along only scientifically trained personnel. Should ethicists or clergy accompany the voyage? Would people trained in the humanities or the social sciences be eligible? People gave varying answers depending upon how they related science and values and whether they accepted the social sciences as truly sciences.

The reporters' background reports engage the television audience until the launch. The digital countdown clock shows T minus two minutes, and with the commercials out of the way, viewers can sit back and comfortably enjoy the historic launch of the first — and perhaps last — Mars colonization project.

※　　　※　　　※

By now perhaps you are wondering when the television journalists will begin to address the outstanding moral issues that this description of the past and projection into the future raises. Surely they should say something about the morality of our treatment of the environment. Is not continued dumping of CO_2 into the atmosphere contrary to the moral law? Is it not immoral to destroy virgin forests and to eliminate species? Do we not have obligations to preserve ecological systems? Given what we have done to the forests, should we now attack the tundra for our profit? Should the colonization of Mars even be attempted, considering how we have plundered and destroyed Earth?

They also should say something about the methods of assisted reproduction envisioned for Mars. Should embryos be taken along on the voyage and then submitted to the EDU? Should embryos, whether human or animal, be subject to experimentation in the first place? Should we put human blastocysts in such danger? Is the envisioned use of women as surrogates immorally treating them merely as means? Will the conditions under which children are generated on Mars separate procreation from parenting and in doing so violate certain moral obligations we have to children?

Should the reporters not address the moral dimensions of the radical genetic engineering program the colonists have in view? Do we have the right to engineer humans and animals? Do we have moral permission to alter species, including humans, as we want to fit the new environment? Are there moral limits on what the colonists should do?

Merely because we have the technology to colonize Mars, should we? What are the dangers that the possession and use of the necessary knowledge pose? Have we not stepped far beyond our bounds when we leave Earth? Is this not also an example of the human hubris that ultimately seeks to play God? Should we have taken knowledge and technology this far?

Perhaps we expect too much in asking the journalists to consider these delicate moral questions. Journalists, like their audience, have difficulty coming to grips with complex and sensitive moral issues. Yes, it is easy to give one's opinion when called upon. But to locate that opinion within a coherent moral system, to show how it relates consistently to other moral judgments, and to be aware of the many implications of that moral judgment — these things are exceedingly more difficult.

To reflect on these and other issues from a coherent ethical perspective is the point of this book. In the chapters that follow we will address first the question of the relation of ethics and morality to science and develop a particular ethic, taking our cue from the Judeo-Christian Scriptures. Following that, we will apply this ethic to the kinds of moral questions that our futuristic scenario raises — to questions about humans and the environment, to what is morally permissible in reproductive technology, to whether we should genetically alter human beings, to human sexuality, and to questions about the dangers of knowing. Our concern throughout will be to advance our understanding of and appreciation for what a Christian stewardship ethic has to say to thoughtful persons entering the twenty-first century.

CHAPTER TWO

Science as a Human Endeavor

S CIENCE BEARS a special aura in our culture. Some of us can recall the government-sponsored initiatives in the late 1950s. The year 1958 was declared the International Geophysical Year. In the same year the U.S. Congress moved to improve science education by passing the National Defense Education Act, authorizing $280 million for science laboratories and textbooks. The showpiece, carried over into the next decade, was the "New Frontier" penetration into space.[1] The achievements of science and technology have been truly dazzling. The space effort generated notable advances in electronics, new materials and metals, astronomy, and cosmology. Physics probed to new depths the atom shattered by the cyclotron, while biology achieved an understanding of living organisms and biomedicine developed technologies and medicines that were before only dreamed of.

At the same time, however, these astounding successes served to bolster a long-held, popular conception of science that often continues to be reinforced, especially by the media: science confirms . . . ; doctors recommend . . . ; recent experiments show . . . ; for the first time we now can. . . . Three important features form part of this popular picture. First, to be scientific is to be objective, free from presuppositions and subjective biases. The results of science can be presented in a logical way that enables rational beings to understand clearly the claims being made and to evaluate their

---

1. Following the Soviet Union's startlingly successful launch of Yuri Gagarin into space, President John F. Kennedy, in a 1961 speech before a joint session of Congress, committed the country to regain world leadership in science and technology by "achieving the goal before this decade is out of landing a man on the moon and returning him safely to earth."

truth without prejudice. Second, to be scientific is to have empirical grounding or evidence for what one says. Scientists talk about matters of either actual or possible experience, not matters extending beyond the observable, natural phenomena that anyone can in principle experience. Through carefully controlled experiments scientists can verify or falsify the claims being made. Ultimately, all competing claims can be resolved by appeal to experience. Third, to be scientific is to be rational. Matters of science are not subject to diverse opinion to be debated endlessly. They are in principle capable of being resolved by human resources to mutual satisfaction, should the proper methods be applied. Scientific matters can be established to be or not be so. In short, science, in contrast to other disciplines, provides an ideal way of knowing and dealing with the world. Its objectivity is unchallenged, its methods clear, its results ascertainable, testable, and provable. It is free from extraneous theories, philosophies, or metaphysical and religious claims. And above all, it is value free; it describes how things are, apart from the idealizations of how things ought to be. Science is the ideal, proper authority.

The Empiricist Tradition

Behind this popular view of science stands a powerful tradition, going back at least as far as Francis Bacon at the turn of the seventeenth century. Though not himself a scientist, he called himself the "trumpeter of his time," reflecting in his writings the program for the beginning of post-Aristotelian, empirical science. For Bacon, the function of science ultimately was to control nature for human benefit. Under the prior guidance of the Aristotelian search for essences, science had not yet begun to realize its potential. Bacon sought to reform and direct it to obtain practical results that would benefit human life. For him no part of nature was off limits to scientists; all were to be studied and (where possible) controlled.

Bacon believed that science is a form of power. But to exercise power over nature we must understand nature and its causes or principles. "Human knowledge and human power meet in one; for where the cause is not known the effect cannot be produced. Nature to be commanded must be obeyed."[2] Bacon suggests that there are, in effect, two parts to

---

2. Francis Bacon, *The Novum Organum,* in *The New Organon and Related Writings* (New York: The Library of Liberal Arts, 1960), I.iii.

the scientific endeavor. One is the control we will have over nature once science advances far enough. Science gives us power to "command nature in action,"[3] to alter it for human betterment. Bacon envisions God granting us the right to exercise this power. Even the possible misuse of this power, he contends, should not be used as grounds to detract from the divine obligation both to know the principles of and to rule over nature.[4]

But to advance we first must understand nature; we must comprehend what is to be obeyed — the causes of particular phenomena and the laws that govern nature. This means that we must not only have a grasp of the data; we also must discover nature's principles, beginning with the more specific and moving progressively and systematically to the more general. Bacon is confident that we can do this because, since nature is directly available to us, we can properly employ our senses to gather objective data about nature. For science to begin we only need to look, for we are in immediate contact with the facts of nature. Bacon claims that he himself is "dwelling purely and constantly among the facts of nature, withdraw[ing] my intellect from them no further than may suffice to let images and rays of natural objects meet in a point."[5]

Bacon thus becomes one of the first to reflect upon science as invoking both knowing and doing. He sees the primary function of science as *doing*, as wielding power over nature; but at the same time he realizes that doing presupposes detailed knowledge about what one is to change. Thus science begins with *knowing*.[6]

Bacon's empirical, scientific method is not one of mere enumeration of the sensory facts. Facts, to become science, must have organization; they must lead to the principles, causes, or axioms that govern nature. To ascertain the principles that lie behind the facts we need induction. After sensing their world and recording what they experience, scientists are able to discern the relevant intermediate principles or axioms and from these arrive at the most general axioms governing the universe. This will be an orderly process. For Bacon the direction is clear: we move from perceiving particular things to understanding their causes and principles. Bacon

---

3. Francis Bacon, *The Great Instauration*, in *The New Organon and Related Writings*, p. 19.

4. Bacon, *The Novum Organum*, I.cxxix. We will further explore this motif in Chapter Three.

5. Bacon, *The Great Instauration*, pp. 13-14.

6. We will return to this theme of relating knowing and doing in Chapter Eight.

claims that "by this means I have established forever a true and lawful marriage between the empirical and the rational faculty."[7] Science effectively uses reason; mere perceptions of particulars will not suffice.

The problem the empiricist faces is that the senses are subject to error. How then can the collection of empirical data constitute a reliable starting point? Though Bacon agrees that sensory data can be misleading, he is supremely confident in this starting point, for the very senses that deceive supply the means to correct their own defects. What is perceived through the senses and mediated through the axioms or principles that the mind derives from that experience can be tested in experiments. Experiments, for Bacon, can *settle* the matter.[8] Thus Bacon holds that objectivity is possible. Uninterpreted data about particulars constitute its objective starting point. Induction provides access to the principles and axioms of nature, while careful experimentation gives a foolproof method to avoid error.

Unfortunately, we often fail to attain true, informative, objective outcomes. We are misled, he suggests, by various idols. For example, *idols of the tribe* make us prone to bring into our investigations what we want to find, whether from our expectations or from our own desires and wants. We adopt these idols when we take sensory experience as providing knowledge without conducting the appropriate experiments. *Idols of the cave* include the viewpoints espoused by persons in the past whom we respect or the prejudices that each of us might bring to the subject matter from our upbringing or training. Our education instills in us certain perspectives, models, or points of view that color what we perceive. The scientist must destroy these idols of preconception. *Idols of the marketplace* are the errors that result from language. Language can mislead us by importing meanings and significances that are not there. Finally, *idols of the theater* are the wrongheaded theories and systems of the past that prevent us from seeing things as they truly are. All of these detract from gaining an objective, true account of nature. But Bacon's optimistic message is that we can begin anew; we can recognize and escape from these idols into an objective, truly inductive science. From the particular, uninterpreted data we can infer axioms or principles, and from these we can establish critical experiments that will test the axioms by verifying or falsifying them.

Bacon thus believed that science gets us to nature itself in an *objective*

7. Bacon, *The Great Instauration*, p. 14.
8. Bacon, *The Novum Organum*, I.xcix.

manner. Through the application of *reason* we can ascertain the principles or axioms by which nature *itself* operates, and through experiments we can *determine* the truth about nature, without interference from the subjective experience of the knower.[9] "God forbid that we should give out a dream from our own imagination for a pattern of the world; rather may He graciously grant us to write an apocalypse or true vision of the footsteps of the Creator imprinted on his creation."[10] We must, however, apply proper methods and safeguards.

Bacon's formulation of inductive procedures — as developed in his Table of Essence and Presence, Table of Deviation or of Absence in Proximity, and Table of Degrees or Comparison[11] — anticipates the inductive logic of the nineteenth-century philosopher John Stuart Mill. Mill more carefully specifies and refines the inductive methods used by science. Induction is "the operation of discovering and proving general propositions."[12] Specifically, it is generalizing from experience. "It consists in inferring from some individual instances in which a phenomenon is observed to occur, that it occurs in all instances of a certain class; namely, in all which *resemble* the former, in what are regarded as the material circumstances."[13]

Mill formulates the canons of induction: the methods of agreement, difference, residues, and concomitant variation. If these are properly used, he argues, we can not only *discover* the causal connections in nature, but we can *demonstrate* them as well. Inductive logic, he holds, "provide[s] the rules and models . . . to which, if inductive arguments conform, those arguments are conclusive, and not otherwise. . . . [T]he four methods are methods of discovery: but even if they were not . . . they are the sole methods of Proof."[14]

Science, for Mill, can arrive inductively at proven conclusions. It is true that induction presupposes the principle of uniformity, according to which similar causes bring about similar effects, but even this principle can be justified inductively. The more uniformities we discover in nature, the more the general principle of uniformity is justified inductively. Thus,

9. These characteristics of the traditional view of science are emphasized by Del Ratzsch, *Philosophy of Science* (Downers Grove, IL: InterVarsity Press, 1986), p. 16.

10. Bacon, *The Great Instauration,* p. 29.

11. Bacon, *The Novum Organum,* II.x-xiii.

12. John Stuart Mill, *A System of Logic* (New York: Longmans, Green and Co., 1925), bk. III, 1, 2.

13. Mill, bk. III, 3, 1.

14. Mill, bk. III, 9, 6.

science need not proceed by invoking a priori methodological principles; all its principles, including those needed to justify everything else, can be themselves justified inductively by experience.

This view of science as the epitome of rationality, objectivity, and experiential learning receives its widest hearing in twentieth-century positivism. The positivists hold that, to insure objectivity, anything that is neither given immediately in experience nor understood in terms of experience must be excluded from scientific learning. The empirical is the key to the meaningful. But positivists take this emphasis on the empirical to more radical extremes than did their predecessors. Knowledge claims that are not empirical are not false but meaningless.

The positivists argue that any statement that purports to be informative must either be what they call a protocol statement (a statement that reports an immediate sense experience — I see rain drops falling before my eyes) or else be translatable into a protocol statement. Failure to meet this condition means that the statement is meaningless. Positivists restrict what can be known to what can be meaningfully said, and what can be meaningfully and informatively said must be empirically verifiable. Science, then, will be strictly empirical, and to accomplish this it must contain only statements that are empirically verifiable (in addition to analytic statements about the necessary relations between ideas).

According to the positivists, nonempirical claims must not be imported into science. Metaphysical presuppositions about God (that he exists, that he created the world), humans (that human existence is a teleological condition for the way the universe has developed, that humans have souls, that they have internal experiences which are accessible only to themselves), or the world (that it is an ordered cosmos) must be excluded, for literally they are nonsense. Grammatically they look like other meaningful synthetic or informative propositions, and hence they are deceptive. But their meaninglessness becomes evident when we attempt to specify what empirical conditions might be relevant to determining whether they are true or false.

Similarly with value statements. Statements about values do not express meaningful propositions; they are not about any possible sense experience we might have. For example, the claims that active euthanasia is immoral or that engineering bacteria for introduction into the environment is morally good cannot be translated into protocol statements about our experience, for although we can experience euthanasia or the engineered bacteria, we cannot experience the immorality or morality of

15

performing euthanasia or introducing the bacteria into the environment. There are no moral properties in the world. Neither are these statements about ourselves; they do not report our own wishes, desires, or approvals. Moral statements merely express our feelings about certain matters in a way that might be persuasive to others. Consequently, matters of morals are meaningless, and as such they must be excluded from science if one is to avoid importing the meaningless into science. Science should be truly value free.

The three features of presuppositionless objectivity, being empirically testable by experiments, and rationality constitute the nub of the traditional view. Science is *objective.* Biases and presuppositions — whether about reality, the supernatural, our own knowing structures, values, or previous theories — must be bracketed from pure science. Metaphysical or religious beliefs about God and humans; moral beliefs about goods, duties, or virtues; epistemological beliefs about what can or cannot be known; and subjective categories can be put aside. Scientists should come to their endeavors unbiased. Science is *empirical.* Data are out there to be discovered and understood on their own terms. Data get us to nature itself: to its structures and principles of operation. To understand the data is to understand the world, not ourselves. Experiments can be set up to test objectively the various generalizations or causal hypotheses to determine which are correct. Finally, science is *rational;* its claims are verifiable or falsifiable by human experience. The basic explanatory principles and correct generalizations and hypotheses emerge inductively out of the data.[15]

The Centrality of Theory

Alas, this popular model, though initially attractive and embodying important understandings of science, is fraught with difficulties. We will focus on three of the most important ones.

(1) What process occurs when we collect data? Bacon does not get it quite right when he claims that data are self-sorting, that they collect themselves into organized patterns independent of the collector's presupposed patterns of sorting.

15. Ratzsch, p. 22.

To the contrary, theory helps prestructure data collection. Thomas Kuhn suggests three ways in which theories do this.[16] First, theories specify what facts are especially significant for understanding the world. There are an indefinite number of facts one could give about any aspect of the world. Think about all the factual claims that could be made about the room in which you are sitting or about one ant colony over time. Which facts are noteworthy enough to be recorded and preserved? Which have the significance to be correlated with other facts? Which should be ignored? Scientists do not merely go out and collect facts.

James Mannoia recounts a fable once told by Karl Popper.

> Once upon a time there lived a man who wished to give his whole life to science. . . . [T]his man sat down with pencil in hand and recorded in a notebook everything he could observe. He included everything, from the weather, the racing results, the levels of cosmic ray bombardment, to the stock market reports and the appearance of all the planets. . . . Our dedicated observer continued this job every day for the rest of his life.
>
> He had compiled, by the time of his death, the most comprehensive record of nature ever made. . . . When he died, he was so certain that his life had been well spent for the cause of science, he donated his notebooks . . . to the American Academy of Science.[17]

But the Academy neither thanked this dedicated observer nor opened his notebooks, for they contained nothing but a jumble of observations. Unlike this recorder of all he observed, the scientist collects data with an object in mind. And when observers select certain facts as worthy of study some theory must be operating, if not consciously then at least subconsciously.

Second, theories predict that if the theory is true, such and such will occur. Hence the scientist will deem significant the discovery (or, on the other hand, failure of discovery) of what is predicted. Often theory pushes the scientist in particular directions in exploring the facts of the world, directions that would not have been anticipated or undertaken without that theory (whether the theory is true or false).

For example, why do most animals reproduce sexually rather than

16. Thomas Kuhn, *The Structure of Scientific Revolutions* (Chicago: University of Chicago Press, 1970), pp. 25-28.

17. V. James Mannoia, *What Is Science?* (Washington, DC: University Press of America, 1980), pp. 14-15.

asexually, when asexual reproduction seems to conform best to the current theory that in natural selection the fittest are those that preserve their genes by passing them on to their progeny? One theory is that sexual reproduction provides the best defense against the rapidly reproducing, infectious invaders that threaten the existence of all organisms. The diversity in the species that results from combining different gene pools favors the survival of those that are sexually reproduced over those that by cloning inherit repetitive genetic similarity. To confirm this theory the biologist Curtis Lively took snails from two lakes in New Zealand separated by 10,000 feet of mountain and subjected each to parasitic worms that live on the snails from one of the lakes as their normal host. As the theory predicted, the parasites attacked the snails with which they were familiar but could not penetrate those with the unfamiliar genotype.[18] But why test these snails with parasitic worms? The experiment with the snails was undertaken because of a theory about reproduction, not simply as an extraneous event of data collection or because the researcher liked New Zealand snails.

Third, a theory specifies what research is needed or what experiments should be conducted to resolve some of the problems and ambiguities to which the theory has drawn attention but which did not play an active role in its formulation. Continuing the above example, why is it, then, that cloning has survived as a method of reproduction, and why are there some cases where cloning is the predominant method of reproduction? If sexual reproduction better preserves the species against biological threats, cloning should have disappeared in animals. To answer this, Robert Vrijenhoek studied minnows in rock pools in the highlands of Sonora, Mexico. During one drought the rock pools dried up as usual. In one pool, however, when sufficient water returned, minnows swam upstream and recolonized the pool. Some of these reproduced by cloning, others sexually. In this case, however, the clones achieved dominance, constituting ninety-five percent of the minnow population. Vrijenhoek theorized that the clones were more capable of rapid population growth in a relatively open pool, whereas those that reproduced sexually had too small and uniform a population to compete. To test this theory he artificially introduced sexually reproducing female minnows into the pool. This infusion of diversity reversed the population prevalence, so that within two years sexually reproducing minnows outnumbered the cloners four to one.[19]

18. JoAnn C. Gutin, "Why Bother?" *Discover* 13, no. 6 (June 1992): 38-39.
19. Gutin, 39.

In short, though one does not theorize without data, one does not collect data without theory. There is a mutuality of influence not recognized in the traditional model of science. But this is not contrary to our normal patterns of knowing. The lines and dots on this page would be perceived by children who have not yet learned the significance of letters, but they would not be meaningful to them. We, however, see the page as containing letters and words written in English. We perceive them as patterned, not as random marks on paper, because of a theory about letters, words, sentences, and meanings. Once we have this theory, the world comes to us structured according to the theory. Even when we encounter words that we do not recognize, we assume that they make some sense in some language unknown to us. The theory may be tacit, but it functions to structure the data and inquiry nonetheless.

(2) A second objection to the traditional theory is that it is not the mere discovery of data that occasions the acceptance or rejection of one theory in favor of another. Unless the datum fits into a theory that explains it, that shows how it connects with other data or how it can be made to work, it is treated as an anomaly to be rejected or ignored. It is theory-fittingness that prepares the way for acceptance of certain data as significant and telling.

Geologists today almost uniformly affirm the continental drift theory, a theory laughed at only thirty-five years ago. The deciding datum that swung geologists to this position was the discovery of symmetrically striped magnetic patterns on the seafloor in the 1960s. These confirmed that the seafloor is slowly spreading by the addition of igneous material emerging through cracks in the mid-ocean ridges. However, it was not the mere discovery of the varying magnetic stripes in the ocean beds that brought about the revolution in geological thinking. Prior to this discovery there was much evidence in support of the continental drift theory. As early as 1885 the noted geologist Edward Suess fit the land forms of the continents of the southern hemisphere into a jigsaw-puzzle supercontinent he called Gondwanaland. Fossil remains in Africa and South America were known to be similar, and some fossils, such as those of a Paleozoic reptile named Mesosaurus, were only to be found in Brazil and in South Africa. Living species that could not conceivably cross the ocean, such as the garden snail, were also similar on the diverse continents.

If significant, prior, supporting data were available,

What was it about the discovery of magnetic stripes in the ocean beds that caused [geologists] to change their minds? After all, any argument

19

that uses the magnetic data to verify drift must proceed by a very roundabout route. One must first assume that the stripes are a record of magnetic reversals which took place millions of years ago. One must conclude, next, that this is evidence for sea-floor spreading. Finally, one must assume that evidence which implies that ocean floors move tells us that continents drift also. . . . The reason [the continental drift theory was now accepted] was that the theory of sea-floor spreading showed how continental drift was possible.[20]

That is, there was a large amount of prior data that showed consistent land forms on continents separated by oceans: for example, mountain ranges in eastern Canada matched those in Scotland and Norway; mountains in Argentina are continuous with the Cape in South Africa; similar layers of sedimentary deposits and life forms, both living and fossil, exist on the two continents. These data had fit into no theory that could explain how the continents could move from some prehistoric, unified land mass to their present location. This is precisely what the theory of seafloor spreading supplied; a theory about the processes *explained* the evidence. Without that theory, despite all the observed correlations between the shape, geological strata, fossils, and animal life of the various continental areas, the theory of continental drift was rejected.

It is not mere data, then, that bring about a shift from one theory to another. The data must be housed in an explanatory theory, a theory that eventually makes better sense of the existent data than any other. This explains why scientists are more ready to accept the existence of quarks than of UFOs.

Moreover, it is often theory that takes precedence in forming attitudes toward the data. Scientists, convinced of the correctness of their theory, usually will hold on to it despite contrary data or anomalies. This can occur when the current theory has worked well in explaining the data so that one perceives a better reason to maintain the old theory in spite of the anomalies than to jettison it in favor of some less robust or satisfactory theory, when the theory provides the structure for the research activities of the scientist, or when there is no compelling alternate theory.

Consider, for example, the Darwinian theory of evolution understood as the gradual development through natural selection of higher

20. Richard Morris, *Dismantling the Universe* (New York: Simon and Schuster, 1983), p. 113.

species by gradual, continuous change from lower species. This generally accepted theory faces anomalies, some known to Darwin himself. For one thing, if the theory is true, one should find paleontological evidence of a gradual transition from one species to another. But transition forms between species are precisely what are missing. As Stephen Gould explains, "The extreme rarity of transitional forms in the fossil record persists as the trade secret of paleontology. The evolutionary trees that adorn our textbooks have data only at the tips and nodes of their branches; the rest is inference, however reasonable, not the evidence of fossils. Yet Darwin was so wedded to gradualism that he wagered his entire theory on a denial of this literal record."[21] Gould goes on to argue that the geological record presents two features at odds with gradualism: sudden appearance, in which the species "appear all at once and 'fully formed'"; and stasis, in which the species remain morphologically stable from their inception until either their modern form or their disappearance from the fossil record.

The problem occurs at the very beginning of the life record with the "Cambrian explosion." There, almost all of the major groups of modern animals appear in the fossil record, fully formed, without prior fossil evidence of complex, multicelled organisms from which they might have reasonably arisen.[22] As Daniel Axelrod puts it,

> One of the major unsolved problems of geology and evolution is the occurrence of diversified, multicellular marine invertebrates in Lower Cambrian rocks on all the continents and their absence in rocks of greater age. . . . Their high degree of organization clearly indicates that a long period of evolution preceded their appearance in the record. However, when we turn to examine the Precambrian rocks for the

21. Stephen Jay Gould, "The Episodic Nature of Evolutionary Change," in *The Panda's Thumb* (New York: W. W. Norton & Co., 1980), p. 181.

22. Stephen J. Gould, *Wonderful Life* (New York: W. W. Norton & Co., 1989), p. 24. Gould notes:

The Precambrian record does contain one fauna of multicellular animals preceding the Cambrian explosion, the Ediacara fauna. . . . But this fauna can offer no comfort to Darwin's expectation for two reasons. First, the Ediacara is barely Precambrian in age. These animals are found exclusively in rocks just predating the explosion, probably no more than 700 million years old and perhaps younger. Second, the Ediacara animals may represent a failed, independent experiment in multicellular life, not a set of simpler ancestors for later creatures with hard parts. (pp. 58-59)

21

forerunners of these Early Cambrian fossils, they are nowhere to be found.[23]

The transition gap from the Precambrian to the Cambrian is not unique. Similar breaks occur, among others, between fish and amphibians, amphibians and reptiles, reptiles and mammals, and angiosperms and pre-Cretaceous plants.[24] "Transitions between major groups are characteristically abrupt."[25] What, then, is to be done with this anomalous fossil record?

Darwin's approach was to aver hopefully that the missing intermediates would be found. He argued that the geological record was woefully incomplete, so that future discoveries would confirm his theory. Yet the anomaly has not disappeared, even with the significant paleontological work done since Darwin. Michael Denton suggests that the fossil record is probably quite complete, in the sense that probably most of the species to be found have been found. Of the 329 living families of terrestrial vertebrates, 261 or 79.1 percent have been found as fossils, and when birds are excluded the percentage rises to 87.8 percent.[26]

Another approach is to deny that the anomaly is that significant. Darwinists point to the discovery, only two years after Darwin published his epic *On the Origin of the Species by Means of Natural Selection,* of a fossil bird named Archaeopteryx that possessed characteristics of both birds and reptiles. Though possessing barbed feathers, the ability to fly, and a wishbone (furcula), it had a vertebrate tail characteristic of reptiles, claws on its wings, no sertum or breastbone, and reptilian teeth. Its metatarsals

23. Daniel Axelrod, "Early Cambrian Marine Fauna," *Science* 128, no. 3314 (4 July 1958): 7. Michael Denton makes a similar point in *Evolution: A Theory in Crisis* (Bethesda, MD: Adler and Adler, 1986), pp. 162-63:

> It is still, as it was in Darwin's day, overwhelmingly true that the first representatives of all the major classes of organisms known to biology are already highly characteristic of their class when they make their initial appearance in the fossil record. This phenomenon is particularly obvious in the case of the invertebrate fossil record. At its first appearance in the ancient paleozoic seas, invertebrate life was already divided into practically all the major groups with which we are familiar today. . . . Robert Barnes summed up the current situation: ". . . the fossil record tells us almost nothing about the evolutionary origin of phyla and classes. Intermediate forms are non-existent, undiscovered, or not recognized."

24. Denton, chap. 8.

25. Stephen Jay Gould, "The Return of the Hopeful Monster," in *The Panda's Thumb,* p. 189.

26. Denton, p. 189.

were between the split variety of reptiles and the fused type of birds.[27] This, many biologists assert, is a prime example of a preserved intermediary species.[28] Yet room for doubt that it was a link between reptile and bird remains. Bird fossils have been found in rocks from the same period (Upper Jurassic), suggesting that the Archaeopteryx was merely another bird and not the ancestor of today's birds. John Osterm remarks that "it is obvious that we must now look for the ancestors of flying birds in a period of time much older than that in which the Archaeopteryx lived."[29]

A third approach is to admit the geological discontinuity and keep the basic evolutionary theory, but to modify the theory to explain the anomalies. Gould, for example, does not abandon Darwinism; he only modifies it. He agrees with Darwin that small, random, undirected changes occurred and were preserved by natural selection. However, he attempts to resolve the anomaly by arguing that in small, isolated sister species groups the changes occurred as "geologically instantaneous events of branching speciation."[30] When their isolation ended, the newly modified organisms were introduced back into the original environment, and since they were better adapted, they prospered, becoming a widely spread, fully formed species readily identifiable in the geological record. The gaps in the geological record, then, are not problematic; they are apparent, not real. Since the evolution occurred in a small offshoot while the main species remained stable (stasis), the changing group, while it was mutating, was not statistically large enough to be preserved in the fossil record.

27. Peter Wellnhofer, "Archaeopteryx," *Scientific American* 262, no. 5 (May 1990): 70-77.
28. Cecie Starr and Ralph Taggart, *Biology: The Unity and Diversity of Life,* 5th ed. (Belmont: Wadsworth Pub. Co., 1989), pp. 31-32.
29. John Osterm, "Bone Bonanza: Early Bird and Mastodon," *Science News* 112 (24 Sept. 1977): 198. A more recent find of a sparrow-sized bird in China prompted Paul Sereno to state that either "modified birds evolved rather quickly after Archaeopteryx, or the avian lineage goes back much farther than we've found" (John Noble Wilford, "Surprising Flight Disclosed 135 Million Years Ago: A Bird," *New York Times* [12 Oct. 1990], A15). However, even this bird evidences a reptilian lineage. For an argument that the Archaeopteryx is an authentic intermediary, see Philip Kitcher, *Abusing Science* (Cambridge, MA: Massachusetts Institute of Technology, 1982), p. 109.
This discussion raises the definitional and philosophical problem of what constitutes an intermediary. Not infrequently, the discussion between diverging viewpoints turns on the absence of any commonly accepted definition of *intermediate*.
30. Stephen Jay Gould, "Darwinism and the Expansion of Evolutionary Theory," *Science* 216, no. 23 (April 1982): 382.

Of course, the problem with Gould's solution to the anomaly is that it cannot be confirmed geologically. The absence of intermediates in the fossil record that punctuated equilibrium theorists' attempt to explain is the very thing they need to confirm their hypothesis. Nevertheless, their hypothesis aptly illustrates how theorists attach priority to the theory (the central Darwinian theory) as they attempt to explain the anomalies.

One might extend the anomalies to include evolutionary theory's difficulty in accounting for the development of complex organs by gradual change. Not only would intermediate states be nonfunctional with respect to their original structure, but only in the rare case could they serve another purpose until the organ is fully developed. Few random organ mutations seem to enhance function. For example, the eye is a complex organ finely tuned to accomplish its task. Gradual changes in the eye by genetic mutation, as suggested by Darwinian evolutionary theory, would make the eye less, not more, effective.[31] In addition, how does one account for organs, such as the human brain, that have capacities far beyond what the animal uses? There would seem to be no reason for its preservation for future use since it serves no immediate function to make the animal more fit to survive.

One response, preserving the Darwinian theory, is simply to deny any anomaly, that is, to deny that complex organisms cannot arise through minute steps. Richard Dawkins asks whether the human eye might have arisen directly from something slightly different from itself, something that we may call X. The answer, he claims, is yes, provided that the "X is defined as sufficiently similar so that the human eye could plausibly have arisen by a single alteration in X."[32] But this kind of claim will not help

---

31. More than eighty years ago the philosopher Henri Bergson appealed to the example of the eye:

> It must not be forgotten that all the parts of an organism are necessarily co-ordinated. . . . One point is certain — the organ will not give selection a hold unless it functions. However the minute structure of the retina may develop and however complicated it may become, such progress, instead of favoring vision, will probably hinder it if the visual centers do not develop at the same time, as well as several parts of the visual organ itself. If the variations are accidental, how can they ever agree to arise in every part of the organ at the same time, in such a way that the organ will continue to perform its function? (*Creative Evolution* [New York: The Modern Library, 1911], p. 72)

32. Richard Dawkins, *The Blind Watchmaker* (New York: W. W. Norton & Co., 1986), p. 77.

to resolve the dispute, for though Dawkins's response is true, it is tautologously true, and hence cannot function to yield knowledge about the evolutionary process itself.

Further, he goes on to ask whether it is plausible that every one of the intermediates worked sufficiently well so that it assisted the survival and reproduction of the organism. Again he responds in the affirmative, for 5 percent vision is better than "no vision at all. So is 1 per cent vision than total blindness. And 6 per cent is better than 5, 7 per cent better than 6, and so on up the gradual, continuous series."[33] But this argument presupposes that an organ that is 5 percent functional is an eye that is functioning at 5 percent. But the intermediate steps leading to the eye are not eyes at all, but protoeyes of various sorts. If I put together pieces of an engine so that I have assembled 5 percent of it, I do not have an engine operating at 5 percent capacity.

A second approach to resolving the anomalies while preserving the theory is to argue that genetic changes, though themselves slight, might result in substantial phenotypic changes if they operate at the embryonic level. Changes of this sort might give the appearance of a kind of saltation. "Small changes early in embryology accumulate through growth to yield profound differences among adults. Prolong the high prenatal rate of brain growth into early childhood and a monkey's brain moves toward human size."[34]

In short, the history of Darwinism well illustrates the primacy of theory. Despite the presence of anomalies it continues to be the dominant perspective, and for good reason. For one thing, it possesses significant "postdictive" power. "To emphasize the predictive power of evolutionary theory is somehow to miss the point. The primary function of the theory is to advance our understanding of past and present organisms, revealing to us how the features of the organic world can be comprehended by recognizing its history."[35] For another thing, if one is to abandon a theory

---

33. Dawkins, p. 81.

34. Gould, "The Return of the Hopeful Monster," p. 192. See also Francisco J. Ayala, "Beyond Darwinism? The Challenge of Macroevolution to the Synthetic Theory of Evolution," in *Philosophy of Biology*, ed. Michael Ruse (New York: Macmillan, 1989), pp. 125-26.

35. Kitcher, p. 81. Questions regarding the role of prediction in what are termed complex adaptive systems (evolution being an example) are raised by developments in the theory of complexity. See E. Mosekilde and L. Mosekilde, *Complexity, Chaos and Biological Evolution* (New York: Plenum, 1991).

because of anomalies, there must be a strong, reasonable alternative possessed of explanatory power. "All theories are revisable, but not all theories are equal. Even though our present evidence does not *prove* that evolutionary biology is true, evolutionary biologists will maintain that the present evidence is overwhelmingly in favor of their theory and over-whelmingly against its supposed rivals."[36]

The primacy of theory over data can occur also at the beginning of a new theory.

> In 1906 the German physicist Walter Kaufmann published the results of a long series of experiments in which he had measured the mass of moving electrons. His results agreed with some theories and disagreed with others. In particular, they failed to substantiate the predictions of the special theory of relativity. . . . But Einstein was not troubled. . . . [H]e wrote, "In my opinion both [of Kaufmann's] theories have a rather small probability because their fundamental assumptions concerning the mass of moving electrons are not explainable in terms of theoretical systems that embrace a greater complex of phenomena." In other words, no matter what the experiments said, the competing theories could not be true because they did not fit into far-reaching, clear-cut theoretical patterns.[37]

Again, this parallels our ordinary ways of knowing. We are reluctant to abandon a view that has served us well to this point. Should there be reports that our spouse is unfaithful, we are reluctant to believe them if the marriage has worked well. Similarly with politics and religion; conversions from allegiance to one party or religious faith and its ideology to another is painful and resisted, even in the face of contrary evidence, because of the history of that party's or faith's adequacy in our experience.

(3) A third objection to the traditional view of science is that the Baconian inductive method does not accurately describe the way in which many scientists arrive at their theories. Rarely do scientists induce theoretical hypotheses from mere raw data. To the contrary, there seems to be no

---

36. Kitcher, p. 34.

37. Morris, p. 67. In a similar vein, Michael Polanyi points to the case of D. C. Miller, who, in his presidential address to the American Physical Society in 1925, presented evidence contrary to Einstein's theory of relativity that was largely ignored by those in his audience, "the evidence being set aside in the hope that it would one day turn out to be wrong" (*Personal Knowledge* [Chicago: University of Chicago Press, 1958], p. 13).

strict logic, no set pattern, to the discovery of theories or hypotheses. "[T]he theories and explanatory principles which arise within science are products of human invention and insight, not logical results of data. There is no *rigorous logical* procedure which accounts for the birth of theories or of the novel concepts and connections which new theories often involve."[38] The data might help to suggest the theory or hypothesis, but it still requires insight, creative imagination, or intuition on the part of the scientist to suggest a fertile theory to be tested.

Some philosophers of science reinforce the absence of a strict empiricist logic of scientific discovery by distinguishing between the reasons for accepting a hypothesis or theory and the reason for suggesting that hypothesis or theory in the first place.[39] The reasons for accepting the hypothesis can be spelled out via the hypothetical-deductive method. Accuracy of prediction, simplicity, breadth or scope of the theory, and compatibility with other already widely accepted theories are reasons we might have for thinking the hypothesis true. But though the hypothetical-deductive method can help us test our theories, it cannot provide an adequate account of the logic of discovery.

Norwood Russell Hanson focuses on the roles of analogy and symmetry in theory discovery.[40] Whereas employing analogies and recognizing symmetries are proper methods for formulating hypotheses, they would not be proper methods for establishing the truth of hypotheses. The employment of analogy is aptly illustrated in Charles Darwin's development of the theory of evolution. For Darwin the crucial analogy came from his reading Malthus's *An Essay on the Principle of Population*.[41] Malthus argued that continued population growth would create an imbalance between people and agricultural resources and ultimately bring famine. In the resulting struggle to survive, the weak, ill, and lazy would be weeded out in favor of the strong, healthy, and productive. The English bourgeois saw that this theory fit well with their laissez-faire economics; an economy

38. Ratzsch, p. 23. Karl Popper agrees: "The initial stage, the act of conceiving or inventing a theory, seems to me neither to call for logical analysis nor be susceptible of it" (*The Logic of Scientific Discovery* [New York: Basic Books, 1959], pp. 31-32).

39. Norwood Russell Hanson, "Is There a Logic of Scientific Discovery?" in *Current Issues in the Philosophy of Science,* ed. Herbert Feigl and Grover Maxwell (New York: Holt, Rinehart, and Winston, 1961), p. 22.

40. Hanson, p. 26.

41. F. Darwin, ed., *The Life and Letters of Charles Darwin,* vol. 1 (London: John Murray, 1888), p. 83.

left alone would favor their own survival. Darwin, in turn, applied the analogy of laissez-faire economics to nature. Nature, left alone, selects out the weak and ill, allowing the fittest to survive. Economics seen through the eyes of the industrial bourgeois provided the analogy for constructing his theory of natural selection.[42] Thus, "in the machine age he established a mechanical conception of organic life. He paralleled the human struggle [in the age of the industrial revolution] with a natural struggle. In an acquisitive hereditary society he stated acquisition and inheritance as the primary means of survival."[43] Jeremy Rifkin comments that "Darwin dressed up nature with an English personality, ascribed to nature English motivations and drives, and even provided nature with the English marketplace and the English form of government. . . . Darwin borrowed from the popular culture the appropriate metaphors and then transposed them to nature."[44]

Michael Polanyi expands the concept of rationality to include what he refers to as "tacit knowledge." Arguing from the principles and results of Gestalt psychology, he contends that "knowledge of the external world is in general acquired by relying on clues which cannot be fully identified, . . . clues that are often elusive and never fully specifiable. Scientific discoveries are likewise based on clues that are never fully specifiable."[45] This does not make scientific discovery unreasonable or irrational. To the contrary, it is closer to the way we attain knowledge generally. But it does prevent stating explicit rules for the logic of discovery.

Two extremes are to be avoided. The one, touted by the traditional empiricist view, is that discovery is a mechanical or purely logical or inductive process unadulterated by theory. The other is that it is pure guesswork, unencumbered by data.[46] The truth lies in the middle. Consequently, the traditional empiricist view of science must be modified to

42. Jeremy Rifkin, *Algeny* (New York: Penguin Books, 1983), p. 80.

43. Geoffrey West, *Autobiography of Charles Darwin: 1809-1882,* ed. Nora Barlow (New York: W. W. Norton, 1958), p. 23.

44. Rifkin, p. 72. Karl Marx, in a sarcastic vein, made the same observation more than a century earlier in his letter of June 18, 1862, to Frederick Engels. See Saul K. Padover, ed., *The Letters of Karl Marx* (Englewood Cliffs, NJ: Prentice-Hall, 1979), p. 157.

45. Michael Polanyi, "Notes on Professor Grunbaum's Observations," in *Current Issues in the Philosophy of Science,* ed. Feigl and Maxwell, p. 54.

46. Mannoia, p. 14. Mannoia uses the term *abduction,* which he borrows from C. S. Peirce, to describe the process of discovery. Abduction involves both "careful attention to facts and a crucial element of intuitive imagination" (p. 16).

acknowledge a dynamic interplay between data and theory. The traditional view is correct in holding out for the importance of data. Theoretical formulation would not take place without data. New theories often arise when anomalous data cannot be adequately accounted for under other theories. But the data we use are embedded in theory. Even the presence of scientific problems or puzzles, which new theories are proffered to resolve, is theory bound.[47] Since science is a human endeavor it becomes evident that theory is in some sense and to some degree determinative of the facts. To what degree is a matter of dispute, one that we need not attempt to resolve in this book.[48] Instead, having established a primary role for theory, we want to explore the significance of this primacy for our understandings about and dealings with the world.

Paradigms

As we have seen, theories play a causal role in science; fact gathering is not theory neutral. Some theories have a broader scope and hence play a larger role than others. These theories become normative when they govern the formation of lower-level theories, suggest ways in which lower-level theories can interrelate, and significantly direct subsequent scientific research, discussion, and training. They are sometimes referred to as paradigms.

The term *paradigm*, in this usage, owes its current popularity to Thomas Kuhn, who in his seminal book *The Structure of Scientific Revolutions* takes *paradigm* to mean "accepted examples of actual scientific

47. Karl Popper writes: "Some rudimentary theory or expectation always comes first; . . . it always precedes observation. . . . Accordingly I assert that we do not start from observations but always from *problems* — either practical problems or from *a theory which has run into difficulties*. . . . Thus we may say that *the growth of knowledge proceeds from old problems to new problems, by means of conjectures and refutations*" (*Objective Knowledge* [London: Oxford University Press, 1972], p. 258).

48. Some contemporary philosophers of science of a more radical bent have gone so far as to claim that scientific truth is what scientists think, accept, or claim to know as true. Theory, not some objective truth, is what determines what we know, for we possess no theoretically neutral access to the world. Indeed, there is no world, free from human interpretation, accessible to us. According to them we cannot escape the human egocentric predicament.

practice — examples which include law, theory, application, and instrumentation together — [that] provide models from which spring particular coherent traditions of scientific research."[49] Paradigms are fruitful ways of looking at the world, ways that determine the categories to be employed in understanding the phenomena, the relations that hold between the entities thought to exist, the problems that need to be addressed, the research methods to be used, and the kinds of outcomes expected from their application. Kuhn gives as examples Ptolemaic and Copernican astronomy, the Aristotelian conception of motion, Newtonian dynamics, corpuscular optics, Maxwell's mathematization of the electromagnetic field, Einsteinian relativity, and quantum mechanics. In the biological sciences we might add the DNA double helix, the Mendelian view of heredity, Darwinian natural selection, sociobiology, and the Linnean classification.

Kuhn claims that all normal science works from within paradigms. Before the establishment of a clear, disciplinary community in some branch of science, scientists belong to schools that have their own theories functioning as paradigms for them. But since their paradigms are not widely shared, there is no common body of data or beliefs or commonly accepted theoretical framework that can structure what the scientist does. Consequently, each scientist or group of scientists begins as it were from scratch and builds on that. With the development of a common paradigm, scientists are freed from beginning anew each time. The controlling paradigm delineates the body of data and theories that are accepted and prompts the data collection and research to be done. After the adoption of a paradigm, science turns to its articulation. It suggests which data are important, the means by which they are to be gathered, the design and employment of the equipment or instruments

---

49. Kuhn, p. 10. Kuhn himself admits that his use of *paradigm* in that book is often ambiguous and at times equivocal. In his 1969 Postscript he distinguishes between two senses of "paradigm": *(a)* "paradigm" as exemplary problem solutions, and *(b)* "paradigm" as the "entire constellation of beliefs, values, techniques, and so on shared by members of a given community," what he refers to as a disciplinary matrix (p. 175; see also Kuhn, *The Essential Tension* [Chicago: University of Chicago, 1977], chap. 12). In his subsequent writings he prefers to limit paradigm to *(a)*. However, there are good grounds for the second usage as well, particularly in his treatment of seeing in chap. 10 of his *Structure of Scientific Revolutions* and in his willingness to extend paradigm (in a broader sense of disciplinary matrix) to model (*The Essential Tension*, p. 297). It is this usage that, as we shall see, begins to approach the concept of a worldview.

to be used, what problems need to be resolved, and the manner by which resolutions might proceed.

Although there are doubts about Kuhn's description of the actual historical process of science, and in particular about his distinction between pre-paradigmatic and normal science, it is generally conceded that he is correct in his emphasis on the central role that paradigms play in science. Scientists, from the outset, make serious theoretical and methodological commitments. By adopting a certain paradigm, they commit to "explicit statements of scientific law and about scientific concepts and theories," to certain types of scientific instrumentation and the way they are to be used, and to certain views of the world.[50] That is, the doing of science involves certain conceptual, theoretical, instrumental, and methodological beliefs. In this respect scientific knowing does not differ from our other experiences of knowing. Science is a *human* endeavor.

This means that, contrary to the traditional empirical view of science, there are no neutral observations, no observer-free data yielding objective data against which we can compare our theories. Rather, all scientific observation is human seeing. It is seeing from a certain perspective, in terms of certain gestalts or patterns that are products of our biology, environment, value structures, and theoretical frameworks. This means that the data bases that scientists use come already theory laden. The scientists' world is "determined jointly by the environment and the particular normal-scientific tradition" in which they operate.[51]

Kuhn goes on to argue that "when paradigms change, the world itself changes with them. . . . [P]aradigm changes do cause scientists to see the world of their research-engagement differently. In so far as their only recourse to that world is through what they see and do, we may want to say that after a revolution scientists are responding to a different world."[52] We are familiar with the drawings that can be seen first one way (as an ascending staircase, a duck, or an old woman), and then, with a perspective switch, another way (as a descending staircase, a rabbit, or a young woman with a flamboyant hat). Similarly, Kuhn suggests, what the student sees as "confused and broken lines [in a bubble-chamber photograph], the physicist [sees as] a record of familiar subnuclear events." For the novice to see what the physicist sees requires initiation into and adoption of a different

50. Kuhn, *The Structure of Scientific Revolutions*, pp. 40-41.
51. Kuhn, *The Structure of Scientific Revolutions*, p. 112.
52. Kuhn, *The Structure of Scientific Revolutions*, p. 111.

perspective, which in this case embodies a certain scientific theory about particles. But once the perspective of this theory is adopted, the photograph now is viewed differently — as is the world that it pictures.

Ambiguity resides in Kuhn's own view. Is the point he is making ontological (the world itself changes) or epistemological (our perception of the world changes) or both? Interpreted epistemologically, his claim is indisputable, aptly illustrated on a more narrow scope by the gestalt pictures noted above. The ontological view — namely, that changes in paradigms change the world itself — is more radical. Kuhn's point seems to be that the world in which we live and which we come to know has a subjective as well as an objective core to it. We give it a structure that organizes it in a certain manner. We put the world together by our lived experience. For example, one might argue that mirages really exist, for no matter how long we stare at the mirage of water in the dip in the road on a sunny day or tell ourselves that it is an illusion, we cannot escape the conclusion that it is there. Only when we move and change our visual perspective — not alter our paradigm — does the mirage disappear.

There lurks in Kuhn's account a strong element of subjectivity, for if the way the world is depends in part on our paradigms, then we have contributed to the constitution of the world. However, Kuhn assures us that this does not make the entire process subjective. There is objectivity as well, for in addition to paradigms there is a structure of events. There is Bacon's nature to be heard, if not obeyed. Knowers seek to establish a relation between themselves and their world. Hence both poles — data and theory — must be taken into account in analyzing the structure of knowledge. What there is in the world provides a significant check on individual theorizing.

We need not pursue Kuhn into the ontological interpretation; our interests lie in matters having to do with knowledge. If Kuhn is correct that the adoption of a paradigm can bring significant changes in the way we see our world, then it is clear that how we operate in knowing, understanding, and dealing with the world depends on the paradigms we use. Scientists' paradigms in part determine the kinds of theories they attend to and accept, the choice of the data they deem significant, the questions they raise, and the methods they use to explore that data. For example, unless one holds to a macroevolutionary account about gradual species development, one would not look for or perhaps even recognize intermediary species. Neither would one mix hydrogen, methane, ammonia, and water in a reaction chamber and bombard the solution with electric charges to see whether one could generate organic molecules. Neither would one try to discover how RNA and proteins could evolve into DNA coding systems.[53]

Paradigms are not strictly individual but communal in nature. Scientists, like theologians, philosophers, historians, and the like, operate in a community. That community has certain standards of acceptance, so that not just any paradigm or theory counts. Paradigms are intersubjective and hence subject to the collective evaluation and wisdom of the community. Hence, communicability is a criterion for their meaningfulness. And there are criteria for assessing their truth, depending in part on the types of claims made by the paradigm and the ascertaining of what evidence is relevant to evaluating the truth claims made. In science, this means that the hypothetical-deductive method has a distinctive role to play in testing the theories that are paradigm influenced. Objectivity and subjectivity in the epistemological realm need not be incompatible or mutually exclusive. Proffered theories have controls for their acceptance, not the least being reality itself and the community of observers. The theory must make sense of and, to some degree, be testable by experiment in that reality, and it must be able to be reasonably communicated and justified to the community of scientists. At the same time, presuppositions about method, about what is there to be experienced, and how one gets to that world are relevant to our knowing that reality.

What we have suggested to this point is that theories play an important role in our scientific understanding. Not all theories function at the same level. Some are simple, governing more restricted matters or particular phenomena; theories at this level are more readily affected by empirical data, and their change over time might be more incremental. Others are more complex; only a large number of significant anomalies are capable of overthrowing belief in them.[54] Theories are of varying scope and complexity, governing applications of varying generality and themselves embodying narrower or more specific theories. Theories of this broader sort we have identified in the broader notion of paradigm, for they function fruitfully to guide scientific research.

53. The recent discovery of enzymatic RNA is held by some to provide the basis for a solution to this anomaly. See Silvia Culp and Philip Kitcher, "Theory Structure/Change in Contemporary Molecular Biology," *The British Journal of the Philosophy of Science* 40 (1989): 478.

54. Ratzsch speaks about mini- and maxi-theories, but makes essentially the same point (pp. 64-66).

Worldviews

Paradigms, too, come with differing degrees of generality. The most general of the paradigms might be termed "macroparadigms" or "worldviews." Worldviews are comprehensive perspectives on reality, frameworks or generally unified sets of fundamental beliefs through which we view, understand, interpret, evaluate, and act on the world.[55]

Worldviews have numerous functions. For one, they provide a larger framework for *describing* what reality is and how it operates. One might liken a worldview to an old rolltop desk, with numerous pigeonholes into which letters and notes can be put. These holes help organize the welter of paperwork by providing a logical structure in which to place the pieces. Missing from the analogy, however, are the interconnections between all the slots. In the desk the slots stand as independent units, whereas the beliefs, concepts, and structures of a person's worldview will intermingle or connect with each other, ideally to provide a unified if not coherent outlook. This analogy also fails to capture the "view" aspect of worldviews. We might liken worldviews to eyeglasses of different colors and strengths by which we see our world. However, the analogue (glasses) lacks the adaptability and flexibility that can characterize worldviews.

Worldviews also provide a framework for indicating how reality *ought* to be. They contain the ideals around which we should structure our behavior, develop relationships, or build society. As normative, they include beliefs about the good: goods to be achieved, goods as character virtues to be developed, and goods as right ways to act.

Finally, worldviews have an existential function. They provide the framework or context in which questions of *meaning* and *purpose* can be raised and answered. They incorporate the individual's beliefs about what is ultimately significant for human life and destiny and about one's place in society and the cosmos.

Worldviews are often prereflective. In this mode they are the lived, unifying perspectives that arise from the influences of family, church, friends, education, media, and culture. They can be conscious in that people are aware of their own ways of knowing, perspectives, values, and purposes; but more likely they are subconscious frameworks that provide the overarching background for our experience. For example, in this pre-

---

55. James H. Olthuis, "On Worldviews," *Christian Scholar's Review* 4, no. 2 (1985): 155.

reflective mode Christian laypersons might believe that the world is created by God but might not come to any conscious, theological formulation of a doctrine of creation, be unable to articulate what difference this is supposed to make in observing or understanding the world, or never have raised questions concerning the relationship of creation to scientific theories about the origin of the universe. However, even in an unformulated state, this belief that God created the world might be formative in their approach to existence. They might find meaning in the world by holding that everything that happens has a divine purpose, but at the same time they may never consciously formulate this until, for example, their child dies unexpectedly in a tragic accident. As they attempt to incorporate this tragedy into their life by finding some meaning or purpose for their child's death, they might suddenly act on a long-simmering plan to establish a memorial scholarship in their child's honor.

One of education's tasks is to make our prereflective worldview reflective. To do this we need to reflect critically on the beliefs we have, the ways in which those beliefs are tied together, and the manner in which they provide meaning and significance to our lives and guide our actions. To construct a reflective worldview, particular attention must be paid to the themes that not only compose but also unify it. What are these themes? Are they empirically adequate for human experience? Are they logically coherent? And are they existentially relevant to our meaning-achieving endeavors?

This kind of reflection often gives rise to the development of, or to the location of oneself within, an "ism." Worldviews frequently assume the character of "isms": for example, naturalism, secular humanism, theism, scientism, environmentalism, Platonism, Catholicism, Hinduism, Marxism, libertarianism, etc.[56] We probably will not agree with all the dogmas, theses, or theories of the "ism" with which we link or identify our worldview; worldviews belong to individuals with idiosyncrasies. But more or less locating our worldview in an "ism" provides us solidarity with and support from a community of generally like-minded individuals. It frees us from the necessity of individually providing rational justification for our beliefs or developing all of their interconnections. Supplying this is an important function of the community.

This suggests that we rarely hold our worldviews in isolation. Rather,

---

56. In this regard, see the first volume in this series, Arthur F. Holmes, *Contours of a World View* (Grand Rapids: William B. Eerdmans, 1983).

we function within a community that holds a similar set of beliefs. Sometimes we work with people who have similar worldviews, though in our increasingly pluralistic society this becomes less and less the case. More frequently, we worship or associate socially or politically with people of like mind. We share the worldview of a particular religious denomination or church congregation, club or social organization, group of supporters of a political party or cause. Whatever the context, that community will share important, central, common theses or paradigms of our worldview. For example, those in the Christian community will have a set of common concepts (e.g., creation, redemption, sacrifice), a set of examples or exemplars to which they appeal (e.g., examples of God's acts in history: the exodus; Jesus' birth, crucifixion, and resurrection), a certain set of stories that embody the standard motifs of the "ism" (Eve and Adam's fall, Abraham's attempted sacrifice of Isaac, David's battle with Goliath, Esther's salvation of her people, the compassion of the good Samaritan, the sower sowing his seed), and even a set of personages who model central values (Joseph, Moses, Ruth, David, Mary, Jesus, Paul, St. Francis, Mother Teresa). These elements make mutual understanding and dialogue possible, for they provide the common background that is assumed in the community's discourse about the way things are, the way things ought to be, and what creates meaning and purpose for life.

Though different worldviews are informed by diverse considerations and hence possess a degree of incommensurability, they often possess many common elements. In fact, since worldviews reflect and speak to the universal human predicament, one would be surprised if there were not many common themes and theses. Further, particular paradigms (including not only theories but methods, practices, patterns of reasoning) also might be compatible with differing worldviews. This is probably the case for many of the paradigms used in science. The Christian and the non-Christian, though differing at certain key points in their respective worldviews, both can commit themselves to certain methodological assumptions about how to study the world (e.g., to the hypothetical-deductive method), to theories that describe processes in that world (electromagnetism, relativity, microevolution), to adoption of certain laws of nature (gravity, laws of genetic inheritance, uniformity), and to accepting certain data as facts about the world.

This means that the doing of science should not necessarily be assumed to be compatible with only one worldview. It can, for example, be an appropriate part of both the naturalistic and Christian worldviews (though one can think of other worldviews, e.g., Hindu Vedantism, where it would fit less well). Nor need one think that the scientific conclusions of scientists working within different worldviews will differ. They might, but whether they do depends on whether the higher-level paradigms with which particular scientists operate are incompatible with central contentions of others' worldviews.

A worldview, then, is a complex of theories, interpretations, models, methodological commitments, symbols and metaphors, and moral and aesthetic ideals that provides the larger context for our experience. It forms, colors, molds, and at times distorts this experience. But for all that, we could not know the world without this overarching background set of beliefs.

Worldviews, Science, and Goods

It is obvious that worldviews, insofar as they address concerns about ideals, include ethical reflections. Worldviews contain descriptions not only of the ideals or conception of goods that we already bring to human experience but also of the goods that we ought to achieve and the evils we ought to avoid. This concern with the way things ought to be constitutes the normative dimension of the worldview and inserts values into our considerations.

The introduction of values raises some controversial issues. Related specifically to the theme of our book, does this mean that, contrary to the positivists and the traditional empiricist view of science, concern for values penetrates down to science itself? Is there a place for values in science, or must science and its methods attempt to remain value free, reserving value considerations to worldviews?

In contrast to the traditional view of science, Kuhn argues that values form part of paradigms. "Usually they are more widely shared among different communities than either symbolic generalizations or models, and they do much to provide a sense of community to natural scientists as a whole."[57] Values, he concludes, operate in the very doing of science. For example, he sees values functioning in science's predictive activities: predictions ought to be accurate and to stay within the margin of permissible

57. Kuhn, *The Structure of Scientific Revolutions*, p. 184.

error. Theoretical considerations also invoke values: in assessing a theory, scientists invoke such values as simplicity, self-consistency, plausibility, compatibility with other theories, accuracy, breadth, and fruitfulness.

It is true that these characteristics are valued, but their inclusion under the umbrella of values suggests how broad — and perhaps thereby how meaningless — the term *value* has become. In effect, anything can be a "value" if and when it is used in an evaluative endeavor or is desirable. Hence, what some would call merely the criteria of an adequate theory here become values.

But values in the guise of theory criteria differ from the values that especially concern the ethicist. If scientists opt for the more complex rather than the simpler theory or are inconsistent in their theories, this would not reflect on their moral character. It might reflect on the way in which they do their science. It might even suggest something about their state of mind, that they have a nonrational or irrational streak. But it leaves untouched important questions about right and wrong, about what we should do and should not do morally in science or technology.

But there are other values — or, better, goods — that enter into the practice of science that have ethical significance. One might argue that scientists ought to have integrity, tell the truth, share the information they gather with their community or the public, be concerned with the effect of experimental procedures on the experimental subjects, not treat their subjects merely as means for their own advancement or the advancement of scientific knowledge, not spoil the environment that they study, and have worthwhile benefits in view for the risks to which they put their subjects. Scientists who lack integrity or who falsify their data are acting wrongly in a way that reflects on their moral character.

Science is not an impersonal affair. As we have argued, science is a *human* way of knowing. Despite what some scientists claim or would like, they cannot abstract themselves in their scientific practice from moral considerations. The concern for goods, both in terms of goods to be achieved and the rightness of acts to be performed, ought to be a part of their concern. Failure in these regards affects the success of the scientific endeavor, for science cannot proceed if the data from experiments are falsified, if false data are introduced (e.g., the Piltdown man), or if there is a refusal to share both the results and the processes of research. Misleading the community of scholars is costly in wasted time and effort. But much more than this, moral failure affects the scientists themselves. It destroys their moral character.

38

The traditional empiricist is incorrect in holding that values do not affect the results of scientific procedures. Scientists, like other human beings, have worldviews. Worldviews contain values, and values in turn will affect the way people practice science. And the practice of science has a direct bearing on what conclusions scientists come to. It affects what issues or problems they consider significant and worthy of their efforts, which experiments they will undertake and which not, how they will go about doing those experiments, how they will see or interpret the results of their experiments or observations, what they will report and how they will report it, and what correlations they draw. Nowhere is the relevance of values to science clearer than in the area of biology, where often the choices made about what experiments to perform and how to perform them reflect on the moral character of the experimenter. The history of human experimentation is replete with atrocities in this regard. Nazi experimentation was perhaps the most unconscionable, but the pain and suffering it imposed on its experimentees were hardly unique. Prisoners and persons in mental institutions also have suffered greatly at the hands of scientists.[58] Soldiers, patients in hospitals, even the general public were unconsenting experimentees in radiation tests conducted by the Energy Department, Defense Department, NASA, Department of Veteran Affairs, and the CIA as recently as the 1960s.[59]

Consequently, it is important that we reflect on the relations between science and morality. To do this we must return to the question of worldviews and reflect on the ethical model or paradigm that will guide our subsequent discussion. Care must be taken about which ethical paradigm to adopt. We will consciously choose a paradigm from within the Christian worldview to structure our discussion of the interface of science and morality. In the next chapter we will articulate that paradigm.

58. M. H. Pappworth, *Human Guinea Pigs: Experimentation on Man* (Boston: Beacon Press, 1967).

59. "Fallouts from Nasty Secrets," *TIME* 142, no. 27 (27 Dec. 1993): 26; Keith Schneider, "U.S. Expands Inquiry into Its Human Radiation Tests," *New York Times* 143 (31 Dec. 1993): A8; Arjun Makhijani, "Energy Enters Guilty Plea," *Bulletin of the Atomic Scientists* 50, no. 2 (March-April 1994): 18-29.

CHAPTER THREE

A Christian Ethic
——— *of Stewardship* ———

OUR DISCUSSION to this point suggests that our worldviews help to shape our understanding of reality. We have seen this with respect to the scientist and can correctly infer that this holds no less true of the ethicist than the scientist, the layperson than the scholar, the Christian than the non-Christian. We have also noted that our worldviews contain paradigms and motifs that help us address particular issues. Hence it is appropriate to inquire concerning the paradigms that are going to be formative of our Christian understanding of the interface between biology and ethics.

In part, the paradigm a person adopts will be personal, since it provides the structure through which experience has meaning for that person. In one sense, it is *that person's* paradigm, a personal perspective on reality. Yet it cannot be strictly personal. First, without communal categories we can neither communicate our ideas to others nor in return understand their ideas. Without commonality, the discourse of others would be the kind of babel we encounter when we pick up an advanced textbook on a completely new subject. The categories used to describe and explain do not connect with our own experience. In short, unless our paradigms are comprehensible by the larger community in which we operate, we cannot escape a solipsism of understanding, where only we can comprehend what we think and experience.

Second, unless the paradigms are communal, there is no way to check their adequacy. This does not mean that our paradigms must be accepted by everyone in the community. The community of thinkers is diverse, and the Christian, for example, need not satisfy totally the demands of non-Christian worldviews or paradigms. Indeed, one would expect that at times there would be significant differences among the varying perspectives. Those who believe that God exists will sometimes take a different approach than nonbelievers to the meaning of what is experienced or interpret differently the larger context of an experience. What is required is that we attempt to make our paradigms intelligible to the larger community and place them on the table for discussion, comment, and evaluation.

To put it another way, we search for a middle ground between Christianity in isolation and Christianity held hostage. In the former, Christians assert a priori the adequacy of their worldview and invite others into it. Here the circle is closed; discourse with nonbelievers can only take place using the categories, motifs, and criteria of the Christian. In the latter, Christians and nonbelievers feel that Christians must satisfy all the epistemic demands of nonbelievers, or else what Christians assert cannot be thought rational or true. As the first is a tyranny of the Christian, the latter is the tyranny of the secularist. The middle ground is one of dialogue and not closed presuppositionalism, of critical awareness but not capitulation, of recognition of differences but not closed-mindedness.

Biblical Basis of Christian Paradigms

The question arises concerning how the paradigms that Christians employ connect with the Bible. At the very minimum, we should adopt paradigms that cohere with explicit biblical paradigms or themes. Coherence is a necessary condition for truth. Inconsistency indicates the need for reevaluation; either one or all of the inconsistent propositions must go, or else the thinker must show that the inconsistency is only apparent.

But coherence is a minimal condition; paradigms that are *central* to Christian understandings also should be at least implicit in Scripture. Paradigms can be consistent with the Christian worldview and yet have nothing especially to do with Christianity. For example, many scientific paradigms, like Mendelian genetics or natural selection, are consistent with the Christian worldview, but since they likewise are consistent with many other non-Christian worldviews, they do not constitute the core of the Christian perspective.

One might argue further that merely being implicit in Scripture, particularly in the important areas of ethics and theology, is also often insufficient; central Christian ethical and theological paradigms should be explicit in Scripture. Being explicit need not be a necessary condition for being accepted, for there is no reason to think that Scripture contains all the paradigms relevant to evaluating and interpreting human existence. However, the importance of a paradigm for a particular worldview will be proportional to how central that paradigm is to that worldview or how directly related it is to other central themes or paradigms. If a given paradigm is tied to a worldview's central themes, it should be found in its core documents. For Christians, Scripture constitutes those core documents.[1]

We want to reiterate that not all paradigms to which Christians appeal will be implicitly or explicitly biblical. This is the case for much of our knowledge that is not distinctively or essentially Christian. We might suggest a kind of sliding scale: the closer a paradigm lies to the core of the worldview and the more it functions in a normative role governing other paradigms or applications, the more necessary it is that it be explicit in the worldview's critical, central documents. Since we want to ask how biology and ethics can be interrelated within the Christian worldview, we will employ an ethical paradigm that is connected with an essential, explicit doctrine of the Christian faith. Our paradigm is biblical in essence and origin.

One might query why we speak about paradigms rather than Scriptural moral injunctions. Why not simply look up the proper answers to moral dilemmas in the Bible? We prefer to speak about biblical paradigms rather than about specific biblical prescriptions or proscriptions because it is often difficult, if not impossible, to find explicit statements regarding many of the moral problems we face. This applies, though not uniquely, to biological ethics. For one thing, reflections about (in contrast to allusions to) biological science are difficult to find, if not wholly absent from, the Bible. Biblical authors invoke biological analogies to illustrate points about the human condition and human experience or about spiritual truths (recall the parable of the sower),[2] but they neither develop a biological taxonomy nor conduct any scientific biological study. Similarly, the Bible

---

1. This reflects the Protestant position; Catholics would want to expand this to include tradition and certain authoritative, infallible pronouncements of the church; Jews would appeal to the Old Testament and Talmud.

2. William Dyrness, for example, points to the biblical connection between obedience to God and barrenness. "Stewardship of the Earth in the Old Testament," in *Tending the Garden,* ed. Wesley Granberg-Michaelson (Grand Rapids: William B. Eerdmans, 1987), pp. 59-60.

never explicitly considers the ethics of many biological issues, such as abortion, in vitro fertilization, genetic engineering, use of pesticides, tropical deforestation, species preservation, fetal and animal research, genetic variability, or the basis for alcoholism, schizophrenia, or sexual orientation. Hence, particular answers to ethical problems relating to biology will not be forthcoming from specific passages in Scripture. For another thing, requiring specific biblical moral dicta often leads to prooftexting. The moral statements presented in Scripture are often stretched beyond their original intent or twisted entirely out of context. To require the moral judgment to be explicitly biblical introduces the danger of making the biblical authors say something they really did not say about a topic they never addressed.

In what follows we will develop a paradigm that provides ways of addressing from a Christian perspective moral issues that arise in biology. Our paradigm is explicitly biblical in that it is grounded in motifs that form the core of the biblical understanding of the world. However, since Scripture never applies this paradigm to biological issues, it is our task to make that application. At the same time, one need not be a Christian to see the force of our paradigm or the truths of its application. That is, though our paradigm has a distinctively Christian basis, the paradigm itself can be understood and applied, at least in part, without invoking that basis.

The Creator and His Stewards

Several paradigms or motifs are central to Christianity. Some of them, such as the fall, atonement, and justification, especially address the fundamental issue of reconciling us to God. The paradigm that will guide our discussion of the relation between ethics and biology is creation. It is the first theme readers encounter in the Bible, and in one form or another (for example, as re-creation) it continues throughout Scripture.

Viewing the world as created means that we cannot take the natural system as completely self-explanatory. Scientific principles and natural laws explain how physical things hold together, break apart, and in general operate, but they cannot explain the existence of the universe itself. Science starts with the universe as a given; it can legitimately neither ask nor answer the question why there is something rather than nothing. Neither can science explain why particular laws or principles rather than others govern the universe. Why, for example, does the universe operate according to the strong and weak nuclear forces rather than according to some other force? The fact that science cannot give an ultimate explanation for the existence of the universe does not count against science; such is not within the purview of scientific methods.

Theists, on the other hand, can provide an answer to the question of why the universe exists and operates according to the principles that it does — that is, by grounding the fundamental principles of nature in the conscious, creative act of God. It is true that theists, too, face limits, for they cannot determine why God chose *these* principles and laws for his creation rather than others. Neither can this claim be tested in experience. Yet despite these limitations, the creationist paradigm adds a meaningful dimension to our understanding of the origin of the universe by locating it in the purposive activity of God.

A second result of viewing the world as created comes from viewing creation, not as a once-for-all event, but as an ongoing process relating God to the world. Larry Rasmussen notes that "strictly speaking, 'creation' is our term, and not in the Old Testament sense. The verb form, 'creating,' is common in the Scriptures. But a noun form used to refer to some vast entity is absent, which only underscores the sense of the Creator's ongoing creating and sustaining — indeed, underscores the unfinished character of the world."[3] Within the Christian tradition, the doctrine — that by his continuous creating God upholds or sustains the universe — means different things to different people. To some it means that, were God not to exist for a moment or were God to withdraw his sustaining activity, the world would simply cease to exist at that point. The world is radically contingent upon God. Others interpret God's sustenance of creation to refer to the fact that there are certain events in the universe that, because they are so extremely unlikely, beg for some explanation beyond (though not excluding) the natural. It is not that one cannot provide a natural explanation (though maybe one cannot provide a completely adequate one); rather, the unlikelihood of the event requires an additional causal or explanatory element — namely, the active influence of a divine being.

Two points at which this is seen are in the creation or emergence of life from the nonliving and in the creation or emergence of intelligent

3. Larry Rasmussen, "Creation, Church, and Christian Responsibility," in *Tending the Garden,* ed. Granberg-Michaelson, p. 116.

consciousness. Regarding the emergence of life, there are aspects that natural selection does not so easily explain. For example, natural selection has difficulty doing more than describing how living things came to be from the nonliving. If one analyzes the elemental chemical-physical components of the universe, one can see that, though life is possible, it is extraordinarily unlikely. Natural selection cannot explain why these components should give rise to *life* rather than to the nonliving. There are many possibilities. What makes these components move toward life-engendering combinations?

> Envisaging how a living cell could have gradually evolved through a sequence of simple protocells seems to pose almost insuperable problems. If [our] estimates are anywhere near the truth then this would undoubtedly mean that the alternative scenario — the possibility of life arising suddenly on earth by chance — is infinitely small. To get a cell by chance would require at least one hundred functional proteins to appear simultaneously in one place. That is one hundred simultaneous events each of an independent probability which could hardly be more than 10^{-20} giving a maximum combined probability of 10^{-2000}.[4]

Regarding the emergence of consciousness, natural selection leaves the processes of the universe directionless. It says only that the fit survive, but it places no requirement or direction on the fit. This means that though natural selection can account for individual cases of organic adaptation, it cannot account for the fact that species do not merely sustain their existence but show evidence of development toward higher and more complex organisms, culminating in rational beings. For some, the doctrine of creation claims that God lies behind this developmental process, directing it to the production of more complex beings who can interact consciously with himself and others.

Our interest in the creation paradigm extends beyond and differs from (though does not exclude) both of the above. To develop how the paradigm works for a biological ethic, let us begin with the Genesis account of creation. Throughout Genesis the author(s) uses the cultural motifs, conceptual schemas, language, and literary genres available at his time and place to portray God's relation to his creation. The author pictures God as a great monarch ruling over his vast realms of creation. For example,

4. Michael Denton, *Evolution: A Theory in Crisis* (Bethesda, MD: Adler and Adler, 1986), p. 323.

Genesis 11 records that Yahweh, apparently having received disturbing reports, comes to view for himself the construction and hubris of those living in Shinar. The author paints a similar scene in Genesis 18, where Yahweh walks incognito through his lands to survey for himself the almost unbelievable moral conditions reported to his courts. In Genesis 28, at a place he names Bethel, Jacob dreams that he sees a stairway stretching between the heavenly court and earth, where he sleeps with a stone for a pillow, with messengers of the divine monarch going back and forth between heaven and earth, apparently reporting to Yahweh, who stands at the summit of the stairway.[5]

Elsewhere in the Old Testament God's kingship extends to everything, from the people of Israel to all the nations, to the heights and depths of creation. John Stek notes that "within the conceptual system of the Old Testament the most common metaphor for the created realm is that of the kingdom of God."[6] He goes on to note that "God is said to be or is depicted as King over all the earth." Psalm 29 sees God enthroned over all creation, powerful of voice to have nature do his bidding. In Psalm 95 the Lord is hailed as king above all gods. All nature, from the depths of the earth and the sea to the tops of the mountains, is his. So, too, are the people of the earth. The next psalm, Psalm 96, echoes the theme. In his courts God is to be praised by the earth and sea, the fields and their inhabitants, and the trees of the forest. His power extends to the forces of nature. "Clouds and thick darkness surround him. . . . Fire goes before him. . . . His lightning lights up the world; the earth sees and trembles. The mountains melt like wax before the LORD . . . of all the earth" (Ps. 97:2-5). In Psalm 47 he ascends to his heavenly court, accompanied by hymns of praises. From there he reigns over all the nations of the earth. The creation is his kingdom, his dominion (Ps. 103:22). His kingdom,

---

5. God's kingship is emphasized throughout the biblical account. Israel's demands to have a king were taken to be a rejection of God as their king. God says to Samuel, "Listen to all that the people are saying to you; it is not you they have rejected as their king, but me" (1 Sam. 8:7). The theme that God is king forever, ruling over all other kings, runs through the Psalms. As the almighty sovereign, he is worthy of praise and the object of petitions. For the prophets, God is the king who delivers (Isa. 33:22; 42:15-16). In the Gospels God's kingship is more veiled, perhaps for political reasons. It comes through the parables (e.g., Matt. 22:1-14), but more so through Jesus' emphasis on the kingdom of God.

6. John H. Stek, "What Says the Scripture?" in Howard J. Van Till et al., *Portraits of Creation* (Grand Rapids: William B. Eerdmans, 1990), p. 250.

which is everlasting, tells about his glory and might (Ps. 145:11-13). The entire creation belongs to him as his kingdom; over it he rules in majesty and splendor (1 Chron. 29:11).

As the earthly monarch rules by decree, so does God. By his decree he established the heavens above, the sun, moon, and stars, the waters in the heavens and in the deep oceans. "Lightning and hail, snow and clouds, stormy winds . . . do his bidding" (Ps. 148:1-8). They are appointed to their places and function by the Lord Almighty (Jer. 31:35; Job 28:26).

It is very likely that this same picture of a most powerful monarch stands behind the two Genesis creation stories. In the first account (Gen. 1) creation is an ordered series of acts by which the territory — the heavens, the earth, its land and water, its vegetation, and its animal inhabitants — is brought into being by God's royal decree and established with its proper function in God's domains. God decrees that there be light and there is light. God separates light from darkness as day from night. God decrees that there be a sky to separate the heavenly waters from the earthly waters, and it is so. God decrees that the oceans and lakes be gathered together so that land might emerge. God commands the land to produce vegetation and to propagate, and the fertile earth produces. God calls into being the celestial bodies to order the heavens, and it happens as he commands. He sets the stars, sun, and moon where he wants them to function. God speaks so that the land, water, and air teem with birds and sea creatures. He speaks again, and the land produces living creatures.

Everything follows from his royal decrees and commands; he announces, and it is accomplished. His word is efficacious, even down to the highlight of the creation, where he proposes to make human beings in his image and does so.

> Before proceeding to the creation of humankind, God announces his momentous intention to do so. Both in itself and in the language employed, this announcement recalls the scene in a royal council chamber in which a king announces his impending action to the members of his court. . . . Moreover, God's assignment of specific provisions for humanity (vv. 29-30) recalls the royal assignment of food at the king's table (cf. Gen. 43:34; 47:22; 2 Sam. 9:7, 13; 19:28 . . .).[7]

Though the fashioned kingdom stands over against the creating Monarch, it is his: its origin lies in his royal commands. Thus by his

7. Stek, p. 233.

creative ordering God stakes his claim to the entire universe. It becomes his own kingdom, over which he rules. That it is his is shown by his naming of what he creates. He calls the light "day" and the darkness "night," the expanse "sky," the dry ground "land," and the bodies of water "seas." In the author's culture, to name something shows either that one possesses it or that one has power over it. To name it is to make it yours; in these stories it is a royal appropriation.

The creation account is thus the narrative of the establishment of a great kingdom by no less than the king of the earth. It is neither historiography nor a scientific account, though nonetheless significant and informative.[8] It is a theological assertion about the origin and true ownership of whatever is. Since everything was created by God, it is his, from light and darkness to humankind. All fits into his dominion and purposes, and he sees that it is good.

But who will administer God's vast domain, his land, his house? Like other people with wealth, monarchs entrusted their households and property to their stewards,[9] so that they could attend to other important matters. Likewise the Creator, in the second creation story (Gen. 2), entrusts his lands to stewards. These stewards are special, for not only will they administer the kingdom, but unlike the rest of creation, they will bear the very image of the Monarch. As oriental kings placed statues of themselves throughout their territory, signifying their claim to that land, God placed those created in his image in the land to represent his interests.[10]

Much debate surrounds discussion of the content of the *imago Dei* (image of God). Diverse interpretations abound.[11] Fortunately here we do not need to determine its precise content. What is of interest is one of its *functions*. Being created in God's image endows people with the authority

---

8. Stek, pp. 229-30, 241-42.

9. They were a common part of the Genesis storyteller's world. As assistant ruler of Egypt, Joseph had his steward (Gen. 44:1), just as he had been one himself to Potiphar (39:4). Even a nomad like Abraham had one (24:2).

10. "Man is placed upon earth in God's image as God's sovereign emblem. He is really only God's representative, summoned to maintain and enforce God's claim to dominion over the earth." Gerhard von Rad, *Genesis* (Philadelphia: Westminster Press, 1972), p. 60.

11. For a brief discussion, see V. Elving Anderson and Bruce R. Reichenbach, "Imaged through the Lens Darkly: Human Personhood and the Sciences," *Journal of the Evangelical Theological Society* 33, no. 2 (June 1990): 197-213.

to act in behalf of the true owner of the world. Humans exercise dominion over the earth as a reflection of God's ultimate lordship.[12]

But what acts are humans to perform on God's behalf? How are they to exercise stewardship? The creation story in the first chapter of Genesis speaks of representatives who are both blessed and commanded to be fruitful and fill the earth, who are to subdue and rule over it (Gen. 1:26-28). Some have seen only a blessing in this verse, for it seems to parallel the blessing in 1:22. "This word addressed to humankind is not a directive or commission. . . . It is a benediction."[13] But Genesis 1:28 contains both a blessing and a command, for as Stek himself notes, humans were created "*so that* they may rule." The statement establishes, on the one hand, that procreation is a blessing, a theme consistently repeated throughout Scripture, not least of all in the climax of Jesus' birth (Luke 1:42). At the same time, however, the statement goes beyond sanctifying fruitfulness to stating an injunction. Whereas animals have no choice in their blessed reproduction, humans do, and hence their conduct is regulated by the command. "The blessing imposed on human fertility is after the model of the one pronounced over the animals before, but there is also a more than subtle difference that disappears in the translation. . . . What he had merely commanded of the beasts becomes a form of conversation when addressed to man."[14]

In the story in Genesis 2, God appoints his stewards to work in and take care of God's special garden, the pleasant place of God's walking (2:15; cf. 3:8). Consequently, in the two creation accounts we have three commands in which God assigns his stewards three administrative functions.

(1) One of these functions is to be fruitful and increase the population to *fill* the land. In the first Genesis creation story, the creation of human beings in God's image is followed by his decree: "Be fruitful and increase in number; fill the earth and subdue it" (1:28). In the second account, though the earth is fruitful, it is not full; though it is pregnant with all kinds of things,

---

12. "Thus the *form* of land management in Israel may be 'like the others,' even with a king and the apparatus commonly associated with that institution, but the *intent* is to be peculiar, as peculiar as Israel and its relationship to the land are peculiar. Other kings inclined to control the land as a possession. This king is to manage the land as a gift entrusted to him but never possessed by him." Walter Brueggemann, *The Land* (Philadelphia: Fortress Press, 1977), p. 75.

13. Stek, p. 251.

14. Bruce Vawter, *On Genesis: A New Reading* (New York: Doubleday and Company, 1977), p. 60.

it is empty of humans. The original man is pictured as alone in the vast garden of Eden, surrounded by creatures he can name and hence dominate, but with whom uniquely human, meaningful relations cannot be established (2:18-20). To alleviate his loneliness, a partner is created out of and for him. Together they are to fulfill the injunction to fill the earth.

Today, when the earth abounds with humans, we have a hard time comprehending any contemporary relevancy of this assigned function. One feels a similar historical ethereality in reading this injunction as when one reads the ancient Chinese philosopher Mo Tzu bemoaning the decrease in China's population brought on by late marriage. This appears to be one administrative function at which God's stewards have succeeded — perhaps too well. The human population doubled to 5 million between 1950 and 1987, and is projected to double again before the end of the twenty-first century.

Increased population puts increased stress on the earth and its resources. For one thing, it provokes more consumption. More water is needed for human use; more land is turned over for agricultural use to feed additional people; more agricultural land is taken over for housing; more lumber is grown and harvested; more fossil fuel is extracted. For another, more pollutants are thrust into the biosphere. Wastes are dumped into rivers, oceans, and landfills; more carbons and nitrous oxides are sent skyward.

Faced with all kinds of ecological disasters projected for the twenty-first century if we proceed apace in populating the earth, we might simply abandon this injunction to fill and concentrate on the other two. Indeed, we might even consider reversing the assigned role to fill by adopting significant steps to curb human population growth to save our environment for future generations.

But there is another way of approaching this injunction. We can interpret it to mean that we now need to pay attention more to the qualitative aspect of our filling the earth than to the quantitative. The earth does not need more humans; that seems clear. But perhaps it needs better humans, humans more disease-resistant, genetically superior, more intelligent, sympathetic, moral, and spiritual, better adjusted to and able to cope with their environment. With our rapidly increasing knowledge about the human microsphere and our developing technology, we stand in a position to improve our progeny. Already we can diagnose and treat diseases before birth and perform fetal surgery. Through an analysis of a couple's genetic load, we can predict the probability that their children will inherit certain genetically determined characteristics and use this information in genetic counseling. We possess the knowledge and ability to determine the genetic structure of embryos in vitro, so that the physician can implant in the uterus only those free of certain genetic defects that would result in painful diseases or life-threatening deformities. If we develop the capacity to perform germ line genetic intervention, we might be able genetically to tailor future generations to certain broad specifications.

A qualitative interpretation of the injunction appears to give us the permission — and perhaps more strongly, since it is a command, the obligation — to change the creation for the better. In the past we have focused on changing the environment for human betterment. Now we have enormous powers to begin to redesign the kinds of human beings we want on earth. But *ought* we to do what we *can* do? Is it moral to change ourselves? What restrictions, if any, are to be placed on the stewards of the earth? As we shall see in the next section, these very same issues arise with other aspects of stewardship.

(2) A second function assigned to the newly created stewards is to *rule* the land as the representatives of the Ultimate Ruler. "Rule over the fish of the sea and the birds of the air and over every living creature that moves on the ground" (Gen. 1:28). The Psalmist writes,

> You made [humankind] ruler over the works of your hands;
>> you put everything under his feet:
> all flocks and herds,
>> and the beasts of the field,
> the birds of the air,
>> and the fish of the sea,
> all that swim the paths of the seas.

(Ps. 8:6-8)

Genesis portrays this dominion over the earth and its inhabitants in several ways. When in chapter one the man and woman are told to subdue and rule the earth, the terms used for "subdue" and "rule" *(kabash* and *radah)* are powerful, dominance terms. "*Kabash* is drawn from a Hebrew word meaning to tread down or bring into bondage, and conveys the image of a conqueror placing his foot on the neck of the conquered. . . . The other verb, *radah,* comes from a word meaning to trample or to prevail against and conveys the image of one treading grapes in a winepress."[15]

15. Loren Wilkinson, ed., *Earthkeeping: Christian Stewardship of Natural Resources* (Grand Rapids: William B. Eerdmans, 1980), p. 209.

Thus the command to subdue "literally . . . implies trampling under one's feet, and it connotes absolute subjugation (cf. Jeremiah 34:11, 16; Zechar- iah 9:15; Nehemiah 5:5; 2 Chronicles 28:10)."[16] As the oriental monarch exercised absolute hegemony over his kingdom, so the stewards are en- dowed by God with similar power.

The writer of Genesis also depicts human control over nature by assigning humans the power to name the animals (2:19-20). As we noted above, to name something was to show either ownership or power over something. To name it was to make it yours. So Adam's act of naming all the living creatures shows that he has taken charge over them.

One should be careful about the true source of this power: it is delegated, not self-derived. The real Owner brings the animals before the namer. Yet God gives him a real charge of stewardship, not a nominal one, for the real Owner of the animals accepts the names the steward assigns.

If the charge to humans is to rule, to gain control over, the ethical question arises how humans are to rule. Supposing that God created them for his purposes, what is to be done with, by, and for the ruled depends on those very purposes. What obligations does the steward have with respect to God's dominions? In the third function assigned to the steward the Genesis author gives us the guidance necessary to begin to construct the requisite ethic of ruling.

(3) The third function is to work with and to care for the property of the Owner. The steward has a *tending* function. "The Lord God took the man and put him in the Garden of Eden to work it and take care of it" (Gen. 2:15). In the word *abad* or "work," sometimes translated "till," one can easily see here an agricultural motif, appropriate to the garden assignment. Yet the story makes no such clear identification of the way we should tend the earth on behalf of the Landlord. True agricultural elements do not appear until the next Genesis stories. Thus there is an openness to the content of this caring, best captured here by translating *abad* as "working" or "serving."

The latter is emphasized by the parallel *shamar* or "keeping." The command here is to watch and preserve, to guard and protect.[17] Caring, hence, is to be understood in the sense of working on behalf of (serving), protecting, and benefiting. This becomes even more poignant if one sees

---

16. Vawter, p. 60.

17. The same word is used by Cain when he questions whether he is his brother's keeper (Gen. 4:9), a task on which he has obviously defaulted.

that creation is an *oikos,* a vast house that holds a public household *(oikia).*[18] Hence we have the responsibility to care for creation in the same way we are obligated to care for our own house and household.

But, one might ask, is there not a sense in which the Landlord himself does not really seem to care for his creation? After all, the Creator is pictured in another Genesis story as being willing to destroy the whole works because his stewards were morally corrupt. All of living creation, God threatens, must go (Gen. 6:7). Yet one must read the entire story. For one thing, all are not destroyed in the catastrophic flood. The hero of the story, Noah, is told to preserve all the kinds of animals for a new start. The floating zoo is full. But even more significantly, at the end of the story the narrator relates a change of attitude in the Monarch. He will not take such radical steps again, no matter how bad his stewards turn out (8:20-22).

One would like to know why God experiences this change of heart. Did he see that mass destruction accomplished very little? It is not long — the end of the same story — before humans are back to immoral business as usual (Gen. 9:20-23). Was it that he was impressed with the sacrifice, with the realization that there would always be a remnant to worship him? The narrator seems both to support this — by placing God's change of mind in the context of Noah's sacrifice — and to reject it — by his claim that God would no longer take such actions, even if there were none to render sacrifice. Perhaps God's regrets have to do with the fact that God really cares about his creation. Yet even this suggestion is puzzling, for almost in the next breath the storyteller relates that God now allows humans to be more than vegetarians, to bring the death they indirectly have brought into the world by the flood directly upon the animal world they rule (9:2-3).

So in what sense does God care about his complete creation? It is true that there is no explicit ecological ethic in Scripture. No one rues the cutting down of Lebanon's cedars, the disappearance of endangered species, or the polluting of Israel's streams. Yet there is little doubt that God has an interest, if not a stake, in his creation. The heavens and earth created by God announce his glory (Ps. 19). The human creatures who are a little lower than the heavenly beings are crowned with glory and honor and charged not only to tell of God's wonders (Ps. 9:1) but also to care for that which tells of God's wonders (Ps. 8).

18. Rasmussen, p. 116.

Jesus' analogy of the sparrows and hairs of our head is telling at this point (Luke 12:6-7). Though of comparatively little worth to humans, the sparrows sold in the market for a song are not forgotten by God. Similarly, though we oftentimes appear to be of little value, yet God has numbered the very hairs of our head. The comparison is at least one of caring, for what is numbered and not forgotten is worth something to the rememberer.

The ruling, then, ought to be done with the spirit of caring for what is ruled. Although a model of the absolute monarch underlies the stories, the type of monarchy appealed to is not one in which only the good of the monarch is in view. "These injunctions are in complete contrast to the concepts of the ancient Near East. Here is an organic rather than a strictly monarchical view of kingship and ruling, one that is further illumined in the life and ministry of Christ, who came not to be served but to serve (Mark 10:45)."[19] There is no justification of rape and pillage of the earth. No selfish abuse is sanctioned. The good not only of the Landlord but also of his lands and creation, his *oikos,* is envisioned. Stewards are servants who, in being commanded to serve *(abad),* benefit that over which they rule. Good kings are to bring prosperity to their people (Ps. 72); good shepherds are to take care of, protect, strengthen, bind up, and recover the sheep (Ezek. 34). "Stewardship is . . . dominion as service."[20]

Filling, subduing, and caring for — these three stewardly functions are of immediate interest to us. In the time of the Old Testament patriarchs, though the acts of carrying out the directive to rule over and take care of this creation were difficult to implement, the ethical issues raised were less profound. For example, the degree to which the relatively few humans could have an impact on the environment was minimal indeed. The environment and its denizens were often as much a threat to humans as vice versa. Frequently the environment exacted its toll on humans in ways that humans could scarcely match in return.

But things differ today. Technology has dramatically increased our power to fill and rule the world. Indeed, some wonder whether we now have the power to escape our bondservant status, to take charge for ourselves over the creation. With our mobility, we can transplant plants and animals from one environment to another, where they have no natural

19. Dyrness, p. 53.
20. Wilkinson, p. 224.

predators to control them. With our machines we can denude the earth of its trees, reshape the landscape, rechannel and tame the rivers,[21] and reclaim land from the sea. With our chemicals we can slice holes in the earth's atmosphere, make our waters undrinkable and our lands uninhabitable, and destroy the very animals Adam was busy naming. And now, with our knowledge of the microsphere, we are busy not only understanding but reshaping organisms themselves. We will soon understand where the genes are located and what functions they have, to the point of using that knowledge to reshape God's creation.

This has been called "playing God."[22] As currently employed, the term has a decidedly negative connotation. To play God is to appropriate to ourselves functions and tasks that properly belong to God, to change what should not be changed. But in light of the functions God assigns to his created stewards, playing God is one of our tasks. This has been part of our assignment from the beginning, so that, contrary to popular opinion, the phrase carries less of a stigma than it seems to at first. By playing God we are not pretending to be God, but rather acting as his representatives on earth. Made like God, we stand in his place, to help fill it, to occupy his territories by subduing and ruling what he has made, and ultimately to care for it.

Perhaps the tricky word in this cliche is *play.* Stewards are not at play in representing our Landlord. It is our life, our business. We are not simulating a divine role; we are carrying out the divine mandate. Stewardship is not leisure, not recreation, but an employment. We are to work, however, not for ourselves but for our Landlord.

Here arises a most significant element of stewardship. Humans in their role as stewards are *accountable* to the Lord for their stewardship. This is the human uniqueness in creation: among all the physical creation only humans are morally accountable for their actions. Some have taken this to be the very essence of the *imago Dei.* God has given us the business of stewarding for him, and he will call us to account for what we have done for him, for ourselves, and for that over which we rule. Stewardship is serious business indeed.

So we are stewards for the Owner of the world, to fill, rule, and care

21. Wesley Granberg-Michaelson notes the use of the military motif of war, applied to nature, in an interesting example involving rechanneling the Mississippi River. *Ecology and Life* (Waco, TX: Word Books, 1988), chap. 3.

22. For example, June Goodfield, *Playing God: Genetic Engineering and the Manipulation of Life* (New York: Random House, 1977).

for creation. We are stand-ins for God. Given this paradigm, the relevant ethical considerations now become clearer. As stewards we are to *fill* the world; we direct our ethical attention to our obligations to change the world in various ways for the better. This involves concern not only for quantity, but also (and especially) for the quality of what we produce or bring about. As stewards we are to *rule over and subdue;* we are given dominion. Ethical attention is directed to the extent of our power and to the relations that should hold between ruler and subjects, between stewards and what we are stewards over. As stewards we are to *care for* the earth; we pay ethical attention to the ways in which we use our powers over nature. We have obligations to God and to the persons and things over whom we are stewards, not only to profit the Landlord but also to benefit and do justice to other stewards and the creation itself. The Christian ethicist will be concerned with specifying the ethical obligations that arise as the paradigm is applied to particular ethical issues.

Tensions in Stewardship

For now, however, we want to focus on perhaps one of the most significant paradoxes of being a steward. Stewards are charged with seemingly contradictory obligations; they are charged to promote the good of the owner through both conservation and change. They are to preserve what the owner already possesses; what is valuable and essential is not to be lost. At the same time stewards are to profit the owner through risking the estate, by making changes in what is entrusted to their care. How can they do both — preserve and change, conserve and risk?

Interestingly enough, there is biblical precedent for this very contrast. In one of Jesus' parables (Matt. 25:14-30), a householder, about to go on a journey, calls his servants and entrusts certain of his capital to them. As stewards, they are faced with the problem of what to do with the property assigned to their care. And here the tension enters. One servant takes seriously the steward's preservative role. He knows that his master is a hard man, and, not willing to risk what was entrusted to him, he hides it in the ground, so that when the master returns nothing that he was given charge over will be lost. The steward expects praise for his preservation of the owner's resources.

But to his shock and dismay, the mere preserver receives condemnation, not praise. Why has he not invested the capital, expanded the owner's possessions, and enriched his master? He has failed as a steward and is immediately relieved of his position. What the owner had entrusted to him is turned over to one of the two stewards who achieved a profitable return on their investments of the owner's property. One point of the story, then, is that stewards who never risk the trust but merely preserve fail in their office. Stewards are obligated to both preserve and profit (change).

We might apply this point to our concern with bioethics. If we are God's stewards of the biosphere, we likewise must both preserve and change. But this raises five important and difficult questions.

(1) As stewards engaged in the task of improving the prospects of the Landlord, *what are we obligated to change?* Though the universe is ordered, the order is not always beneficial either to humans or to their counterparts in nature. The environment that makes agriculture possible can also threaten its bounty. The earth-watering rain floods; the life-giving sun bakes the crops or in excess creates conditions ripe for the multiplication of devouring insects. Viruses threaten the quality and length of human, animal, and vegetative life. Though we are not faced with the unmitigated disasters pictured by Voltaire in *Candide,* neither do we have Leibniz's best of all possible worlds. Happiness and pain, the beneficial and the discordant, mix.

It would seem, then, that change appropriately ought to begin with what negatively affects human existence. We ought to search for preventions of and cures for ravaging diseases. We ought to explore ways to prevent or mitigate dysfunctions that threaten newborns. The environment should be tamed and transformed to be more inhabitable and hospitable: rivers should be dammed to prevent life-threatening floods and to generate electricity; mountains should be moved to facilitate transportation; lakes should be created to provide adequate water resources; forests should be harvested to furnish lumber for human habitations and daily use. The Green Revolution is legitimate because it prevents malnutrition and starvation. It would seem, then, that we are obligated to change that which either does not benefit us or has a potential to harm us, in order to make our planet a better place for us to live.

But why should we change the environment to benefit *humans?* Is this not a kind of hubris, contradicting the basic intent of the original model according to which the steward was to profit the Landowner and not himself? In the Genesis story, *God* sees that what he creates is good. The good is for God as the valuing Landowner. But now we speak of changing the creation to benefit human beings, the stewards. Have we not unwarrantedly switched the valuer?

Not really, for both creation stories indicate that God meant creation to be not only for himself but for humans as well. In the first story, the creation of humans caps the long creative process before God's rest. In God's final creative act, he forms something to bear his image and to rule over the rest of creation. True, the humans are made, as are the other creatures, from the dust of the earth. In this respect, we are like the other animals; we come from the earth and return to it. Yet we experience a transcendent aspect as well, for only of us does the author say that we are made in God's image and likeness. In the second story, God specially breathes into the man his divine breath. He places him in the fecund park, gives him the fruit of the trees for food, and even marshals the animals for him to name. It is not hubris, then, to hold that change should benefit humans as well as the Landlord. The precedent is found in our paradigm.

But is it always better to change the environment for human benefit? Would it not be better at times to change humans, where possible, to fit the environment rather than vice versa? It is often the case that when humans are benefited by changing the environment, more harm than good results overall. We so alter the environment that it cannot recover: gone are the white and red pine forests of the northern Midwest and their ecology, turned into houses and railroad ties; gone is the prairie and its ecology, converted into endless corn and wheat fields; the hills of Pennsylvania and West Virginia lie buried under slag heaps, mined for their energy-producing coal; we have polluted the pristine lakes of New York and Scandinavia with the mercury and acids of our industries; lost are an unknown number of species of plants, animals, and other organisms. Would it not be more effective, if not more right, to begin to exercise our new power, where appropriate, to alter humans themselves to adapt to their environment, rather than always vice versa? That is, does not the obligation to change extend beyond changing the environment to include altering human beings themselves to adapt better to their environment?

In the past, change was directed almost exclusively to the environment; alterations of humans were minor. Yet it would appear that we should alter ourselves as well. But in what ways? Perhaps the most obvious way has to do with our life-styles. This injunction can be directed to a variety of aspects.[23] *(a)* It is easy to confuse wants with needs. Yet a life of simplicity, where the basic needs are met but many desires are creatively

23. One can find a number of concrete, helpful suggestions in Wilkinson, pp. 261-62.

rechanneled, can be both satisfying and ecologically sound.[24] In place of our emphasis on individual ownership, we can begin to share our resources with our neighbors, so that not so many possessions are required. *(b)* We can alter our current pattern of energy consumption to move away from those resources that are most polluting and destructive to extract and ultimately to use less energy and create energy in more environmentally friendly ways (such as through solar or wind power). To employ more environmentally safe energy practices will certainly cost more initially, but this cost is part of what is involved in changing ourselves. *(c)* We can alter our habits and behavior — in how we drive or plan our trips, what and how much we buy, where we live, and what kind of living space we have. For example, what do we eat? Can we substitute more ecologically sound items in our diet? By eating more grains and vegetables we can reduce the need for cattle, which are hard on their environment. The recent emphasis on recycling illustrates the changes we can make to our ways of doing things. We can alter the way we use materials, what materials we use, and what we do with them after we are done.

(2) The reader may think that it is enough to hold that changing the environment and human persons is *permissible;* do we want to argue that it goes beyond permissibility to being *obligatory?* Ethics makes a distinction between what is permitted and what is obligatory. For example, a thoughtful defender of capital punishment would maintain that capital punishment ought to be permitted without holding that it is mandatory. Or again, risking my life to attempt to save someone from drowning is permitted but might not be obligatory, especially if I possess marginal swimming skills. This distinction is embodied in discussions of the *lex talionis:* Does the injunction of an eye for an eye state what ought to be exacted, or does it merely delineate the maximum one is permitted to exact?

Which of these — permission or obligation — applies to our acts of changing our environment and ourselves? If one takes the injunctions of Genesis and Jesus' parable seriously, it would seem that we are not just permitted to change things for the better; this is our obligation. The steward in Jesus' parable was not merely permitted to invest his trust; he was obligated to do so, and failing to do this, he was roundly condemned. As stewards, investing in change is our mission, our business, something

---

24. Ron Sider, *Rich Christians in an Age of Hunger* (Downers Grove, IL: InterVarsity, 1977).

we should be about. There is a moral injunction on us to be about the business of improving our world.

But what about changing ourselves? Above we spoke about changing our desires, actions, and habits. It is appropriate likewise to stress the obligatory nature of these changes. Our normal reaction to the obligation to improve is to change something else to fit our own needs, wants, and behavioral patterns. This, it is argued, is less risky, at least to us, not to mention being often cheaper and easier. But this normal reaction often runs counter to our stewardship ethic, for we have obligations not only to benefit the Landowner but also to benefit that over which we rule. And to change what we rule to benefit ourselves might violate this trust. Thus we reject the easy, nonsacrificial route of always changing what is around us instead of directing attention to changing ourselves, our habits, and our behavior.

(3) This raises a third question, namely, *for what purpose are we to change creation?* We have the injunction to rule over the trust for the profit of the Landlord. But how will the Landlord gain from our investments? What is God's profit?

This is a difficult question, for the classical approach to creation has been to argue that God is perfect, lacking nothing. He created the world out of his plenitude, not out of any need. Had he created it from need, it would mean that he lacked something, which would be inconsistent with his absolute perfection. This view originated as early as the third century with the philosopher Plotinus, who was himself inspired by the Greek philosopher Plato and was an intellectual mentor to the Christian theologian Augustine. Plotinus held that the world comes into being by radiating or emanating from the One. The emanations do not arise from the will of the One but out of necessity. They derive from the perfection of the One, not out of any need or lack that the One has. The superior has no need of the inferior, though the inferior needs the superior.

Two of these three Neoplatonic ideas are rejected in classical Christianity. First, the world does not emanate from God but is created by him *ex nihilo* (out of nothing), apart from himself. Second, God creates freely; his act of creating is not necessary, though his act of creating must be consistent with his character and essential nature. The only act that is necessary to God is that he seek to promote his own good. The end of creation, then, is the goodness of God. But the third idea — that God creates not out of need but out of fullness — is preserved. Lacking nothing, God does not create to acquire something. "He intends only to communicate His perfection, which is His goodness. . . . Therefore divine goodness is the end of all things"[25] — not in the sense that things add to God's goodness, but in that the good they achieve emulates God's goodness.

On this interpretation, God's relation to the creation is unilateral, not reciprocal. He contributes to it; indeed, it would not exist without him and his continuous, sustaining power. But since nothing can add to his infinity, all that takes place ultimately has no effect on him; it neither fills a lack in him nor adds anything to him. But if creation makes no difference to God and his existence, one has to wonder what the significance of the creative act was from God's perspective. Perhaps the Hindu notion, expressed by Sankara and Ramanuja, that God created out of no motive but only for sport, is closer to the truth, given these parameters.[26]

Because of this difficulty, among others, some contemporary theologians go on to reject the third Neoplatonic characteristic as well. The relation between God and his creation is reciprocal; each contributes in significant ways to the other. God's own self-realization is found, in part, in the realization of his creation, just as the realization of the creation is found in achieving the good for it initiated by God. Our response to God has significance for his life, just as his response to us has significance for ours.

Returning to our original question, we ask, what profit ought we to bring to God? On the classical view, God as the Landlord gets from his creation nothing that he needs; nothing can contribute to him and his existence. He receives praise or glory, but the praise, as it were, only adds to an already overflowing cup. One cannot add meaningfully to the infinite. God does not benefit from changing the creation.

On our revised view, however, the Landlord gains a great deal from his creation. The creation contributes to God's ongoing life. His good is achieved, in part, by and through the goods realized in the universe. Some of these goods will be moral and spiritual goods, realized through the right actions, motives, and virtues of his created stewards. Others will be goods of self-realization, where the creation and the stewards realize the potentialities that fulfill them. They will also include aesthetic goods — goods of order, harmony, and beauty. In sum, we should change what detracts from meaningful existence and overall goodness and beauty, as well as that

25. Thomas Aquinas, *Summa Theologica* I, Q44, a4.
26. Bruce R. Reichenbach, *The Law of Karma* (London: Macmillan, 1990), pp. 74-75.

which has been affected by our sinful acts, for the good that we achieve contributes both to God's good and to ours.[27]

In a backhanded way, the obligation to change the world for the good not only legitimizes both science and technology; it also saddles their practitioners with moral obligations, for in knowing and changing the world they are faced with the obligation to realize the good. Their endeavors have, as their end, the profit of the Landlord, his stewards, and the earth that they care for. Likewise they must have concerns about the appropriate way in which to realize these goods (the means). Put another way, the steward paradigm invokes both teleological and deontological concerns.

Science and technology, according to this model, are not value neutral. They must address concerns about what is worth knowing in light of human needs, about what is worth doing in light of potential human good, and about what scientists should be doing in light of the development of their own moral character. In all of this they answer not merely to themselves but to the God who charged them to care for the world. If we are God's stewards, we are not our own, but have obligations to bring about God's good. Scientists, then, are held to the same accounting as all other stewards, while at the same time, because of their unique links with the earth, they have a special accounting in terms of what they have done with their knowledge and technical abilities.

(4) If we are to change, *what are the limits of the change?* Are we permitted to change absolutely everything? Or are there things that must be preserved, left unchanged? To answer this let us return to our biblical paradigm. In the first Genesis creation story, we are told six times that God views his creation and sees that it is good. One way of understanding goodness in this story is to connect it with human happiness and well-being. In God's creation there was no evil, either moral or natural. Moral evil, in the form of sin against God, is introduced in a subsequent story, where humans, seeking to be like God, violate the divine command. Natural evil enters as a punishment for human disobedience. The disobe-

---

27. The obligation to change what we have destroyed by our sin is consistent with the redemptive motif that runs through Scripture. The apostle Paul speaks about creation, along with us, groaning, waiting for redemption (Rom. 8:22-23). Together we desire God's redemptive and renewing acts. We are not contending that humans merely by themselves are able to provide creation's needed redemption. However, in our stewardly role we can be representatives of God as he works out his redemptive plan for creation.

dient now suffer pain in childbearing, difficulties in farming recalcitrant land, and mortality. Goodness, according to this view, was there in the beginning; it is a reflection of the goodness of God.[28] Not even with the creation of finite creatures was evil introduced, only its possibility. Only when humans act in defiance of God does evil actually begin.

Another, more likely interpretation of the goodness in the story is that the creation, seen as good, is capable of doing what God intended it to do. The universe is ordered and functional. The light is good because it functions to separate day and night. The separation of the waters from the land is good because it allows the development of vegetation on land and creatures both in sea·and on land. All in all, there is a fit harmony of means and ends, of order and purpose. Everything has value, then, because it has a place in and contributes to the whole.

What would happen, for instance, if all individuals were capable of interbreeding, if there were no protected gene pools to constitute species? Put another way, what if there were no species, only individuals? Then an organism that had successfully adapted to its environment could breed with an organism suited to another environment, producing offspring suited to neither. The superior genotype for that particular environmental niche would be lost, as it combined with an entirely different one. Under this scenario there could be no environmental stability for the organisms.[29]

Because it is ordered, nature is also intelligible. If it were an indeterminate chaos, it would be impenetrable to human reason. The concepts through which we know the world are universals covering numerous particulars that are alike in relevant ways. Knowledge of biology depends on knowledge of such things as cells, chromosomes, strands of DNA, genes, nucleotides, viruses, amino acids, and proteins, and of such processes as cell division, gene expression, and cloning. It would be humanly impossible to know these objects and processes if each thing or event were unique and uncategorizable. If such were the case, to know something about one object or process would tell us nothing about other things or processes, making generalization, induction, and hypothesis and theory formation impossible. Order and repetition, in terms of both things and processes, are necessary for knowledge, scientific and otherwise. Inso-

28. Wilkinson (pp. 205-6) particularly sees the goodness manifested in the freedom that God grants to the creatures he has made.

29. Ernst Mayr, *Populations, Species, and Evolution* (Cambridge, MA: Harvard University Press, 1963), pp. 19-20.

far as it is knowable, the created world is a "fit object for human inquiry, understanding, and control."[30]

So what, then, is to be preserved? If we are to contribute to human wholeness and benefit what exists in the universe, the structures that already serve to foster these possibilities ought to be preserved. This includes, among other things, preserving genetic stability as well as genetic variability, the integrity of the environment, balances in nature that facilitate its operations, the viability and integrity of human life, human faculties that make possible the enjoyment of quality of life, freedom of choice that makes for moral responsibility, the conditions necessary for attaining spiritual growth and maintaining a relationship with God and other humans, and the order and balance necessary for aesthetic ends.

Undoubtedly the limits of the conservator will be the sorts of things that scientists continually bump up against. Just as scientists continually push at the borders of the *knowable* and the *doable,* they will also be pushing at the borders of the *ought to be known* and *the ought to be done.*

The debate in the 1970s over whether scientists should proceed with certain kinds of recombinant DNA experiments illustrates the tension. In July 1974, the Committee on Recombinant DNA Molecules of the Natural Academy of Sciences published a letter in *Science* and *Nature* that called their fellow scientists to a voluntary moratorium on certain recombinant DNA experiments.[31] The ground for the moratorium was the uncertainties about what could result from accidents in the laboratories. What would result if *E. coli* bacteria, combined with the genes from a tumor virus, became carcinogenic and were carried inadvertently beyond the laboratories? Could they colonize human intestines and convey cancer to humans? Or would we have the means to control a humanly manufactured scourge that, made resistant to certain antibiotics, escaped from the controlled laboratory environment? With the moratorium, "oughts," for the moment, controlled the "cans."

Yet the self-imposed moratorium lasted less than a year. Safety guide-

---

30. Langdon Gilkey, *Maker of Heaven and Earth* (New York: Anchor Books, 1965), p. 121.

31. A moratorium was suggested on two types of experiments: "the construction of new recombinant DNA molecules carrying combinations of antibiotic resistance not known to occur in nature," and "the construction of hybrid molecules containing DNA from animal viruses." *Bioscience* 25 (Apr. 1975): 240.

lines for continuing experiments with recombinant DNA, developed at the Asilomar Conference in February 1975, replaced the moratorium; the desire to know and do could not be restrained. As one researcher put it, "If you're dedicated to the truth, you have to say that there are no truths not worth seeking."[32] Today research proceeds apace,[33] accompanied by legal challenges.[34]

(5) Being a landowner and entrusting the estate to stewards involves risks. *What are the risks?* And are there limits to the risks God takes?

According to classical theology God is not a risk taker. Even before he created the world he knew everything that would happen to it. Indeed, some theologians hold that God knew not only what would happen (the actual) and what could happen (the possible), but also what would happen if certain other things happened that never did (called middle knowledge). That is, God knew that I would get up this morning at 6:20 (the actual), that I could have arisen earlier (the possible), and that had I chosen to arise at 6:13, I would have chosen hot rather than cold cereal for breakfast (middle knowledge). That is, God knew ahead of time every possible scenario and what would happen if that scenario came to pass. Hence he was in a position to create those beings that, by their already known choices, best suited his purposes.

Reformed theology goes even further. God did not create on the basis of his knowledge; rather, he actually foreordained the existence of every creature and the performance of every action.[35] The world, its contents, and all its events are predestined by God.

In these theologies, God risks nothing in creation. For some theologians there are no surprises because God knows all possibilities, all actualities, and what would have happened if things had been different. And for other theologians there are no surprises because all is foreordained by God; he controls everything.

Yet neither of these positions captures the drama of the biblical narrative. A profound sense of disappointment is expressed in the questions

---

32. "Politics and Genes," *Newsweek* 87 (12 Jan. 1976): 51.

33. The report (*Science* 338 [20 Apr. 1989]: 607) that seven researchers at the Pasteur Institute in Paris had been diagnosed with rare forms of cancer suggests that safeguards were inadequate. All seven worked concurrently on the same floor and had been involved with research using recombinant DNA.

34. W. French Anderson, "Big Changes at the RAC!" *Human Gene Therapy* 5 (Nov. 1994): 1309.

35. John Calvin, *Institutes of the Christian Religion* 1.16.2; 3.21.5.

God poses to the first sinners. To the hiding Adam, whom he commonly encountered as he walked in the garden in the cool of the day, he asks, "Where are you?" Of the murderer Cain he inquires, "Where is your brother?" The wickedness of the recalcitrant humans becomes so great that God is sorry he created them and threatens to destroy them all. Saving only Noah and his family, he tries his experiment with humans a second time, with a similar, unsatisfactory outcome. Later on, he chooses Saul as the first king of Israel, and then when Saul disappoints him, he repents of having chosen him and has to restart the kingly lineage with David.

We could go on, but the evidence is clear: God *is* a risk taker. The creation, especially insofar as it includes free human beings who can choose to obey or to disobey, to better or to destroy the world, is fraught with risk. But if he is a risk taker, what does the Landlord give to his stewards to venture, to risk?

One answer is that God has given to us everything within our reach to risk. Whatever we can touch we can affect, for good or ill. We have the power to create and to destroy, to benefit and to despoil. Yet the biblical story suggests that there are divinely imposed limits, that there are things we cannot reach. In our paradigmatic Genesis story, the man and the woman, following their disobedience, are driven from the divine park, Eden. Although they attain knowledge, immortality is forbidden to them. This suggests that there are certain things that God withholds, things that are his to give and that are not attainable by human technology.

It is difficult to discern what those limits on our power are. Yet even with limits, much (though certainly not all) of the earth is at risk. Even the great natural catastrophes of the past, though they destroyed many species, left the earth with life. However, the extent of our power means that science is serious business. This is not a reason for not engaging in science and technology. To the contrary, we have already seen that science is legitimized by the injunction for change. But it does mean that in doing what we do, we must be single-mindedly concerned with the good that benefits the earth we are to care for, the stewards for whom the earth was created, and God who created both the earth and its stewards.

Does Nature Need Stewards?

It should not be thought that our stewardship paradigm lacks critics or stands problem free. One objection in particular is posed and must be addressed.[36] Simply put, why should one think that nature needs stewards, that it has to be taken care of at all? If it has existed for eons, for hundreds of millions of years, without human supervision, cannot it continue to take care of itself without humans?

Stephen Jay Gould puts the objection this way:

> The views that we live on a fragile planet now subject to permanent derailment and disruption by human intervention [and] that humans must learn to act as stewards for this threatened world, . . . however well intentioned, are rooted in the old sin of pride and exaggerated self-importance. We are one among millions of species, stewards of nothing. By what argument could we, arising just a geological microsecond ago, become responsible for the affairs of a world 4.5 billion years old, teeming with life that has been evolving and diversifying for at least three-quarters of that immense span? Nature does not exist for us, had no idea we were coming, and doesn't give a damn about us. . . . We are virtually powerless over the earth at our planet's own geological time scale. . . . On geological scales, our planet will take good care of itself and let time clear the impact of any human malfeasance.[37]

The objection poses a direct challenge to our paradigm and an indirect one to our original analogy, which likened God's employment of stewards to that of a secular monarch. Without proper supervision and administration, the secular monarch's kingdom would go to wrack and ruin. The laws have to be enforced; taxes collected; revenues disbursed; infrastructure constructed and maintained; disputes settled; criminals caught, prosecuted, and punished; the citizenry protected and defended against both internal and external threats. Since the kingdom cannot run effectively on its own and since he cannot be everywhere, the potentate places his representatives or stewards throughout his vast empire to administer it and guard against disruption, upheaval, and social turmoil.

But the earth is different. Its kingdom existed long before any poten-

---

36. This objection was suggested to me by Anne Hall.

37. Stephen Jay Gould, "The Golden Rule — A Proper Scale for Our Environmental Crisis," *Natural History* 99 (Sept. 1990): 30.

tate placed stewards throughout the land. It functioned well on its own for many geological eras; through stasis and catastrophe, life persisted and evolved. Why should one think that if God had not recently placed his stewards throughout his kingdom, the place would have gone to wrack and ruin? Without stewards, nature has taken care of itself and can continue to do so. Organisms can procreate, thrive, adapt, develop, and evolve. Of course, not every natural thing thrives. But not only is continual thriving not required in nature; it is not desirable. Without death there is no birth or life; without decay there is no renewal.

Even the powers of the alleged stewards in the analogy are grossly disparate. The potentate's stewards might indeed be capable of either preserving or annihilating his kingdom. Their wisdom, perseverance, and selfless concern for the people might furnish the glue that holds the kingdom together; their inept economic policies, lack of military preparedness, dishonesty, or selfish greed could spell its quick demise. But not so for any paltry human stewards of the earth; the immense residual powers of nature dwarf their power. "All the megatonnage in our nuclear arsenals yield but one ten-thousandth the power of the asteroid that might have triggered the Cretaceous mass extinction. Yet the earth survived that large shock and, in wiping out dinosaurs, paved the road for the evolution of large mammals, including humans."[38] Even if we initiate a catastrophic nuclear war, though humans may not survive, some forms of life will survive to continue nature's evolutionary process.

As we said, this poses a serious objection. To construct a response to this question — namely, does a stewardship paradigm entail that nature cannot take good care of itself? — it is important to inquire concerning the view of nature presupposed and to note the context in which the question is raised.

First of all, the objection assumes that nature has operated and continues to operate on its own. No omnipotent potentate reigns over nature; no other being guided its fantastic development from primeval soup to conscious animals or was involved in preserving it through its catastrophic collisions with asteroids. In such a view, the concept of stewards who act on behalf of someone greater than nature would not connect with any other theses. Gould is no theist.

Yet our stewardship paradigm assumes from the outset that there is a God who is the original and continuing source of nature, its creative and

38. Gould, p. 30.

sustaining cause, a being who can act and has acted in geological as well as in human history. As omnipotent, God can act directly on or indirectly through his creation. This claim forms the heart of our theistic worldview. Thus Gould's presuppositional claim that nature has been on its own over the geologic eons reveals a critical difference in our worldviews.

Admittedly ours is not a scientific hypothesis. It can, however, be argued for on the scientific grounds that the conditions of the universe necessary for the existence of living, knowing, valuing beings are so unlikely as to be staggering. For example, had the "Big Bang expanded at a different rate, life would not have evolved. *A reduction by one part in a million million* at an initial stage would have led to recollapse before temperatures could fall below ten thousand degrees. An early *increase* by one part in a million would have prevented the growth of galaxies, stars and planets."[39] Or again, if the gravitational force had been slightly greater, all the stars would have been blue giants, whose life span is too short to allow the evolution of intelligent life. However, if it had been slightly less, the universe would have been devoid of many elements essential to life. In effect, what is a priori unlikely is necessary not only for what exists now, but for the very possibility of seeing its purposefulness. At the same time, this theistic thesis cannot be empirically tested, for we do not know the manner in which or the conditions under which God acts. For the Christian, it is a belief engendered by acceptance of revelation, which is itself the result of divine interaction with creation.

A second consideration that must be raised in responding to the objection is this: What is meant by "take good care of itself"? The objection presupposes that nature is a complex system composed of many individuals or parts. In the geological "long run" the parts are insignificant; whether any particular organism, species, or ecosystem survives or thrives is irrelevant.[40] What matters is the "survival" or continuity of nature, with

39. John Leslie, "The Anthropic Principle, World Ensemble, Design," *American Philosophical Quarterly* 19 (April 1982): 141.

40. Gould takes pains in his article to show that the long-run view need not apply in the short run. In the short run, which encompasses the human time scale, it is reasonable to try "to preserve populations because the comfort and decency of our present lives, and those of fellow species that share our planet, depend upon such stability" (p. 26). In the case of the Mount Graham red squirrel, the local subspecies whose existence, threatened by the proposed building of an astronomical observatory, occasioned the article, their survival might interest "students of evolution" studying "peripheral populations, living in marginal habitats."

earth's "prosperity," reasonable "stability," diversity, and the possibility of "paving the road" for the development of more complex forms of life.

Given this interpretation of "take good care of itself," the answer the stewardship paradigm might give to the question posed depends upon the context in which the question is being raised. If it is raised outside the context of the existence of human beings, then clearly the answer is no; a stewardship paradigm does not entail that nature cannot take good care of itself. Before humans existed on earth, nature could procreate, continue, achieve ecological stability and diversity, and develop, as the biologic-geological history of the world shows. At the same time, we do not affirm that nature can take good care of itself apart from God. However, this is part of the creation, not the stewardship, paradigm.

What does follow from our negative answer to the question posed about nature in this prehuman context is that the primary reason why God created human beings cannot be that God needed stewards to preserve the environment. It is not as though nature, failing in its purpose or structure, in a state of chaos and destruction, required that God create stewards to enable it to survive and prosper or to restore order.

The fundamental reason why God made us is well stated in the first question and answer of the Westminster Catechism: our "chief end" is "to glorify God, and to enjoy him forever." People were created to return love and glory to their Creator in a way that supersedes the response to God that nature can make. Nature gives glory to God by doing what it does (Ps. 19); people do it by their free, loving response to the love God has shown them. Just as we desire the free return of love from those whom we procreate, and just as we would prefer this love over any forced or necessary responses of a mechanism or robot programmed to respond that it loves us whenever we ask it, so God desires that we freely return love to him. The first and greatest commandment, that we should love the Lord our God with all our heart, soul, mind, and strength (Mark 12:30), provides the reason for our being.

Since stewardship does not provide the primary *raison d'être* for the existence of humans, the stewardship commands must be seen in the context of the prior existence of human beings, who are created by God, whose very existence is valued by him, and whose free love is sought in return. In the context either of the existence of humans or of God's intention that there be humans, the answer to our question — does the stewardship paradigm presuppose that nature cannot take good care of itself? — is yes, for with the introduction of powerful, conscious beings who can radically affect their environment, nature itself lies in jeopardy. Its stability, diversity, and ability to evolve and to prosper is under possible attack. To return to our kingdom analogy, it is as if the very creation of stewards created the problems that the kingdom faces. The stewards are the possible catalyst for societal disruption. At the same time, the existence of people is so valuable that it is worth risking the entire kingdom to bring them into being. Hence, stewardship becomes one of their primary charges.

One might liken the situation to commanding my son to keep his room orderly. Does my stewardship command presuppose that his room cannot be ordered by itself? No. Without him, things would stay as they are. Hence I did not bring him into the world in order to keep his room orderly. At the same time, having my son is more valuable than not having him. But having him means that there will now be disorder in his room; his very creation means that there will be the disorder that must be addressed. Hence, it is meaningful to give him the injunction to keep his room orderly; this becomes one of his assigned tasks.

As Gould correctly notes, we might not be capable of jeopardizing the ultimate existence of nature, but God's plans for nature might well extend beyond nature's mere survival of any humanly contrived holocaust. Gould sees nature not merely surviving but prospering and providing a base for future evolution. But the theist can well hold that there is more to nature's purpose than this. Nature is so important that God seeks to renew and re-create it.

The New Testament speaks about creation groaning in the pains of childbirth, waiting for its redemption. The human fall had a negative impact on nature. Nature, "subjected to frustration," in "bondage to decay," waits in "eager expectation" for its liberation (Rom. 8:18-25). This liberation began with the death and resurrection of Christ and will be fulfilled in the eschaton with his return. What this theological affirmation means in practical, ecological terms, we admit we do not know. Theological descriptions of the renewal are exceedingly vague and at times biologically naive.[41] For example, some suggest that in the new kingdom there will be no more death in nature. "This abolition of the present world form will show itself negatively in that the rigid, fundamental law, to which all life in the present world is subject . . . will be cancelled — the biological principle that life can only increase and multiply by a process in which other life is suffocated and

41. For a brief review of the literature, see David S. Wise, "A Review of Environmental Stewardship Literature and the New Testament," in *The Environment and the Christian*, ed. Calvin B. DeWitt (Grand Rapids: Baker Book House, 1991), pp. 117-34.

destroyed with pain and deadly torture."[42] But since death and decay were present in the natural world long before humans came on the scene and were placed in the garden, and since death — whether of plants or animals or other organisms — is necessary for life, the changes envisioned for nature must encompass something other than death's elimination if the future kingdom is to be continuous at all with this one.

Similar things might be said about those who see the fall of humans as creating a widespread ecological disorder that needs repair in the eschaton. Genesis knows no such scenario. The punishment for Adam and Eve's disobedience applies to human agricultural activities. Sweat and toil will not produce the abundant fruits the original pair were accustomed to. In their place will grow thorns and thistles.

What the redemption of nature will be like, precisely how nature will be made new, how it will find itself in a new harmony with God, and how the curse brought about by Adam's sin is reversed (Rom. 5:12-19) remain some of the many divine mysteries. There is the same openness about what will result from God's re-creative and restorative acts as there was before the original creation. None beforehand could have predicted or surmised the outcome of God's creative acts. Yet as the New Testament affirms, just as God in Christ made the universe (John 1:3; Heb. 1:1-3), so in Christ he will renew heaven and earth (Rev. 21:1-5; cf. Isa. 65:17). In Revelation, "the garden has become the city, but the city is reminiscent of the garden."[43] But we cannot speculate about the ecology of either.

Because our presence puts the kingdom at risk, the commands to stewardship take on importance and urgency. In this new context, the stewardship commands to fill, rule over, and care for creation attain special significance. Precisely what this significance is remains to be seen; it forms the substance of the remaining chapters in this book.

We have provided a general outline of our Christian ethical paradigm. We must now apply it to specific biological issues. In doing so, we will achieve a greater understanding of the ethical significance of the three divine commands and the role they can play to guide us in making moral decisions.

---

42. Louis H. Taylor, *The New Creation* (New York: Pagent, 1958), quoted in Wise, p. 127.

43. Ronald Manahan, "Christ as the Second Adam," in *The Environment and the Christian,* ed. DeWitt, p. 46. See also Gordon Zerbe, "The Kingdom of God and Stewardship of Creation" in the same volume (chap. 4).

CHAPTER FOUR

Stewardship
——— of the Environment ———

IN 1970 the northeastern region of Brazil experienced its worst drought in thirty years. This, combined with skyrocketing population growth, worsened the economic outlook for the already beleaguered inhabitants of this underdeveloped and impoverished yet heavily populated region. After visiting the area, Brazilian president Emílio Garrastazu Médici announced plans to alleviate some of the pressures on the Northeast by opening the Amazon basin to agricultural development. The emphasis would be on the resettlement of one million landless poor, 75 percent of them from the Northeast, on small farms in the interior.

The centerpieces of the Brazilian resettlement program were to be the Transamazonian and the Cuiabá-Santarém highways. The Transamazonian Highway was announced, contracts let within two weeks, and building commenced less than three months later.[1] Planners in Brasília envisioned small, indigenously farmed agricultural plots lining the highway and feeder roads. At regular, ten-kilometer intervals settlers would construct *agrovilas,* small settlements with a store and elementary school. Towns *(agrópolis)* every fifty kilometers would link this network of villages, and every one hundred fifty kilometers, larger urban centers of fifty thousand people *(rurópolis)* would serve as regional agricultural centers for distribution and education.

Following an initial six-month subsidy, the colonist farmers rapidly

---

1. Goodland and Irwin note that as late as 1969 no such roads existed in the development plans for Amazonia. R. J. A. Goodland and H. S. Irwin, *Amazon Jungle: Green Hell to Red Desert?* (New York: Elsevier Scientific Publishing Company, 1975), p. 10.

were to become self-sufficient on their fertile, one hundred hectare plots. Their initial crops of rice, manioc, corn, and beans were to be sufficient to feed their families. Eventually cash crops would be shipped to markets by trucks plying the highway, providing the settlers with additional disposable income. The Amazon would be settled, the poor given land to farm, and the economy stimulated.

Unfortunately, things turned out differently than the military government planned. The projected urban areas along the Transamazonian Highway either were malaria infested or had poor soils. Instead of the anticipated one million settlers, by 1977 less than eight thousand families had been resettled in three model colonies.[2] Much of the best available land was quickly taken by the mixed white and Indian population that already scratched for a living in the region. For many of the colonists the promised agricultural education was never made available. The crops the farmers planted were annuals that quickly depleted the fertility of the soil; at best only one or two years of good harvest was possible. Since the large amounts of fertilizer necessary for agricultural success were economically inaccessible to them, they had to move on to other sites.

When this small-farmer development project faltered, the government turned to more economically promising schemes. The foundation for these had been laid already in 1965, when the military president of Brazil, Castello Branco, declared "Operation Amazonia," a program to open the Amazon for development. The governmental program, expanded in 1968 under SUDAM (Superintendency for the Development of Amazonia), aimed at attracting investors for large-scale agricultural development (including ranching) and industry. But the increasing economic pressures of a rising foreign debt in the mid-1970s, caused in part by the world oil crisis, moved Brazilian thinking on Amazonian development significantly in the direction of large-scale commercial enterprises. After economic stability had briefly returned to Brazil's economy, the search for multinational development began apace. SUDAM approved hundreds of big cattle projects, so that

2. Roger D. Stone, *Dreams of Amazonia* (New York: Viking Penguin, 1985), p. 90. Even had the original plans been followed, the project would not have accomplished one of its major tasks, that of relieving the poverty of the Northeast. The population of the Northeast, at twenty-three million people, is growing at a rate of 2.8 percent, which means that the opening of the Amazon to settlement, three-fourths of which was to come from the north, would allow the resettlement of less than one year's population growth. A solution to population growth by providing an outlet it was not.

some 95 percent of the new landholdings in the Amazon were of 10,000 hectares or more. Unlike coffee plantations, the large new cattle estates and soy bean and sugar cane farms employed little labor. Many of the dispossessed or discharged tenants and rural laborers migrated to the cities, swelling the ranks of the unemployed and underemployed and aggravating all the urban social problems. Thousands of others drifted to the Amazon frontier, becoming *posseiros* (squatters) who raise subsistence crops.[3]

The shipping magnate Daniel K. Ludwig inaugurated one of the largest efforts along the Jari River. This was not the first time a multimillionaire entrepreneur attempted large-scale development of the Amazon. In 1927 Henry Ford, in order to break the South Asian rubber cartel, had created over five thousand square miles of rubber plantation at what became known as Fordlandia. Technology and mass production techniques were to be applied to the production of rubber. However, a combination of factors — infestation of the rubber trees by the South American leaf blight *(Dothidella ulei)*, poor planting techniques, personnel conflicts between Brazilian workers and American bosses, conflict with a Brazilian government newly protective of its politically vulnerable interior, and transportation problems created by seasonal low water on the river — conspired to bring the project to a halt by 1945, after an investment of more than $20 million.

Ludwig's attempt to grow rice and pulpwood on a plantation twice the size of Ford's met a similar fate. Insects attacked the gmelina trees he imported from Asia, while leaf-cutter ants went after the more toxic Caribbean pine. The long-grained rice grown along the floodplain succumbed to armies of insects and various diseases. By 1982 the pulp mill that was to process lumber for world consumption was feeding on its own prematurely cut pulpwood just to keep going. Jari was about to go the way of Fordlandia, despite the millions in capital expended. Again the Amazon basin won a reprieve, though only temporarily.

Just looking at the Amazon, one would think that the soils must be exceedingly rich to support such lush growth. The nineteenth-century Amazonian biologist Alfred Wallace wrote, "nature and climate are nowhere so favourable to the laborer and I fearlessly assert that here, the 'primeval' forest can be converted into rich pasture and meadow land, into

3. Benjamin Keen and Mark Wasserman, *A History of Latin America* (Boston: Houghton Mifflin, 1988), p. 389.

cultivated fields, gardens and orchards, containing every variety of produce, with half the labor and, what is of more importance, in less than half the time that would be required at home, even though there we had clear, instead of forest ground to commence upon."[4]

But appearances are deceiving; the truth is precisely the opposite. Much of the forest litter of leaves and branches never reaches the ground but is trapped on branches or in crotches of trees where it enters the nutrient cycle. The leaf litter that does reach the forest floor is either quickly encircled by tree roots and absorbed into the living trees or fed on by fungi that in turn feed the nutrients to the tree roots. Thus the nutrient cycle of the forest plants is practically closed. Very few nutrients ever get into the soil and water, which means that turning the forest into agricultural land is often doomed from the start, except where there are small pockets of fertile soil or land suitable for small-scale, slash-and-burn enterprises.[5]

Destruction of the Rain Forest

The Amazon basin's story is not unique; it can be repeated for the tropical forests around the world. Until the twentieth century, the more than fifteen million square kilometers of tropical forest between the Tropics of Cancer and Capricorn stood largely untouched, except for symbiotic human existence. The forest yielded reluctantly to the temporary destruction caused by slash-and-burn agriculture but was never threatened in the way that the forests of the eastern and midwestern United States were threatened and in large measure destroyed during the eighteenth and nineteenth centuries. In the twentieth century, however, the tropical forests, too, have come under attack. Half of the tropical forests are already gone, and the destruction continues at a rate of more than 250,000 square kilometers each year.[6] Madagascar provides a bleak example; 95 percent of its original 590,000 square kilometers

4. Quoted in Stone, p. 64.
5. Goodland and Irwin, p. 28. "Some studies have shown that as little as one-tenth of 1 percent of forest nutrients ever penetrate below the first 5 centimeters (2 inches) of soil in tropical rain forest." Adrian Forsyth and Ken Miyata, *Tropical Nature* (New York: Charles Scribner's Sons, 1984); quoted in Roger Thompson, "Requiem for Rain Forests?" *Editorial Research Reports,* 20 Dec. 1985, pp. 953-54.
6. Ronald B. Nigh and James D. Nations, "Tropical Rainforests," *Bulletin of the Atomic Scientists* 13 (March 1980): 12.

of rain forest has been cleared for lumber and the raising of coffee, spices, sugarcane, and cattle. Rain forests have largely disappeared from India, Bangladesh, and Sri Lanka, while other Asian countries like Malaysia, Thailand, and the Philippines have only patchy forests remaining. Countries like Sierra Leone and the Ivory Coast in West Africa face similar situations, while in Central America two-thirds of the rain forest is gone and the rest is rapidly disappearing.[7] In a dramatic Landsat photo, the border between Mexico and Guatemala is clearly delineated, as on a map, between the tilled, destroyed forest in Mexico and the preserved forest in Guatemala.[8]

The causes of the devastation in the Amazon basin are typical of the worldwide problem. One of these causes is the combination of the pressures brought by the burgeoning numbers of peasant farmers, who cut down forests to establish subsistence farms and who use agricultural practices that are appropriate to temperate-zone agriculture but inappropriate to rain forests. Though the population explosion alone has not stimulated the invasion, it is a major factor. "In just 30 years the population of tropical nations is projected to double."[9] To the population explosion one must add the social pressures caused by poverty[10] and the intensification of mechanized, export-oriented agribusiness that turns tenant farmers into part-time laborers and forces the landless, displaced poor to seek whatever land is available to feed their families. These pressures in turn have forced governments to allow and in some cases encourage the landless to settle in more marginal agricultural areas, including the virgin tropical forests.

For example, the Brazilian government has facilitated settlement of the tropical forests in the far western part of the country along the Peruvian border for the peasants moving north and west from southern Brazil. Brazil's creation of a road from Cuiabá to Pôrto Velho opened up the remote state of Rondônia, first to a trickle and then to a flood of immigrant farmers. The population increased from 111,000 in 1975 to more than one million in 1988. Those living there, increasing at the annual rate of 15.8 percent, already have had a significant impact on the environment.[11]

---

7. Nigh and Nations, p. 13.
8. Wilbur E. Garrett, "La Ruta Maya," *National Geographic* 176, no. 4 (Oct. 1989): 474-75.
9. Thompson, p. 952.
10. In Brazil alone, the most populous country in South America, 70 percent of the people have subsistence-level incomes.
11. Stone, p. 94. For satellite photographs showing the change in just one region, see Norman Myers, *The Primary Source* (New York: W. W. Norton, 1984), pp. 182-83.

Whereas in 1975, 1,250 square kilometers of forest had been cleared, by 1988, 60,000 square kilometers (20 percent of the total) had been cleared and 180,000 had been degraded.[12]

Large, invading populations and the establishment of permanent deeds and titles to the land makes traditional slash-and-burn farming impossible. Restricted to their designated plot, the new farmers must make do with land they have acquired. But the attempt to establish permanent farms on tropical soils, without massive use of fertilizers and insecticides, generally is doomed from the outset. The nutrients that sustain the forest are not in the soils but in the forest growth. The burning of the forest captures the nutrients in the ash, but this resource is depleted after the first harvest or two, and the land then ceases to yield even a subsistence harvest.

Another cause of the devastation of the forests is likewise agricultural in nature, but on a more extensive scale. Whereas peasant farmers affect small plots that, once abandoned and left alone, in time can recuperate, the bulldozers used on large monoculture plantations can destroy four to five hectares a day, which leads to denuding thousands of acres at a time. In addition, many types of monoculture — cotton, tobacco, sugarcane — are destructive when carved out of rain forests. Frequently the plantations displace the local peasant populace, forcing them into ever more agriculturally marginal areas.

The most troublesome type of monoculture in recent decades is cattle ranching. Like Brazil, Costa Rica in the 1960s and 1970s developed large cattle ranches in marginal areas. The extensive lands needed for grazing, if taken from nonproductive forest, are relatively cheap, and the cattle, left to forage for themselves, require little maintenance or oversight. Indeed, it is estimated that "over ⅔ of the agriculturally productive land in Central America and Panama is devoted to this pattern of extensive livestock production."[13]

The beef produced, however, is not for domestic consumption. In fact, Latin American domestic consumption of beef has declined during the very period in which the cattle industry has significantly expanded. The benefits of the cattle industry do not reach the local masses but become part of the profits of the transnational corporations based abroad. The

12. Norman Myers, "Tropical-forest Species: Going, Going, Going. . . ." *Scientific American* 259 (Dec. 1988): 132.
13. Nigh and Nations, p. 14.

beef, too lean for use in the more profitable table cuts, is deboned and shipped in frozen packages to the Northern Hemisphere, where it either ends up in the fast-food restaurants or on the tables of Japanese, United States, and Western European consumers as hamburger or processed meats or is turned into pet food.

As a result of the ranching, not only are large tracts of land deforested, but the soil is trampled and compacted, leading in some areas to lateriza-tion (a process whereby the earth turns almost as hard as cement). The planting of grasses to sustain the cattle is only a short-term solution, as they quickly deplete the soil of whatever nutrients remain after the burnoff. The result is often a quasi-desert, where the rains quickly run off, erosion occurs, and the density of cattle per acre diminishes. "Tropical South and Central America support over 180 million cattle on 297 million hectares of pasture land cleared from forest. This represents a density of around 0.6 animals per hectare, extremely low by any standards."[14] After ten or fifteen years, their fertility gone, the pastures are abandoned, and the ranchers have to find other land.

A third cause of the deforestation is the harvesting of the forest timber. The large-scale commercial harvesting of tropical hardwoods is mainly for export to the developed world. The developing countries see this as a quick solution to their burdensome balance of payments.[15] However, the methods used result in the destruction of surrounding vegetation (and hence of the ecology of the forest) and in the compaction of the earth, which prevents the forest from regenerating naturally. At the same time there is little concern for reforestation to maintain a sustainable resource. The result has been that some countries, like Thailand and Nigeria, which were once exporters of tropical hardwoods, are now im-porters. At the current rate of harvesting without reforestation, countries like Cameroon, Gabon, Ivory Coast, Malaysia, and the Philippines also will soon join that group.[16]

However, by far the greatest immediate threat to forests stems from the fact that more than two-thirds of the people in developing countries

14. Nigh and Nations, pp. 13-14.

15. This should not be seen as unique to tropical forests. The current harvesting of Siberian timber, which is being done to finance international debt, replays the scenario.

16. "In Ghana, for example, exports have fallen from a high of 124 million cubic meters in 1973 to 11 million cubic meters in 1982." *Tropical Forests: A Call for Action*, vol. 1 (Washington, DC: World Resources Institute, 1985), p. 10.

depend primarily on their surrounding forests for cooking and heating fuel. The result is that "more than 80% of the wood harvested in developing countries is used as fuelwood."[17] And of the two billion harvesters, three-fourths of them are cutting fuelwood around their homes faster than it can be replenished. The constant pressure of the need for wood puts devastating pressures on the ecology surrounding their villages. With the removal of the forest vegetation, the water is no longer held in the soil by the tree roots, and instead of percolating into the soil to replenish wells, it runs off with its load of soil to the rivers. This, in turn, results in massive soil erosion, desertification, and water pollution.

Effects of Tropical Deforestation

But why should we worry? Is not tropical forest destruction merely a by-product of the inevitable agricultural and economic development of that region? Have not tobacco and cotton farms and the fertile corn belt in the United States successfully replaced the eastern and midwestern forests? To the contrary, the radical deforestation of the tropical forests has several significant and potentially long-lasting effects.

(1) Deforestation on a large scale will bring about at an unprecedented pace the extinction of many species of plants and animals. To date scientists have documented 1.7 million of the estimated five to ten (some suggest thirty) million species that inhabit our planet.[18] This means that most species are yet to be discovered, analyzed, and documented as to their niche in the environment. Of these species, an estimated two-thirds to as many as 90 percent live in the tropics.

In many cases the destruction of their habitats ensures the destruction of large numbers of species. Some have suggested that as many as one hundred species are disappearing each day.[19] In Madagascar, for example, 90 percent of the original vegetation has disappeared, destroying much of

17. *Tropical Forests,* vol. 1, p. 5.

18. A caveat in any discussion of tropical species concerns what species are and how species are to be distinguished from other species on the one hand and from subspecies on the other. The geneticist R. J. Berry suggested to me that since genetic structures of species vary, species delineation in tropical plants and animals at times approximates subspecies determination elsewhere on the planet.

19. Eugene Linden, "The Death of Birth," *Time* 133 (2 Jan. 1989): 32.

its unique flora and fauna. Furthermore, even if some species survive extinction, the destruction of their diverse populations greatly reduces their genetic variability.

In the tropics the individual members of many species are scattered over a wide area, while the types of species in one area may differ widely from those nearby. As Norman Myers explains,

> These distribution patterns mean that many tropical forest plants are comparatively rare, which makes some of them liable to local extinction when a forest is subject to intensive exploitation. So extremely rare are the individuals of certain plant species in these forests, notably orchids and legumes, that we know a few of them from only a single encounter several decades ago. In the Malay Peninsula, a begonia was discovered during the 1940s, and, because of its ornamental appearance, it has become a popular cultivated plant in Europe, with millions of specimens in gardens, nurseries, and so forth. But it has not been found in the wild again. In Amazonia, numerous tree species were identified between 50 and 100 years ago that have not been located since.[20]

It is true that species come and go. One estimate of what is termed background extinction ("defined as the steady rate of species turnover") for the Tertiary Period (65 to 2.5 million years ago) is two families per million years. Much more serious were the mass extinctions that are held to have occurred at least five times in the history of the world. For example, during the mass extinction at the end of the Permian Period, perhaps half of the families disappeared. In the four others (at the end of the Ordovician, Devonian, Triassic, and Cretaceous Periods) perhaps as many as 11 to 14 percent of marine animal families were affected.[21] Nevertheless, the unprecedented pace and scale of the contemporary destruction are troubling. None of the geologically recorded mass extinctions would approach what would happen should humans rapaciously destroy so many other species.

Furthermore, there is the possibility that what is destroyed might have benefited us. One loss would be genetic variability. Hybridization has narrowly specialized our crops, dramatically improving their yields. Yet this productivity has a cost, for in making crops uniform across the

20. Myers, pp. 30-31.
21. Cecie Starr and Ralph Taggart, *Biology* (Belmont, CA: Wadsworth, 1989), pp. 566-67.

world their ability to respond to various diseases, pests, and changes in weather is diminished. It was germ plasm from long-established genetic lines that contributed to ending the 1970 leaf fungus blight affecting the U.S. corn crop. Five of the thirteen crops that supply 90 percent of the world's protein and calories come from the tropical rain forests.[22] Should we lose genes from wild varieties that enable them to adapt to different pests and diseases, the future production of these crops could be jeopardized. Genes from wild species also might be capable of enhancing our domestic crops. For example, the biotechnological firm NBI transplanted genes from tropical tomatoes to increase the density of U.S. tomatoes.[23]

Other losses might take place in the area of medicine. Wild plants contribute significantly to modern medicine by providing drugs for a variety of illnesses. As many as half of our current drugs derive from wild plants,[24] and half of these derive from the tropics.[25] For example, quinine, derived from cinchona bark, made possible the treatment of malaria. The Squibb Corporation uses venom of the Brazilian pit viper to develop the drug for high blood pressure called Capoten.[26] Chemicals derived from the Madagascar rosy periwinkle plant now help fight lymphocytic leukemia; chemicals from the Indian snakeroot plant form the base for tranquilizer products; and diosgenin from the tropical Mexican yam is used for manufacturing oral contraceptives.[27]

The point is that so much is unknown about what we would be destroying. Hence we have no way of knowing what possible benefits might be lost in the process of providing hamburger and tropical hardwoods for the world's wealthy countries. Genetic capital once destroyed can never be reclaimed.

The potential for genetic diversity in a host of undiscovered plants, animals, fungi, and other organisms has begun to entice commercial medical and agricultural companies back into the tropics to retrieve what remains. The new hitch is that many countries are following the lead of Costa Rica and demanding a certain percentage of the royalties from the

22. These five are rice, peanuts, sorghum, yams, and coconut. Thompson, p. 952.
23. Linden, p. 33.
24. Myers, p. 132.
25. Myers, p. 7.
26. Linden, p. 33.
27. Thompson, p. 951.

production of medicines or crops that contain the genes derived from species in their countries. Their argument is that genes are also part of their natural resources, and if the developed countries expect them to preserve their natural environments, the developed countries should be willing to share the profits made from what is discovered in them. The U.S. government and many companies have resisted making such deals, both because many of the developing nations often do not respect patents and copyrights and because businesses find it difficult to factor the cost when what might or might not be discovered is unknown.

(2) Destruction of the rain forests leads to the dramatic destruction of the land itself. The tropical sun bakes the soil, while the rain causes serious erosion. Erosion from cultivated fields is many times higher than that from land with natural vegetation. "In the Amazon basin a hectare of land under forest was found to lose only one kilogram of topsoil per year, while the same area cleared of forest may lose up to 34 tons per year."[28] Not only does the land itself become unusable, but runoff threatens waterways and the dams associated with them. Some observers have warned, for example, that deforestation of the watershed around the Panama Canal threatens it with silt buildup.[29]

(3) The forests contribute substantially to the hydrological cycle of a region. In the Amazon basin, for example, more than half of the rainfall originates from within the basin.[30] Water enters the atmosphere by evapotranspiration from the trees. Destruction of the trees breaks the cycle, leading to flooding (where the soil and vegetation cannot absorb the great amounts of rain water and it runs off quickly) and drought (where the vegetation is not putting back into the atmosphere enough moisture to return to earth).

(4) Forests also moderate climatic changes. Their water cools the air, moderating the heat that is concentrated over the tropics. Equally important is their contribution to absorbing carbon dioxide. World carbon dioxide levels have increased significantly in the last century, from 280-290 parts per million (ppm) a hundred years ago to 350 ppm today. With the increasing burning of fossil fuels, it is projected that carbon dioxide levels could reach 500-700 ppm by the year 2050.[31] Carbon dioxide molecules trap the earth's heat, preventing it from escaping into the atmosphere. This

28. Nigh and Nations, p. 15.
29. Nigh and Nations, p. 15.
30. Stone, p. 115.
31. Linden, p. 38.

so-called greenhouse effect means that eventually the temperature of the atmosphere will increase. "If current predictions prove correct, the rapidly increasing concentration of atmospheric carbon dioxide will raise the average global temperature by about one degree by the year 2000 and by two degrees within 70 years."[32] Should this occur, it could melt the polar ice caps and cause the deserts to expand. The resulting melt water would flood the coastal areas, which on every continent contain major population centers, while currently productive land like the North American plains would become dust bowls unsuitable for agriculture. Currently, tropical forests absorb a substantial amount of carbon dioxide, thus contributing to the stabilization of the earth's climate.[33]

(5) Finally, there are all the unknown "people" effects. Outside encroachment on the rain forests has contributed to the decline and even extinction of the native populations. In Brazil alone the native American population has declined from more than one million (some say it was more like five million) at the time of the Portuguese settlement to fifty thousand today.[34] And even the native Americans who remain are not safe. Goodland and Irwin, in their early report on the effects of the Transamazonian Highway, quote the director of the governmental organization established to protect the interests of the native Americans as saying, "Indian programs shall obstruct neither national development nor the axes of penetration for the integration of Amazonia."[35]

There are also the possible long-term consequences that an exploitive economy and a destroyed ecology can have upon those who remain in the area. Currently the Food and Agriculture Organization of the United Nations estimates that eighty-eight million people live in rain forests.[36] If the above prognostications are correct, then the poor who remain will be mired in their poverty, in an area denuded of resources, unable to sustain their livelihood, and threatened by floods and drought.

32. Nigh and Nations, p. 16.

33. Scientists debate whether tropical forests absorb more carbon dioxide than they release, or whether a balance between release and consumption exists. There is also debate on whether the destruction of forests might actually stimulate other plant growth, which in turn would compensate for loss of forests in absorbing carbon dioxide from the atmosphere. Cheryl Simon Silver, *One Earth, One Future* (Washington, DC: National Academy Press, 1990), p. 122.

34. Goodland and Irwin, p. 64.

35. Goodland and Irwin, p. 58.

36. Thompson, p. 953.

Of course, not all is this bleak. Brazil's election of Fernando Collar de Mello as president in the last week of 1989 was taken as a sign of a change in Brazil's concern with its environment. He vowed to reverse exploitation of the Amazon and to drive out those who were illegally prospecting on Indian land. He also created the new position of environmental secretary and named a leading environmentalist to the post. Destruction of the forest declined from a rate of 27 to 20 percent between 1989 and 1990. Yet problems persist. The environmental secretary was forced to resign in 1992; and President Collar himself, tainted with scandal, was impeached and resigned late in 1992. Further, some argued that the decline in forest destruction was not as hopeful a sign as it seemed, that in fact it was due to a decline in the world economy, not to conservation. Brazil's debt remains large, its economy stalled, and the debate rages both within and outside Brazil how best to administer the forests of the Amazon.

A more hopeful story can be found in Costa Rica, which contains an estimated 5 to 7 percent of the world's biodiversity.[37] Variance in altitude (from sea level to over 12,000 feet) and in rainfall creates twelve distinct ecological zones. Costa Rica has intentionally slowed the process of tropical forest destruction by designating 13 percent of the country as national parks, wildlife preserves, and protected areas. Surrounding these areas are buffer zones of forest preserves and private wildlife areas that further extend the amount protected.

Preservation has not been easy. When a large banana plantation closed in the early 1980s, the unemployed workers migrated into Corcovado Park on the Pacific coast and began panning for gold. This created high levels of ecologically disruptive sediment in the river and introduced toxic mercury into the water. Though the government eventually drove the eight hundred squatters out, the event revealed that merely designating areas for conservation was insufficient.[38] Nor could all the parks be adequately policed. Conservation had to be coupled with social policies and actions that provided economic opportunities for the landless and jobless and for the ever-increasing peasant population whose marginal farms would quickly give out. Without economic assistance, peasants would be driven into the preserves looking for new land to cultivate. To achieve

37. Leslie Roberts, "Chemical Prospecting: Hope for Vanishing Ecosystems?" *Science* 256 (22 May 1992): 1142.
38. Katrina Brandon and Alvaro Umana, "Rooting for Costa Rica's Megaparks," *Americas* 43, no. 3 (May-June 1991): 22.

both conservation and social development, the government is courting governments, environmental organizations, and businesses around the world. Governments and environmental organizations are asked to swap Costa Rica's foreign debt for conservation of its tropical forest and for investment in making these areas contribute to Costa Rica's economy through such things as ecotourism. Multinational companies are invited to prospect genetically in the parks and preserves for a fee and future royalties on what is discovered.

A further hopeful sign can be seen in the agreements signed by both developed and developing countries at the 1992 Earth Summit on economic development and the environment at Rio de Janeiro. The conference generated much North-South debate, a significant hesitancy on the part of the United States' Republican administration to participate and make concessions, and many recommendations. The United Nations was authorized to begin negotiations on an international forest treaty. A climate treaty was signed for the reduction of the emission of greenhouse gases, a treaty that, though presently toothless, lays the basis for future action if the current increasing trend continues. Though no target date was set, the industrialized Northern countries agreed to move toward committing 0.7 percent of their economic output toward environmental concerns and global cleanup. The conference signals a new intent, if not yet much in substance, by developed and developing nations to cooperate in economic development and environmental preservation.

The Ethic of Stewardship: Ruling the Earth

The story of the causes and potential effects of the destruction of the tropical rain forests takes us to our broader concern. What moral obligations do we have to the environment? To answer this we will apply our ethical model of stewardship to environmental considerations. What we have to say about environmental ethics in general we can apply to our case study of the tropical forests.

In Chapter Three we saw that our stewardship on behalf of the Landowner involved three injunctions: to rule over, fill, and tend the earth. Let us begin with the injunction to rule over the earth.

It is easy to misconstrue this injunction to sanction human selfishness and rapaciousness. The tropical forest, its exploiters argue, is there to be subdued. It is filled with prime timber that can be cut, burned for fuel, or exported and used for construction, paneling, furniture, or pulp. Its colorful birds and unique mammals substantially profit traders who sell them to zoos or for pets. Its land, though marginal, still can support some peasants who, while not becoming wealthy, can at least farm in order not to starve. And large tracts can be ranched, even though sparsely, to provide meat for the developing world either to consume or to export to satisfy its balance of payments. More generally, the environment — any environment — is to be used for human benefit, as a means to human fulfillment. Nature serves a utilitarian function. If we are to rule the land, then we should be about the business of getting from it the most that we can.

But even such a narrow, human-centered approach to ruling over the earth need not exclude considerations of the future. If we are going to treat the earth in a utilitarian fashion, we must weigh long-term as well as short-term costs and benefits. We have an obligation to evaluate what impact our exploitation of the environment will have down the road for us and for succeeding generations. What will we leave for our children to use? What burdens will the changes we make in our environment impose upon them?

Mere short-term exploitation of the environment often yields only slight benefit for a few persons. In the tropical forests, the agricultural benefit the peasant farmer can derive from one plot is short lived. Though the slashing and burning of the forest converts some of the plant nutrients into ash, these are absorbed by the first crops or are leached down too deep in the soil for crops to recover, while other nutrients like nitrogen are lost. With nothing left to sustain yields, the farmer must soon search for another area to cultivate. The burned, used, and then abandoned field will not become fertile again for three to fifteen years, depending on the soil and environment.

Neither will the conversion of marginal land to sustained agriculture generally benefit that farmer. Without lime, fertilizer, herbicides, and pesticides, the land soon will have to be abandoned as unproductive. And if chemicals are provided for sustained use, questions must be raised about their long-term effect on marginal soils. In particular, to what extent will they permanently poison the soils and waters into which they enter, and how will they affect the humans who till the soil and drink the water?

Large ranches might benefit large cattle corporations, their stockholders, and customers in a more long-term fashion, but ultimately they will experience the same fate. The profits will not be sustainable, for eventually the land will produce less and less vegetation for the roaming cattle. As we noted above, after ten to fifteen years, new ranch land must be carved out of the primary forest. "In Brazil nearly all the ranches established before 1978 have been abandoned."[39] For the consumers, the destruction of what might, in the long run, be able to contribute substantially to their health (recall that one-fourth of our drugs come from tropical flora) seems a very high price to pay for cheaper hamburgers. For the exporting countries, exploitation of the immense natural resources in their tropical forests will be likewise short lived if it is done solely in an extractive way. Temporarily it may provide some relief from the pressures of foreign debt, as well as wealth for a few of the privileged, but once the resources are exhausted they cannot be reclaimed, and without resources the nation will be in even greater danger of adding to its foreign debt simply to sustain itself and of slipping further into national poverty. There is a lesson to be learned from nations like Nigeria and Thailand, which once were lumber-exporting nations but now must import lumber themselves.

One could proceed with this kind of utilitarian, anthropologically centered consideration to argue that we have ethical obligations not to rape the land and plunder its forests and minerals. In this view, the ethical obligations are not to the land and forest but to ourselves and our children. When we preserve the environment by careful management, it is not for the benefit of its nonhuman inhabitants; it is for the present and future human good. It is in our long-term interest to find ways to manage it successfully, to preserve what fertility it has and to increase it where possible, and to restore usages that will eventually benefit humans, whether in terms of foodstuffs, timber, medicine, or overall contributions to making the planet healthier and more fit for human habitation. "Although hunting, fishing, butterfly collecting, and wildflower picking provide immediate recreation, these species must also be managed with regard to their recreational use by future generations of humans."[40]

An Environmental Ethic

This anthropologically centered, utilitarian calculation is superficial, though not unimportant. To develop a true environmental ethic, we must go beyond

---

39. Thompson, p. 956.

40. R. Dale Guthrie, "The Ethical Relationship between Humans and Other Organisms," *Perspectives in Biology and Medicine* 11, no. 1 (Autumn 1967): 55.

considering the environment as a mere instrument for the realization of human happiness and well-being. We thus distinguish between an ethic of environmental management and a true environmental ethic.[41] In a management ethic the environment is treated strictly as an instrument for human benefit. Its moral use is to maximize human well-being, both short- and long-term. Hence the environment's proper place in moral evaluation lies in the means part of the means-end calculation. Since the environment has no moral standing in its own right — that is, since it is not part of the end for whose good we act — actions or effects are not weighed morally in terms of how they affect the environment or anything in it. Consequently, the environment's moral position is continuously precarious, depending on how humans evaluate its potential for human use.

Aldo Leopold gives the example of songbirds. "At the beginning of the century songbirds were supposed to be disappearing. Ornithologists jumped to the rescue with some distinctly shaky evidence to the effect that insects would eat us up if birds failed to control them. The evidence had to be economic in order to be valid."[42] From the management perspective, individuals and ecosystems can teeter on the brink, while humans debate their economic, recreational, pleasure-producing, aesthetic, or spiritual value for humans; and they can disappear should the calculation change.

In an environmental ethic, however, some things in the environment possess moral standing (possess moral worth independent of being means to another's good). As such, there will be times when the environment and its contents will be part of the "for whom" aspect of moral calculation. Actions and effects will be weighed not solely in terms of human advantage but also from the perspective of things in the environment that have moral standing. Actions will tend toward being good if propitious for them, bad if unpropitious. (We say "tend toward" because, as we shall see shortly, human interests and prima facie obligations must also be taken into account in deciding the morality of an action that affects the environment.) In a true environmental ethic, then, some things in the environment must be considered to be things that can be benefited or harmed and not merely means to *human* satisfaction.

41. Tom Regan, *All That Dwell Therein* (Berkeley: University of California, 1982), pp. 187-88.
42. Aldo Leopold, *A Sand County Almanac* (New York: Oxford University Press, 1949), p. 226.

Something has moral standing if it deserves moral consideration. It is frequently noted that we can determine which things have moral standing by noting what can have interests. However, "having interests" is an ambiguous phrase. One can have interests in the sense of being able to *take an interest* in things. I have an interest in my house because I have wants or desires with regard to it that I attempt to satisfy: I want it to have a new roof, a paneled study, a different colored carpet. Understood in this sense, having interests is a feature only of conscious beings, and perhaps only of self-conscious beings capable of forming beliefs. The emphasis on beliefs arises because to desire something requires that desirers have beliefs both about what they want and about how to attain it. This sense of interest is important when we consider the scope of our moral actions. But at most this sense of interest constitutes a sufficient, not a necessary, condition for ascribing moral standing; that is, if something self-consciously entertains beliefs, this means that we must give it moral standing. However, being able to take an interest does not appear to constitute an adequate limit or restriction to what has moral value; it is not a necessary condition. For example, we consider neonatal infants as having moral standing, even though they have no conscious awareness of what is in their interest and apparently can entertain no beliefs. One might reply that we ascribe to them moral standing because they have the potentiality for conscious awareness. However, we also give moral standing to comatose patients for whom there are no such prospects. For example, despite their condition it is immoral to abuse or mistreat them.

Accordingly, "taking an interest in" is not the sense of having interest that we use to ascribe moral standing. Rather, we employ "having interests" in our environmental ethic in the sense of *being able to be benefited or harmed.*[43] That which has interests can be improved or ruined by something that is done to it, though it need not be conscious of what is done. For example, just as we say that a neonatal infant has interests that must be protected because the infant can be benefited by actions taken regarding it (for example, surgery), so a tree can be benefited by, for example, the removal of strangling mistletoe. Though neither is aware of nor can consciously appreciate the beneficial action, each is benefited by the action. In both cases the benefit can only be appreciated consciously by an observer or proxy for that being.

These two senses of having an interest are obviously independent.

---

43. Kenneth Goodpaster, "On Being Morally Considerable," *Journal of Philosophy* 75 (1978): 308-25.

The interests we have (what we desire) might not in fact be in our best interests (as when a child desires to touch a hot stove). Similarly, what is in our best interests might not be what we desire (as when someone recommends that we stop smoking or overeating). And nonconscious things can have interests (can be benefited or harmed) quite apart from their realizing it.

Now though most (if not all) things in a given environment do not have interests in the first sense (if we are correct that having desires requires beliefs), many things can have interests in the second sense. For example, we can improve the health of a plant or a dog by removing insects from it; we can threaten the plant's health by neglecting to water it. It is not merely because the removal of ticks from a dog makes things better for *us* that removing ticks is good (though of course having our dog free from ticks might be a good for us); it is also good because it is better for the dog. The dog's life is improved by being made tick free.[44] Similarly with trees. Pruning a tree can benefit us because it makes the tree more productive; but it can also benefit the tree in that the removal of dead or diseased limbs keeps it healthy. Even particular ecosystems can have moral standing, insofar as actions can be taken to benefit or harm them. In short, humans are not the only things for which there are objective goods. Since actions can be taken regarding things in the environment that make life for them better or worse, they have moral standing and must be taken into account in making moral decisions.

This is not a new insight. Aristotle long ago argued that things in nature have their natural ends just as humans do. Indeed, according to him they achieve their ends, not consciously, but in terms of their formal and final causes. The rejection of the idea of formal and final causes, along with the mechanization of nature in the Enlightenment, resulted in the devaluation of nature and the corresponding adoption of an instrumental approach to it. Hence, in a real sense our environmental ethic returns to very ancient roots, though not accompanied by all the ancients' metaphysical baggage.

Unfortunately, moral appeals for restraining the rapaciousness by which we subdue or rule over the earth often are based on little more than management ethics. We are told not to destroy the trees because if

---

44. At the same time, of course, by removing the tick from its source of food we have not benefited the tick. This aptly illustrates the difficulties and ambiguities that lie in wait for those discussing how to intervene to preserve something's interests.

we do then future generations cannot use them or because once a country exhausts its resources it will be impoverished. Or we are told to preserve our wilderness because it meets our needs. One environmentalist writes, "We not only value the wilderness because of its own superlative values but because our experience in the wilderness meets fundamental human needs. These needs are not only recreational and spiritual, but also educational and scientific, not only personal but cultural."[45] But *mere* utilitarian appeals run counter to the concept of stewardship we introduced in Chapter Three. As we argued there, ruling or subduing the earth has to be done not only for the benefit of the Landlord and human beings but also for the benefit of what is ruled. When we consider the ruled to be an essential part of the stewardship ethic, we concede that it has interests that must be taken into account in making moral evaluations. Its benefit becomes part of the total ethical calculation. In short, with its tripartite structure, a stewardship ethic is truly an environmental ethic.

Why Worry about Interests?

Why, it might be asked, should we worry about the interests of plants and animals and ecosystems over and above our interests or when they conflict with our interests? The ruled are only sometimes conscious, and even when they are conscious, the issue of whether they are self-conscious and can appreciate the goods in which they have an interest is disputed. Why not simply adopt a management ethic?

Nontheists have difficulty establishing why it is that things in nature should be treated as having moral standing, as having value other than as a means to something else's good. One response is that they have moral standing because we give them moral standing. But this will hardly do, since without any objective grounds for such attribution it is open to the grossest arbitrariness. We can give moral standing to whatever we like and conversely (and perhaps perversely) withhold it from anything we like.

A second, more reasoned response is that things in nature have moral standing because they possess certain properties. But what properties must something possess in order for it to have moral standing? It is charged that

45. Howard Zahniser, "Wilderness Forever," in *Wilderness: America's Living Heritage,* ed. David Brower (San Francisco: Sierra Club, 1961), pp. 156-57.

often there is a kind of human egocentrism about any such list we might propose. Something has moral standing only if it has properties like those that I, the valuer, possess and that give me moral standing. Suppose one suggests that to have moral standing one must be capable of being self-conscious, appreciating and taking reasoned action to achieve its interests, and evaluating whether these interests are attained. On these criteria, only humans (and divine beings) would have moral standing. Animal rights activists, however, accuse those who make rational consciousness a necessary condition of moral standing of "speciesism";[46] they suggest that only beings who are conscious can have moral standing, since only they can act to achieve or avoid what is in their interests. Plant rights activists in turn accuse this view of "sentientism";[47] what gives something moral standing, they argue, is the fact that it is alive. But others might suggest that this last view smacks of a "vitalism" that overlooks the moral standing of inanimate nature, of the land itself; they would argue that what we need is a "land ethic." Beyond name-calling, the problem is to isolate the properties relevant to moral standing and to justify their selection. The appeal to rational consciousness seems to have some merit because the establishment of values requires a valuer, and only rational beings can take an interest in and hence value what is good for them. But as we have already seen, an environmental ethic appeals to the weaker sense of "having interests," that is, of being benefited by actions. Only if the stronger sense of "having interests" is used can rationality provide the criterion for ascribing moral standing.

Others suggest that moral standing is determined in terms of inherent goodness.[48] But why, it might be asked, is something inherently good? Proponents of this view admit that they have no satisfactory reply.[49] If x is inherently good but not y, there must be something about x that gives it its inherent goodness. But such an approach in effect reduces inherent goodness to the previous position — namely, that some property makes x good — and consequently it falls prey to the same objections. To specify any particular intrinsic property as necessary to moral standing would be arbitrary.

46. Peter Singer, ed., *In Defense of Animals* (London: Basil Blackwell, 1985), pp. 4-6.

47. Regan, p. 184.

48. James D. Heffernan, "The Land Ethic: A Critical Appraisal," *Environmental Ethics* 4, no. 3 (Fall 1982): 239.

49. Regan, p. 202.

One advantage of our ethical paradigm, then, is that it provides a basis for moral standing. The Christian argues that neither humans nor nature are inherently good. Value or worth is something that is not inherent in creation but is not merely utilitarian either. Value or worth — moral standing — is connected to God. Something is good because God, who is the ground of values, values it.

In the Genesis creation account God looks at creation — the land and seas, vegetation, heavenly bodies, and animals — and sees that it is good (1:4, 10, 12, 18, 21, 25). Indeed, everything is very good (1:31). There is no hint that these things are good only for humans, for most of God's declarations of goodness occur before God creates humans.

Psalm 104 similarly sees that the goodness of things in some cases can be independent of human valuation. There the Psalmist praises the God who "makes springs pour water into the ravines" (v. 10) so as to provide water to the beasts of the field, the wild donkeys, and the birds. The trees are given by God as a home for the birds and storks, whereas the "high mountains belong to the wild goats" (v. 18). "The lions roar for their prey and seek their food from God" (v. 21). Even the wild animals have moral standing.

However, the claim that things derive their value or moral standing from God is not unproblematic.[50] It reintroduces the ancient Socratic dilemma developed by Plato in his *Euthyphro:* Is something good because God commands or determines its goodness, or does God value it because it is good? Is goodness determined by God, or does God value what he knows to be good?

To opt for the former saddles us with the problem of divine arbitrariness, for if goodness or moral standing is determined by God's will alone, if it is merely a product of his proclamation, then there is no basis for his valuation. Indeed, there could not be; otherwise we have invoked an autonomous principle of moral valuation. To opt for the latter — that God values something because it is good — seems to return us to the problems noted above, namely, determining what properties provide for goodness or moral standing.

To address this dilemma, we must first look at the Genesis story. Careful reading does not indicate that God's valuation is a baseless pronouncement. It is not by a mere act of God's will that creation is seen to be good. Rather, the story says that "God saw that it was good." So what

---

50. The following objection was raised by Eric Hauer.

94

did he see that allowed him to determine that it was good? Unfortunately, the Genesis account does not tell us. But the issue is worth exploring, especially in light of twentieth-century philosophy's rejection of an ethical naturalism that holds that values can be determined solely from an analysis of natural properties.

Creation's goodness cannot lie merely in the fact that God created it. It is not as if God created something and by virtue of his act he decided that it is good. Such would be suspect, for what is there about being a maker that allows or provides for the maker to assign goodness? I might paint a prairie landscape and determine or assert upon its completion that it is good. But my bold assertion bears little weight unless I can point to some objective properties that provide the basis for ascribing worth to the painting.

Furthermore, created things have no set of common physical properties in virtue of which they would be good or have moral standing. Rationality, sentience, or life does not suffice, for the created earth, seas, and heavenly bodies that God saw were good lack these features.

One possible basis for God's ascription of goodness has to do with the functions and functioning of the created world. It is reasonable to claim that at the end of each creative period, God saw creation's goodness in the way each thing was performing its assigned or proper functions within the economy of creation. As omniscient, God knew the natural order and what its functions were, and in the case of biological organisms he aligned creation so that the functions were fulfilled. Goodness in the biological sphere relates to the created world realizing its functions.

It is important to note that this view of goodness differs from the above problematic account of goodness as a property. "Goodness is not a property of things in the usual sense of property at all. To say that a thing is good is not like saying that it is round or square, or pink or blue, or late or early, or above or below. . . . Instead, it seems to be what some . . . call a consequential or supervenient property." That is to say, it is a property that a thing has in virtue of other properties that it possesses. "The properties of a thing can be the sources of its goodness or value . . . just insofar as they are properties that evidence the perfection or complete actuality of the thing in question."[51] Goodness, then, is a supervenient property that a thing has insofar as it realizes its potential or achieves its

51. Henry B. Veatch, *The Ontology of Morals* (Evanston: Northwestern University Press, 1971), p. 109.

function. In this sense goodness is not arbitrary but instead is grounded objectively in the teleological character of biological nature.

This view does not put goodness independent of God, for it is God who created objects with the particular functions that they have. Goodness is thus neither simply intrinsic to things nor completely autonomous of God. There is a subjective element in determining goodness and moral standing, for God sees creation as good and values it (sees it is desirable) insofar as creation functions according to his design and purposes. Since he established the design and purpose, nature satisfies his desires. At the same time, however, the account provides an objective ground for goodness insofar as the functions of things in nature are determinable and really satisfy the ends desired by God.

This account should not be taken to mean that organic functions cannot change or evolve over time, as the environment changes. We do not understand creation in a static way that precludes adaptation and change. God's dynamic relation to nature is progressive, not punctual.

To return to and summarize our main argument, if one is going to develop a stewardship ethic that includes ruling, it will be one that considers not only the good or benefit of the rulers and the One on whose behalf they rule but also the good or benefit of the ruled, for the ruled, like the steward, has value bestowed on it. The steward is to act in the interests of all three parties: God, human beings, and nature. It is illegitimate to appeal to the biblical injunction to subdue the earth to justify raping it. Here we take issue with those who argue that being a Christian commits one to holding that "God planned all of this explicitly for man's benefit and rule: no item in the physical creation had any purpose save to serve man's purposes. . . . Christianity . . . insisted that it is God's will that man exploit nature for his proper ends."[52] To the contrary, ruling is to take into account the interests of the Landlord, ruler, and ruled; hence mere exploitation is immoral. Subduing must also be seen in light of benefiting the ruled, for it, too, has interests. Its interest is not *only* to serve human interests.

---

52. Lynn White, Jr., "The Historical Roots of Our Ecological Crisis," *Science* 155, no. 3767 (10 Mar. 1967): 1205.

The Ethic of Stewardship: Caring for the Earth

According to the Genesis story, God put the man in the royal garden to work and tend (care for) it. The command to rule thus involves the injunction to *care for* or tend.

Some might think that from our contention that the ruled (namely, creation) has moral standing it follows that the best way to care for the environment is to leave it alone. The environmentalist Tom Regan, for example, contends that if an object or thing has moral standing, then the proper attitude toward it should be respect: "to treat it *merely* as a means to human ends is to mistreat it. Such treatment shows a lack of respect for its being something that has value independently of these ends."[53] This seems true enough.

But what follows from not treating something merely as a means to our own gratification? Regan goes on to derive what he terms the "preservation principle," namely, that if we respect something, then our approach to it should be one of "nondestruction, noninterference, and generally, nonmeddling."[54] Regan's preservationist approach would respect the tropical forests by removing humans, or at least all civilized humans, from the endangered areas. There would be no hunting or fishing except by the few indigenous tribespeople. Mining, timbering, and agriculture would be prohibited. Dams, airports, highways, and hotels would not be constructed. People would enter the preserve only as passive, noninterfering observers of the tropical bounty. Preservation is for Regan a moral imperative. He does not claim that interference in nature is never justified, but interference cannot be justified simply in terms of human interests.

One should be clear here. This type of preservationist is not arguing that we should take steps to preserve each of the species or even all of the ecosystems. To the contrary, the preservationist approach is that we take no steps at all, that we withdraw our steps as much as possible from nature. We are to tread so lightly on nature that we leave minimal imprints. This preservationist approach is a noninterfering approach; nature and evolution are to resume their courses unimpeded by humans.[55] Species and

---

53. Regan, p. 200. The idea of respect traces back to Leopold, p. 220.
54. Regan, p. 200.
55. Regan appears to ride the fence on whether the preservation principle is prima facie or absolute, though he tends in this passage (pp. 200-201) toward the prima facie view when he suggests that it would be appropriate to interfere with a river that is silting

ecosystems will naturally disappear and evolve. The catalyst, however, will not be human agency but rather the very same agents that brought about the present diversity in our natural ecosystems.

Why should one think that *noninterference* rather than *conservation* devolves from the respect that one owes to things that have moral standing? Our stewardship ethic considers the benefit for both the ruled (here, nature) and the steward. Hence, though we are not to treat nature *merely* as a means for human satisfaction, nature can be treated as a means to benefit humans. Utilitarian considerations are not triumphant, but they are relevant.

When we bring in considerations related to caring for creation to benefit human beings, we find that merely leaving nature alone — mere preservation — is unjustified. Rather, the proper approach is conservation, which involves a thoughtful balance between the preservation of nature and its resources and the careful, appropriate use of the environment's resources for human benefit in ways that leave the environment ecologically sound, biologically diverse, and aesthetically pleasing.

There is, of course, a pragmatic argument for conservation-minded intervention as opposed to mere noninterference. Total bans on the use of things in the environment are ultimately unworkable. In the matter of the tropical forests, the pressures of human survival, international debts, and corporation drives for quick returns on investment would mean that any general ban on cutting trees would be ignored. One cannot tell a starving peasant that he should not burn another patch of forest to plant maize to feed his family, even if it contains the last specimen of a certain tree species. The hungry subsistence fisherman will not spare the manatee he has just captured, nor will the lifelong logger give up his livelihood for the northern spotted owl, even if it is the last one on earth. Humans have always had an impact on their environment, and they always will.

But mere unworkability should not determine our ethic. The justification for our intervening in nature must derive from our obligation to benefit human beings as well as nature. Since nature, including the tropical forests, produces things that benefit humans, nature appropriately can and should be used for human benefit. This might sound like the old utilitarian justification for exploitation, but it is not, for our

---

up naturally, to "preserve or increase what is inherently valuable in nature." But, one might ask, why is a "wild and free" river more intrinsically valuable than a muddy, lethargic river?

use of nature must be accompanied by the respect for nature that derives from its moral standing. A mixture of both preservation and conservation is called for. Nature's resources must be harvested in such a way that where they need not be tapped to meet basic human needs, or where their extraction will cause irreparable harm to nature, they are preserved; where they are tapped but restorable, they are restored; and where they are tapped but not restorable, they are used wisely, and the place from which they are taken is restored either to its natural state or to some other aesthetically acceptable, productive use with its own sustainable ecosystems.

Conservation produces both immediate and long-term benefits for future generations. It also functions to value nature, for it rules out destroying the environment merely for human benefit; it affirms that nature has moral standing, and hence the inroads we make on it should be as nondestructive of ecosystems as possible. This is not always possible; to turn prairie into farmland will replace one ecosystem with another. A conservationist approach, however, will find a place and role for both.

Inequality of Humans and Other Species

It might be objected that if nature has moral standing, we cannot use it at all for our benefit. From this preservationist or noninterference viewpoint, everything that has moral standing has equal standing. So if prairie has equal standing with farmland, the northern spotted owl with the jobs of the Northwest loggers, there are no grounds for destroying the former to create or preserve the latter.

Dave Foreman, founder of Earth First!, puts this view radically and graphically. "An individual human life has no more intrinsic value than does an individual Grizzly Bear life. Human suffering resulting from drought and famine in Ethiopia is tragic, yes, but the destruction there of other creatures and habitat is even more tragic." Consequently, we are justified in not taking action to prevent Ethiopians from starving, for though their plight is lamentable, to save them might wreak even worse havoc on their already fragile environment. Even "such apparent enemies as 'disease' (e.g., malaria) and 'pests' (e.g., mosquitoes) [are] not manifestations of evil to be overcome but rather [are] vital and necessary components of a complex and vibrant biosphere. . . . The preservation of wildness and native diversity is *the* most important issue. Issues directly affecting only humans pale in comparison."[56]

To opt for the equality of humans with ecosystems, to argue that "we must return to being animal," stands contrary to our paradigm. The stewardship ethic will not argue for equal standing. The doctrine of *imago Dei* places both unique position (privilege) and unique responsibilities on human beings. Only humans are made in God's image, and hence there is a prima facie case for human preference. Thus, where there is conflict between human needs and environmental moral standing, human needs should predominate — to the extent that genuine needs are represented.

Determination of genuine needs has proved a thorny issue, especially as it relates to the matter of scarce resources. At the very least, genuine needs consist of what is necessary for human survival. Heffernan sets forth this principle: "The survival interests of human beings ought to outweigh those of the rest of the biotic community and the survival interests of the rest of the biotic community ought to outweigh the nonsurvival interests of human beings."[57] Though clearly stated, his restriction of needs to mere "survival interests" makes the principle too narrow, for we have genuine needs beyond those having to do with mere survival. It is true that we need nutritious food, clean water, and adequate clothing and housing. But we have spiritual and aesthetic needs as well. In assessing human needs, it is important not to reduce them to the merely quantitative and material.

We must be careful, however, not to confuse needs with wants. Coats are necessary for those living in the colder climes. But coats made from the white skins of Harp seal pups are not a necessity. Housing is essential for humans, but it need not require expensive redwood decking, the procurement of which will bring about the demise of the remaining old-growth redwood ecosystems along the Pacific coast.

In short, while our stewardship ethic allows us to use nature to meet our needs, the *imago Dei* confers unique responsibilities upon us. We have the ability to alter the environment more drastically than any other organism. This is where conservation (including preservation) becomes an obligation. In changing the world to benefit us, we must exercise the greatest care and caution to conserve and preserve what exists, in terms of both individuals of various species and (more importantly) ecosystems.

56. Dave Foreman, "Putting Earth First," *Confessions of an Eco-Warrior* (New York: Harmony Books, 1991), in *Environmental Ethics,* ed. Armstrong and Botzler, pp. 422-23.
57. Heffernan, p. 246.

The needs to which we appeal to justify environmental change must be genuine and substantial. Mere profit does not qualify. The use of nature as the means to satisfy those needs must be carried out with the highest respect paid to nature. This means that neither human nor ecological interests can completely exclude the other, but though human interests might have preference, ways to create balance must be sought. In short, we opt for a middle ground between mere noninterference (which treats all creation from an egalitarian perspective) and instrumentalism (which treats creation as a mere means to further human ends).

Solomonic Balance

It goes without saying that theory is easier to delineate than to apply. An example of the difficulty of balancing concerns is the seemingly intractable problem of the conflict between the interests of the northern spotted owl (and the ecology of the old growth forests it inhabits) and those of the Pacific Northwest lumber industry.

At the same time that Congress passed the Endangered Species Act in the mid-1970s, the northwest lumber industry began to harvest timber by clear cutting. This method affected the old-growth forests' ecosystems, as well as soil and water quality. Erosion from the land polluted the rivers and estuaries, threatening their wildlife. Plummeting salmon populations threw the long-established fishing industry into chaos.

One of the apparent casualties of clear cutting was the northern spotted owl. In the mid-1980s, biologists estimated that less than two thousand pairs survived in the Pacific Northwest's virgin timber. They argued that the owl's habitats are restricted to old-growth forests, that it requires a large hunting area, and that its reproductive rate is low. Consequently, continued encroachment on its territory by the lumber industry had put its survival in jeopardy.

Environmentalists took the northern spotted owl to symbolize the degradation of the Pacific Northwest's entire old-growth forests, which were going the way of old-growth forests across the country (see figure 1). Invoking the Endangered Species Act, environmentalists appealed to the Bush administration to prohibit harvesting old-growth forests in the Northwest to save the owl. The timber industry responded that to accede to such a demand would bring dire economic consequences on the region. Reducing by one-third the amount of timber cut on federal lands would

FIGURE I. The Diminishing Forest

KEY ■ **Remaining old-growth forest**
▨ **Younger regenerated forest**

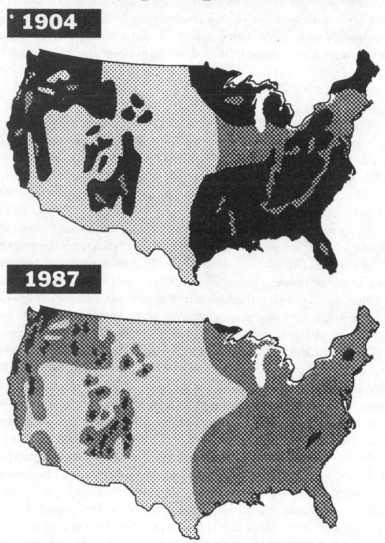

1904

1987

jeopardize the jobs of thirty thousand loggers, significantly lower real estate prices in the region, and devastate both the local and state economies, which are heavily dependent on revenue from the timber industry.

The Bush administration offered a plan that set aside more than seven million acres of federal lands to preserve the owl (this was later reduced to 5.7 million acres, permitting logging in some areas of the owl's habitat to preserve lumbering jobs). But because the plan failed to protect the spotted owl's habitat, it violated the Endangered Species Act. The matter was taken to court, and in 1991 Judge William L. Dwyer imposed a harvesting moratorium on 24 million acres of federal land. This resulted in doubling timber prices in the U.S. and significant job loss in the affected region.

When he took office, President Clinton attempted to fulfill a campaign pledge to achieve a compromise that would satisfy both sides of the debate. He sought to reduce the amount of old-growth forest lumbered to 1.1 million board feet each year, 20 percent of the peak harvest in the 1980s. To compensate for the 9,500 jobs the government estimated would be lost, he infused over a third of a billion dollars in aid yearly to the region, for everything from job retraining to tourism development. Yet this compromise, too, proved unsatisfactory to both sides, who in turn sued. The plan was enough, however, to get Judge Dwyer to lift the logging ban on June 6, 1994.

Clinton's mediation attempt points to the need, in this case as in others, to achieve a Solomonic balance between competing interests. Several questions arise.

(1) Are the human needs genuine? To some extent they are, since limits on logging jeopardize the livelihoods of the loggers and those dependent upon them. Yet lumbering in which more trees are cut than replaced is a short-term occupation. Indeed, it is estimated that at the current pace of harvesting, much of this timber resource will be gone in the next decade, in which case the industry will have done to itself what any government action to preserve the owl would have done. And in fact many jobs in lumber processing have already disappeared, for much of the timber was shipped to the Orient as unprocessed logs.

(2) Is the timber being put to good use, whether in the United States or abroad? Merely because it is there and can be cut and sold provides no grounds for cutting it. Even the fact that the trees will die and rot (go to waste) provides no justification, for waste is itself an important part of the ecosystem.

(3) What effects would destroying all the old-growth timber have on future quality of human, animal, and plant life in the region? The current old-growth ecosystem regulates water levels and quality, cleans the air in a region that is becoming increasingly polluted, enriches and stabilizes the soil in areas subject to substantial erosion, and supports a rich ecosystem of plants and animals. It also protects the rivers as fish habitats, thus contributing to other human industries such as fishing.

(4) Are the claims about the decline of the indigenous ecology, including those about the owl, accurate? Gregg Easterbrook argues that the claims about the owl's status are exaggerated. Recent population estimates are much higher than earlier ones; further, owls are found to be thriving in new-growth forests in Northern California, where small mammals such as wood rats proliferate in the heavy undergrowth.[58] At the same time, there is little dispute that the larger ecosystem of the region is being significantly affected by the lumbering methods used. Any ultimate loss of this ecosystem, as has occurred elsewhere, would be a blemish on our stewardship.

(5) This raises the question whether there might be other ways of resolving the problem. Easterbrook suggests that selection logging and shelter-cutting (where trees are removed in clumps so that the forest canopy remains) could replace clear cutting. Another option would be to accommodate two distinct types of forest areas, managed forests for timber production and old-growth or natural forests. Both options would preserve wildlife habitats and create sustained timber income, while reducing pressure on the old-growth forests. The debate would concern not only the size of each and how they would be related geographically to each other but also how the management could be accomplished in economically sound ways.

(6) Our own timber resources need to be developed and managed in ways that also remove some of the pressure to seek timber elsewhere, particularly in Third World countries such as Indonesia, Malaysia, and New Guinea, which are also environmentally sensitive but which, because of their development needs, manifest fewer ecological concerns and less stringent laws.[59]

---

58. Gregg Easterbrook, *The New Republic* 210, no. 13 (28 March, 1994), 22-23. This argument was made earlier by James Owen Rice, "Where Many an Owl Is Spotted," *National Review* 44 (2 March 1992): 41-42.

59. Philip Shenon, "Isolated Papua New Guineans Fall Prey to Foreign Bulldozers," *New York Times* 143 (5 June 1994): A1, 6.

Our goal here is not to suggest some specific amount of acreage to be set aside or a specific management plan; rather, it is to illustrate that while mere preservaiton is not what our ethic calls for, neither will our ethic allow us to overlook the good of that over which we are stewards. It calls for stewardship of the forest's resources, its unique ecosystems, and its individual members in ways that benefit the ecosystem and what lives in it, as well as present and future human beings. Thus creative ways must be found to balance the needs of the loggers (including ways to help them find other, meaningful occupations), of those dependent on forest resources (including we consumers), and of things in the environment and their ecosystems. In short, changing nature must be done with two eyes open: one cast on the humans for whose benefit we make changes and whose real needs are to be met, and one cast on the nature that we change and that we can benefit or harm.

This larger issue relates to questions with which we began this chapter, namely, the relation between developing and developed countries. The developing countries want development; the developed countries have interests (at least in part) in the developing countries' preserving their resources. How can the two goals be reconciled? Preservation alone would maintain the present disparity among earth's peoples; exploitation alone fails to recognize the interests of the environment.

One response, proposed by a 1989 G7 (the group of leading industrial nations that includes the United States, Britain, France, Germany, Canada, Japan, and Italy) conference called to address biocthical issues, stressed sustainability, stewardship, and quality of life. The proposal recommends

> *Sustainable development,* defined not as a fixed state of harmony, but as a process of change in which the exploitation of resources, the direction of investments, the orientation of technological development, and institutional change are constantly readjusted to reconcile present and future needs. Sustainable development is a process of social and economic betterment that satisfies the needs and values of all interest groups, while maintaining future options and conserving natural resources and diversity.[60]

60. "A Code of Environment Practice," Sixth Economic Summit Nations Conference on Bio-Ethics, Brussels, 10-12 May 1989; printed in *Environmental Dilemmas: Ethics and Decisions,* ed. R. J. Berry (London: Chapman & Hall, 1992), p. 256. For a discussion

The document goes on to recommend "*Full accounting of costs* to ensure that the stock of renewable resources is maintained constant, with the amount of waste or pollution kept below the assimilative capacity of the environment; and non-renewable resources either substituted or harvested minimally."

The proposal is laudatory, for it seeks to encourage both development that fulfills socioeconomic needs and conservation accompanied by minimal exploitation of nonrenewable natural resources and diversity. It does so while holding both anthropocentric and biosphere concerns — though, as one would expect in a document prepared for a summit sponsored by an economic organization, anthropocentric concerns predominate.[61] And interestingly enough, it proceeds from a secular perspective that invokes a stewardship model.[62] The document contains many specific and important recommendations for stewardly caring with which we would concur.

The Ethic of Stewardship: Filling the Earth

Above we questioned why we should preserve species. One reason might be that there is a value to diversity. This brings in the third stewardly injunction, to *fill* the land. As we suggested in Chapter Three, although

---

of applied sustainable development, see Francesca Bray, "Agriculture for Developing Nations," *Scientific American* 271, no. 1 (July 1994): 30-37.

61. To be accurate, the document hints at rather than overtly develops the nonutilitarian value of nature. For example, from the claim that "since we are dependent for survival, health and psychological well-being on the physical integrity of the biosphere . . . , an environmental ethic that encourages responsibility to the human use of natural resources is an instrument rather than an opponent of our self-interest," the document concludes with the non sequitur, "Thus, environment ethics are more than self-interest" (pp. 254-55). The rest of the paragraph addresses not the "more than," but another issue, namely, sharing responsibility. Even the existence values of things in nature are given an economic twist (p. 256).

62. Unfortunately the document does not address the secular stewardship model itself. For example, to whom is one accountable on this model? Is it to other humans or future generations? This would seem to leave us with an anthropocentrically based environmental ethic. Is it to nature? But then what would account for the document's strong anthropocentric tone? It is worth noting that this document and its stewardship ethic were Christianized by being made the basis of the 1991 Church of England's statement on environmental ethics.

the concern in the Genesis accounts was quantitative in terms of producing mere numbers, there are other ways of looking at filling. Here we might suggest that filling can be seen from the perspective of diversity.

Diversity benefits nature itself. Diversity of species is important in an instrumental sense, for distinct species often contribute uniquely to the overall ecology. Certain plant species in the tropical forests are dependent upon particular species for reproduction. "The fig genus, the most distinctive and widespread of plant genera in the tropics, comprises more than 900 species, *each* of which is pollinated by its own species of wasp."[63] Similarly, the durian tree of the Malay Peninsula is pollinated by only one species of bat, which in turn finds its primary source of nectar in those trees. Diversity within a species is also necessary for its survival. When, for example, one strain of plant is attacked by a pest, a different strain with a slightly different genetic structure might survive. Consequently, preserving species diversity is in the interest of and benefits the environment and its diverse organisms.

Diversity benefits humans as well. The benefit might be physiological, for as we noted above, half of our pharmaceuticals have originated from wild plants. We also can use the genetic resources of wild plant varieties to make our domesticated crops more disease resistant or productive. But diversity also has a psychological benefit, insofar as it both produces aesthetic enjoyment and provokes the wonder that often occasions the very activity of science.

There is also within Western thought the tradition that diversity is desired by God. Thomas Aquinas argued that God, in loving himself, wants to see himself multiplied. That is, since God is goodness itself, God wants goodness to be multiplied. But there can only be one divine essence. Consequently, the divine essence must be multiplied as best it can in the creation that is like or replicates God. God, then, in loving himself and willing the good, wills that the universe be diverse to manifest this goodness.[64] This diversity is more than quantitative; it applies to more than mere individuals. Because it is finite, each created thing can emulate him only in part, imperfectly. How each thing emulates God depends on its particular nature or structure. Hence, for God to have a creation that maximally emulates his goodness, there must be a diversity of types of things — species, if you like.[65] In short, the plenitude of species arises from the very goodness of God, and

63. Myers, p. 85.
64. Thomas Aquinas, *Summa Contra Gentiles* 1.75.3.
65. Aquinas, 2.45.2 and 6.

each in some way emulates the divine. Because it emulates the divine, showing forth his goodness, diversity of species is deemed good by God, and this in turn provides the basis for the moral standing that we have accorded creation.

It would be presumptuous to suggest that God must have only one reason for diversity. The reasons might be complex, including instrumental reasons (diversity is necessary for nature to satisfy diverse human biological needs), ecological reasons (diversity is necessary for development, stability, integrity, and success of organisms in their ecosystems), religious reasons (diversity is part of the creation's testimony to the goodness and fullness of the Creator), and aesthetic reasons (diversity is necessary for beauty and human epistemic challenge). Indeed, the diversity might provide for God's own aesthetic enjoyment.[66]

Conclusion

What shall we do about the tropical forests? About the old-growth forests in northern California, the Pacific Northwest, even Siberia? We might side with the individual who recommended that we "have a moratorium on destruction, like a nuclear test ban treaty, until we have time to figure out how to use the forest constructively."[67] This sounds reminiscent of the moratorium called for by those involved in recombinant DNA research in 1974, and it would provide a temporary solution at best.

The action taken with regard to the forests must be in accord with our three injunctions. The forests are to be *ruled,* not as mere servants of human need and whim, but as things that have a unique value given to them by God, as things that have interests and can be benefited and consequently deserve both respect and appropriate human action for their benefit. Forests are to be *cared for,* both for their own good and for ours. In this respect, conservation provides the moral middle ground between rapaciousness (which fails to acknowledge the moral standing of the forests and their inhabitants) and mere preservation (which fails to see that

66. The report "Christians and the Environment," adopted by the Church of England in 1991, speaks about "the pleasure God finds in creation at large and divine concern for its well-being embrac[ing] both wild and tame creatures, whether they are attractive or repulsive, valuable or apparently useless for us."

67. Thompson, p. 962.

morally the forests can provide for human benefit as well). Stewardship means that humans have a responsibility to see to it that forests survive in their diversity and that they survive in ways that benefit present and future generations of organisms and of human beings — especially (but not restricted to) those persons who live proximate to those ecosystems. Some balance must be struck between the changes necessary to make the forests productive for humans and the preservation of those forests as an intrinsic part of nature's economy and ecology. And finally, forests are to be *filled* with the diversity that is necessary for their own ecological survival, for human survival and enjoyment, and for divine fulfillment.

Our study of the causes of the problems shows something beyond all this, namely, the fundamental realization that environmental ethics cannot be separated from ethical reflection about other areas of human life. As we have seen, the root causes of the destruction of the tropical forests, as indeed of much of our environment, include poverty, population explosion, social structures that move peasant farmers onto poor land, the demands of developed countries for luxury goods or cheap goods, greed leading to exploitation, and wastefulness. Conservation of the tropical forests must include measures to control human population growth, to find family-sustaining jobs for the disenfranchised rural poor, and to curb the developed world's exploitation of the developing world's resources. It must address the spiritual problems of human rapaciousness and greed, which seek personal short-term benefits at the expense of others and the environment.[68] It must address problems of justice between individuals and between nations, so that some nations do not have to destroy their natural resources in an effort to satiate the ever-hungry international debt. In short, no environmental ethic can hope to succeed without tying into a broader ethic that considers social, economic, political, and spiritual problems and obligations, and that assesses other relevant moral goods.

---

68. Lynn White puts it well: "Since the roots of our trouble are so largely religious, the remedy must also be essentially religious, whether we call it that or not. We must rethink and refeel our nature and destiny" (p. 1207).

CHAPTER FIVE

Stewardship and Assisted Reproduction:
——— Motives and Rights ———

I N 1969 the English physiologists Robert Edwards and F. D. Bavister
and gynecologist Patrick Steptoe performed what is believed to be the
first successful fertilization of a human ovum outside a woman's body.[1]
The process required minor surgery on a woman to remove a mature egg
from one of her ovaries. The surgeons placed the egg in a nutrient solution
biochemically similar to the environment found in the fallopian tubes.
They then introduced into the solution sperm that had undergone capaci-
tation (the removal from the sperm's surface of the chemical inhibitors
that prevent the sperm from penetrating the ovum). About a day after
fertilization, the egg began to divide.

But fertilization was only half of the process needed to procreate
humans using assisted reproductive techniques. The fertilized egg also had
to be implanted successfully in the uterus, there to develop into a viable
fetus. To do this Edwards and Steptoe injected the prospective mother
with hormones to prepare her uterus to receive the implanted egg. They
then picked up the morula, the size of the period at the end of this sentence,
with a hollow tube and carefully inserted it through the cervix into the
uterus, where they hoped its outer cells would attach to the uterine wall.

---

1. R. G. Edwards, R. D. Bavister, P. C. Steptoe, "Early States of Fertilization *in
vitro* of Human Oocytes Matured *in vitro*," *Nature* 221 (15 Feb. 1969): 632-35. Prior
claims were made in the 1940s by the gynecologist John Rock and in the 1950s by Landrum
Shettles and also by the Italian scientist Daniele Petrucci. The last two were never independently confirmed, and their claims are generally rejected. There is also scepticism
regarding whether fertilization rather than mere segmentation took place in the first case.

Though Edwards and Steptoe had tried as early as 1975 to implant pre-embryos[2] in women incapable of conceiving naturally, no attempt culminated in a birth until 1978. Suffering from blocked fallopian tubes, Lesley Brown underwent an unsuccessful operation to correct the condition in 1970. After she suffered through years of infertility, the remains of her tubes were cut off in 1977. She learned of the work of Edwards and Steptoe and contacted them to see if they could help her. In November 1977, Steptoe gave her fertility hormones to stimulate her ovaries. Then, employing a laparoscope inserted through an incision in her abdomen, he removed an ovum with a suction needle and fertilized it in the laboratory with her husband's sperm. Three days later he implanted the eight-cell pre-embryo in her uterus. The implantation was successful, and on July 25, 1978, Lesley Brown gave birth to five-pound twelve-ounce Louise Joy Brown, the first recorded human birth from in vitro fertilization.[3]

Motives for Doing Assisted Reproduction

The dizzying pace of the development of reproductive technologies since the birth of Louise Brown attests to the desire of prospective parents to have children and to the scientific community's desire to accommodate them. Though satisfying this desire is the primary motive for many researchers, the motivational network is often complex.

2. There are various terms for the stages of human development. The term *morula* is used to refer to the fertilized egg after it has differentiated into a solid ball of about eight to sixteen cells, while *blastocyst* refers to a ball of cells with a surface layer and an inner cell mass. *Embryo* refers to the stage of development from two to eight weeks, while the developmental stage after the eighth week is termed *fetus*. Though technically inaccurate, the term *embryo* is often employed in ethical and scientific discussions to cover the prefetal stages. This gives the impression that the development of what is implanted is farther along than an eight-, sixteen-, or thirty-two-celled organism. To avoid this false impression, we will refer generally to what exists in the period between fertilization and two weeks using the term *pre-embryo*.

3. P. C. Steptoe and E. G. Edwards, "Birth after the Reimplantation of a Human Embryo," *The Lancet* 2 (12 Aug. 1978): 366.

The Motive of Notoriety

For some researchers, the primary motive is to attain renown. To be the first to bring about a birth by using artificial means provides opportunity for both fame and fortune. Those who succeed receive the attention of the media and the accolades (as well as initial scepticism) of their peers. In the case of Louise Brown, reporters lined the hospital's waiting room and hounded doctors, nurses, and parents alike in their search for "the story." Even prior to Louise's birth, the Browns had signed an exclusive contract to tell their story for more than a half million dollars. Parent, newborn, and physicians alike shared the limelight.

Edwards and Steptoe were not the first to report success in the race to bring about conception by in vitro fertilization (IVF). In 1974 the English scientist Douglas Bavis claimed that he had implanted pre-embryos in three women who had each given birth to healthy children. However, no names were provided, and his alleged achievements were never confirmed.

News of Steptoe's success brought laments of lost fame from the American side of the Atlantic. Researchers in the U.S. protested that the race to beget children by IVF was not entirely fair,[4] for in 1975, under a federal order, the Department of Health, Education, and Welfare (HEW) was prohibited from funding IVF experiments unless they were approved by a national ethics advisory board to be appointed by the HEW secretary. This board was not established until January of the year Louise Brown was born, and even then it never approved any IVF proposal prior to its dissolution by President Carter in 1979.

But other firsts were still possible: the first successful conception using frozen sperm, the first surrogate mother artificially inseminated, the first implantation of frozen pre-embryos, the first surrogate mother to bear a child having no genetic relation to her, and the first cloned child (yet to be done).[5]

4. *Time* magazine reported that U.C.L.A. obstetrician Jaroslave Marik bitterly noted, prior to Louise Brown's birth, that "if all the pull and pressures had not been applied, there might be an American woman now about to deliver" an artificially conceived baby (*Time* 112 [31 July 1978]: 62).

5. David Rorvik's claim that it has been accomplished (*In His Image: The Cloning of a Man* [Philadelphia: Lippincott, 1977]) is disregarded as a hoax.

The Profit and Eugenic Motives

The motives of others are more explicitly monetary. For these, assisted reproduction provides a ready means for profit. Investors have sought to capitalize on the desire of infertile couples to have children by establishing sperm banks from which women can select sperm for artificial insemination. Idant established the first sperm bank in Manhattan at the end of 1971. As of 1989, through both its own artificial insemination program and the sperm that it shipped out to physicians and clinics across the country, it assisted in creating more than 30,000 pregnancies. As of 1985 there were more than fifty sperm banks in the U.S., and in 1987 there were more than 23,000 inseminations using donor sperm.[6]

Guidelines for sperm collection and dissemination, developed in 1979, are promulgated by the American Fertility Society, though following the guidelines is voluntary. As of 1991, only New York, Michigan, and Illinois inspected sperm banks. For their legal protection, sperm banks usually carefully screen donors to guard against the transmission of certain diseases. When potential donors come to the sperm bank, they complete a questionnaire asking about their family health history, sexual life, drug and alcohol usage, the amount of exercise they get, and diet, as well as details about education, hobbies, talents, and interests. Once the sperm is collected, it is analyzed for sperm count, shape, and motility and for sexually transmitted diseases and hepatitis. More rarely the sperm is tested for serious genetic diseases like Tay Sachs. The five to twenty vials collected per ejaculation are then frozen to -180° in nitrogen vapor. Two days later one vial is unfrozen to see how the particular donor's sperm have survived. As a safety precaution the donors are asked to return six months later for another HIV test, since HIV can lie dormant for six months before showing up on tests. After all is said and done, BioGenetics, a New Jersey firm, rejects 90 percent of the men who want to be donors on a variety of grounds, from life-style to poor sperm quality.[7] The data collected from the information sheets that sperm donors complete are computerized and put into a catalogue, which is disseminated to clinics, doctors, or individuals who request it. BioGenetics sends out 1,500 to 1,800 vials of sperm per month at $120 each.

At fertility clinics women are given a form on which to select donors according to race; ethnic background; physical characteristics such as eye

6. Benedict Carey, "Sperm, Inc.," *In Health* 5, no. 4 (July/Aug. 1991): 52-53.
7. Carey, p. 53.

and hair color, weight, and height; and social characteristics such as education (Idant accepts only college-educated donors), hobbies, and talents. Women thereby can attempt to control the kinds of characteristics they want their children to exhibit.

Some sperm banks operate with more eugenic motives in view. In the late 1970s Robert K. Graham established a sperm bank in California called the Hermann J. Muller Repository for Germinal Choice. His goal was to carry out in a small way the eugenic program of the late Nobel geneticist after whom it was named. Graham recruited Nobel laureates to contribute sperm to his sperm bank, and at least five complied. He then advertised for women under thirty-five with high IQs who were willing to become inseminated with the sperm of one of these elite. A condition of acceptance was their promise to send to Graham regular reports on the health and IQ of the children conceived.

There is no guarantee that by using the sperm of a Nobel laureate a woman will bear a child with the same high intelligence or drive, or that a person conceived with the sperm of a handsome, musically talented, six-foot four-inch surfer will emulate those characteristics. For one thing, the line separating genetic and environmental determinants of human development cannot be (and probably never will be) clearly drawn. Nor is it entirely clear what or to what degree behavioral or personality traits have a genetic basis and are inheritable. At the same time, if genetic breeding in plants and animals provides any indication of the results for humans, it is possible, if not very probable, that superior physiological and psychological strains generally yield superior progeny.

Profound ethical issues surround eugenics. Without much thought we engage in eugenics in plant and animal breeding and find it not only acceptable but desirable. The seed signs planted in all the Midwestern corn fields amply testify to the prevalent practice of hybridization. There is little food produced in the U.S., whether it be grains, vegetables, fruit, or meat, that has not been genetically modified. But if eugenics can be practiced without moral qualms on the nonhuman world, why should we object to using it on humans? We take up this issue in Chapter Seven.

The Third-Party Altruistic Motive

Another possible motive would be to use assisted reproduction to help third-party individuals not directly involved in the reproductive activity.

114

Recently researchers have become interested in how transplanted fetal tissues can aid people suffering from various debilitating and deadly diseases, including Parkinson's disease, diabetes, and Hurler's syndrome. Not only are fetal cells undifferentiated (in that they can grow into many types of cells) and can join currently existing tissue where they are inserted, but they also grow rapidly and generally are accepted by the recipient's immune system. Though experiments to date have used tissues from normally conceived fetuses, it does not stretch the imagination to see that the same techniques could be applied to fetuses conceived by assisted reproductive techniques.

In 1987 a team of researchers and clinicians in Mexico led by Ignacio Medrazo reported that by grafting fetal brain cells into the brains of two people suffering from Parkinson's disease they were able to partially control the patients' disease symptoms.[8] The fetuses used were neither conceived nor aborted with this transplant use in mind; the tissues from spontaneous abortions were made available for transplantation. Yet following these reports, some people who had Parkinson's sufferers in their family hinted that they would be willing to conceive a fetus to be aborted for the therapeutic use of a family member.[9]

To conceive, let alone to use reproductive technology to assist conception, for this purpose is morally abhorrent because it treats the procreated individual merely as the means to satisfy the health demands of another. The benefit to others might be substantial, as in the case of Parkinson's disease sufferers, yet it comes at a high cost to the potential child, for the acquisition of the fetal tissues requires the death of the fetus. As we argued in Chapter Four, it is morally wrong to treat living things *merely* as means to others' ends, for as God's creation living things have value derived from God. This particularly applies to human beings, who are made in God's image. Creating beings who have the potential to be persons, with the sole purpose of sacrificing their total good for another, fails to fulfill our obligations of stewardship toward them, for it neither gives them a choice in the matter nor respects their interests.[10]

8. Ignacio Medrazo et al., "Transplantation of Fetal Substantia Negra and Adrenal Medulla to the Caudate Nucleus in Two Patients with Parkinson's Disease," *New England Journal of Medicine* 318 (7 Jan. 1988): 51.

9. Temar Lewin, "Medical Use of Fetal Tissue Spurs New Abortion Debate," *New York Times,* 16 Aug. 1987, p. A1, 30. See also Mary Anne Warren, Daniel Maguire, and Carol Levine, "Can the Fetus Be an Organ Farm?" *The Hastings Center Report* 8, no. 5 (Oct. 1978): 23-25.

10. For an advocate of conception for this purpose, see John A. Robertson, "Rights,

One cannot get around this argument merely by denying the personhood of the fetus. No matter what status one gives to the fetus — actually human, potentially human, developing toward full personhood — it is a living organism with interests and an inherent potential to become a human person with a meaningful life of its own; its creation and subsequent termination solely for the good of a third party, without any consideration of its own potential good, is treating it merely as a means and hence is immoral. What it can and possibly would be in its own right is ignored.

Literature that insensitively reports medical research for public consumption often encourages the fear that researchers will misuse fetuses by treating them only as means to others' ends. For example, *Newsweek,* in a report on the use of fetal tissue, entitles a black box illustration showing an eight- to nine-week-old human fetus "The Sum of Its Parts." It goes on to list the potential uses of the fetus, connecting by lines the respective uses to parts of the pictured fetus. The accompanying article notes that "the ovaries of a 2-month-old fetus contain a few million eggs. . . . If such a transplant can give a woman born without ovaries a chance to conceive, the abortion of one fetus could give life to a whole family of babies."[11] In such a sensational context it is easy to overlook or even be sceptical about the accompanying statement by Health and Human Services (HHS) secretary Donna Shalala: "HHS would make sure that the new frontier of fetal research and therapy did not become an excuse to encourage more abortions or cheapen fetal life."[12]

On the other hand, we cannot support the position of those who hold that fetal tissue procured from elective abortions should be prohibited from being used in research and transplantation. In 1985 President Reagan banned the National Institutes of Health (NIH) from using fetal tissue from induced abortions in transplantation procedures; in May 1992 President Bush continued this ban with a presidential veto. Only in 1993, on the third day after he took office, did President Clinton lift the ban by executive order. We agree that abortion must not be conducted for this end. Yet responsible stewardship also makes it morally unacceptable to

---

Symbolism, and Public Policy in Fetal Tissue Transplants," *Hastings Center Report* 18, no. 6 (Dec. 1988): 8.

11. Sharon Begley et al., "Cures from the Womb," *Newsweek* 121, no. 8 (22 Feb. 1993): 51.

12. Begley, p. 49.

waste human tissues that could significantly benefit people who are suffering from neurological and other diseases. Such waste shows a callous disregard for those who have been entrusted to us.

At the same time, this procedure stands upon a thin moral line that must be carefully patrolled, for it begins at the point of a tragedy — the taking of a life. We can guard against, though unfortunately not guarantee prevention of, abuse in using cadaveric fetal tissues by following standard practices regarding taking transplantable organs or tissue from deceased persons.[13] (1) Health care providers must separate any decision related to the acquisition of the tissue (e.g., whether or not the woman wants an abortion, or whether an abortion ought to be performed and how it is to be performed) from any decision regarding its use in transplants (e.g., whether it will be transplanted; optimal gestation age for fetal tissue; who is to get the tissue). (2) It is important to separate those who perform the abortion from those engaged in the subsequent use of the donated (fetal) tissues. In the case of fetal tissue transplants, this helps to prevent taking undue risks that might endanger the woman in order to obtain a fetus whose organ tissues are viable for transplant. (3) The buying and selling of organs (here fetal tissues) should be prevented, including the transfer of funds to support the abortion clinic or process. At most there should be reimbursement of the cost (not profit) for obtaining, storing, transporting, and implanting the fetal tissue. (4) All fetal tissues should be dispensed by a national fetal tissue bank. This would deter profit seekers from capitalizing on the misfortunes of the fetus and prevent the intentional conception or termination of fetuses for transplant into particular patients. Safeguards such as these can preserve the moral integrity of those using aborted fetuses, whether the abortion was induced or spontaneous, to benefit others. This approach properly separates the question of the use of fetal tissues from the ethical question of the morality of abortion.

Critics who are at the same time opponents of abortion object to the use of cadaveric fetal tissue on two grounds.[14] First, they contend that

13. See the recommendations of the 1988 NIH Fetal Tissue Transplantation Research Panel, which concluded (though not unanimously) that the federal government should, with careful monitoring, lift the moratorium on federal funding for the use of fetal tissue in research. *Human Fetal Tissue Transplantation Research* (Bethesda, MD, 14 Dec. 1988).

14. We want to distinguish the case where the fetus is healthy and "viable" from the case where the fetus could not come to term and hence is intrinsically at risk, so that an abortion is necessary to protect the life of the woman or to prevent unnecessary

using another's tissue requires consent, and that is precisely what is impossible in this case. Consent is deemed necessary to protect the dignity of the one whose body will be used. Though generally it is proper and necessary to get either the prior consent of the donor or the present consent of the donor's family as a proxy before organs can be taken for transplant, in the case of fetal tissue the consent of the fetus cannot be obtained. Some suggest that in lieu of this the woman be asked to give proxy consent. She "remains the proper decisionmaker about the disposition and transfer of fetal remains," for she still has a special connection with the fetus.[15]

But in consenting to the elective abortion, has she not voluntarily severed that special connection and thereby forfeited her role as proxy decision maker? "When a [potential] parent resolves to destroy her unborn fetus she has abdicated her office and duty as the guardian of her offspring, and thereby forfeits her tutelary powers."[16] Furthermore, contrary to the NIH panel's contention that "the dead fetus has no interests that the pregnant woman's donation would violate," the fetus is not dead while the woman is pregnant. Hence, to obtain proxy consent from the woman or family prior to the abortion might jeopardize the dignity of the fetus one is hoping to protect. To obtain consent before the woman determines that she wants an abortion might significantly color or influence her decision. It might put the still-living fetus at additional risk in that it might make it easier for the woman to decide to have an abortion, believing that at least some good might come out of it. Even obtaining consent after she has determined that she wants an abortion might pose a risk to the fetus, for it is still alive and potentially unaborted when the consent is sought. It is always possible for the woman to change her mind and for the fetus to be saved.[17] In short,

---

complications. In this case, proxy consent would be appropriate, for the fetus is not at risk simply on account of the decision of the woman or family but is intrinsically or contextually at risk.

15. James F. Childress, in his statement to the Advisory Committee to the Director, NIH, *Human Fetal Tissue Transplantation Research,* C7.

16. James Tunstead Burtchaell, "University Policy on Experimental Use of Aborted Fetal Tissue," *IRB: A Review of Human Subjects Research* 10, no. 4 (July/Aug. 1988): 7-11; reprinted in *Taking Sides,* 5th ed., ed. Carol Levine (Guilford, CT: Dushkin Publishing Group, 1993), p. 273.

17. The NIH advisory panel also stated that those involved in the transplantation of fetal tissue should "accord human fetal tissue the same respect accorded other cadaveric human tissues." The recommendation of respect is ironic, considering that before the not-at-risk fetuses were intentionally aborted, they were given so little respect that their

the proxy consent of the woman's family or of the woman who voluntarily terminates the life of a not-at-risk fetus should be not required or even sought.

Neither should the health care providers involved be petitioned to grant consent, for they possess no special standing with respect to the fetus that would morally confer proxy power over the fetus's disposal. That leaves the state to decide. However, Burtchaell argues that the state cannot grant permission either, for "if the State agrees . . . to consign to research the remains of only those fetuses who have perished under the ultimate abuse, that inevitably places the State in a position of patronage toward their destruction. The State would . . . also be implicitly derelict in its protective powers." In effect, he concludes, one cannot proceed to use fetal tissue in research or for transplants since there is no one who can rightfully grant consent and since consent is necessary to overcome the presumption that a person's body should remain inviolate at death. "How we treat human remains is both a function and a cause of our bond with human persons. . . . If we honor a person while she is living we have no choice but to honor her body after death. To confiscate it discredits all ostensible dignity we accorded that person *in vivo* and orients us to treat still other persons with contempt."[18]

Burtchaell's argument clearly presumes that the fetus is a human being, a claim with which some will disagree. But suppose it has this standing, what kind of disposal of the fetal remains will give the fetus dignity? Mere disposal in refuse containers, as is the current practice, does not ensure this. But would formal burial with appropriate rites do anything more? Would not a greater dignity be found in turning what is, in many cases, an evil into a greater good? Since there is, on Burtchaell's analysis, no one in a position to grant permission, and since to waste the valuable tissues is itself an indignity against the value of humans, it would seem consonant with dignity to use the tissues to meet the basic needs of others.[19]

---

life could be taken virtually at the pregnant woman's will. At the same time, this obligation to respect fetuses clearly follows from our stewardship ethic, though, we would argue, it goes beyond merely treating their tissue with dignity. Kenneth L. Woodward et al., "A Search for Limits," *Newsweek* 121, no. 8 (22 Feb. 1993): 53.

18. Burtchaell, pp. 273-74.

19. Some might see in this the basis for a general argument in favor of obtaining organs from the deceased where no objection is registered, a practice the reverse of what we do currently when we require consent. We would not oppose such an application, though the practice must be surrounded by carefully crafted procedural guidelines and an attitude of respect, not vulturism.

The language used here can be of critical importance. Burtchaell speaks of confiscating or requisitioning someone's body after death. This language is negative and contrary to human dignity. But one can equally speak of dignifying the remains by seeing that good comes from their use, of reclaiming the fetus's loss through the life given to others. This language cannot be used to justify abortion, but it does serve to provide another understanding of preserving dignity. Clearly, the answer to the question of whether such use is contrary to the fetus's dignity turns on what postmortem acts are consonant with dignity — disposal and deterioration, or careful use for the good of others. Our stewardship ethic suggests the latter.[20]

The second argument put forward by critics is that the use of fetal tissue is immoral because someone is benefiting from the immoral practice of abortion and from the death of another, innocent individual, and to benefit from an evil act is to be a party to it. Burtchaell, for example, argues that one can be held morally accountable for an action by complicity, even if one has not performed it. He notes four different types of moral complicity in evil: (1) complicity by active collaboration in the deed, as when one drives the getaway car in a bank robbery; (2) complicity by indirect association that implies approval, as when "an economist secured admission to the workings of a red-lining real estate operation in order to study the racial discrimination at work there" (here the activities were done without the consent of the victims of the redlining); (3) complicity by failure to prevent the evil when possible, as "when an employer turns a blind eye upon careless safety compliance by workers"; and (4) complicity by shielding the perpetrator from penalty.[21] Burtchaell sees the use of aborted fetuses for research or transplantation (he does not address transplantation specifically but would apply the same argument) as an instance of complicity by indirect association implying approval. The other three do not apply because the researcher did not actively collaborate in the deed, is not in a position to prevent abortions, and is not in a position to shield the person doing the abortion from penalty since currently no penalties accrue for conducting legal abortions in the United States.

This is an important argument, for on the one hand it raises the serious question about moral complicity in evil acts and about the degree

20. It is important to stress that we are speaking about using dead fetuses only, not sustaining for these purposes the life of fetuses scheduled for abortion.

21. Burtchaell, pp. 274-76.

120

to which one should be able to benefit or bring good from others' evil actions. Though there are cases that satisfy Burtchaell's second type of complicity (his own example of the economist is not clearly one), there are also cases in which one does benefit from the evil acts of others but no complicity can be alleged. For example, there are no good grounds for thinking that to accept the transplanted heart of a murder victim is to be guilty of complicity with his assailant. Yet to reject the heart would be to refuse to bring good out of evil. What is difficult in cases of alleged complicity is to decide which evils we should bring good out of while maintaining independent moral judgment and which evils leave so tainted a product that they irreparably taint the user and his or her judgment.[22]

Why should one think that use of aborted fetal tissue in transplantation is a case of complicity? Burtchaell claims that it is because the person who transplants fetal tissue "resorts to the abortionist as a ready supplier of tissue from unborn humans who have been purposely destroyed."[23] The physician does not make a one-time use of the person conducting the abortion but becomes that person's ready customer. Physicians transplanting fetal tissue could not continue to function without aborted tissue.

Yet this is true of people who perform transplants in general. They could not be in the business without a steady supply from cadavers. Yet this does not suggest that they approve of the death of the people whose organs they procured or of the way in which they died. Were this a general principle, the mortician would be in a most precarious moral position.

Burtchaell is mistaken in claiming that the person who uses the results of immoral acts, even on a regular basis, thereby gives implied approval and hence is guilty of complicity and is an accessory to the immoral act. One is a party to the evil in cases of indirect association only if the action one takes in turn compromises one's judgments; if, as Burtchaell well puts it, "my association with the evil causes me to be corrupted alongside the principal agent." But even the regular use of tissues from elective abortions does not necessarily suggest or imply that the user approves of abortion or that the user's judgments or character have been

---

22. A current example of this concerns the debate over whether the results of Nazi experiments can be used, even if they could possibly be beneficial. Kristine Moe, "Should the Nazi Research Data Be Cited?" *The Hastings Center Report* 14, no. 6 (1984): 5-7; Mark Sheldon et al., "Nazi Data: Dissociation from Evil," *The Hastings Center Report* 19, no. 4 (July/Aug. 1989): 16-18. See also Arthur L. Caplan, *When Medicine Went Mad: Bioethics and the Holocaust* (Totowa, NJ: Humana Press, 1992).

23. Burtchaell, p. 277.

compromised. It does show that the user is willing regularly to bring good out of evil, to benefit others in need.

The Motive of Benefiting the Infertile

Returning to the question of motives, the most important motive behind the development of methods for artificially assisting human reproduction is the sincere desire to assist couples who are incapable of having children. According to the Public Health Service, there are 2.3 million married couples in the U.S. with women of childbearing age who are infertile (have been unable to conceive after twelve months of intercourse). "During the period 1981-1982, physicians reported a record 2 million visits for infertility problems, twice as many as for comparable two-year periods. Also infertility among younger women seems to be on the increase. In 1965 infertility in women aged 20 to 26 was estimated to be 4%; in 1982 the figure was 11%."[24] This means that a large potential clientele exists that can benefit from this research and developed technology. It is this motive that will inform the rest of our discussion.

Methods of Assistance

Modern medicine has developed a number of ways to assist women to become pregnant, of which we shall mention only a few. Undoubtedly the least ethically controversial involves administration of hormone therapy to either the male or the female, though it carries with it the risk of multiple births. Surgery to correct physiological abnormalities is likewise ethically acceptable. For example, surgery employing fiber-optic technology can be used in some cases to open obstructions in a woman's fallopian tubes.

But for some couples these procedures still do not result in a pregnancy. For example, the biochemical structure of the woman's vaginal lining might be overly toxic to sperm; or her uterus might be abnormally structured, making it difficult for the sperm to reach the fallopian tubes. Or for some reason the male might not be able either to produce or to

24. Ronald Munson, *Intervention and Reflection,* 4th ed. (Belmont, CA: Wadsworth, 1992), p. 469.

deliver enough sperm to impregnate the woman. In such cases, artificial insemination, with or without controlled stimulation of the ovaries, provides a way of overcoming the difficulties. Artificial insemination may use the husband's sperm (AIH), or the sperm of a donor (AID), or a combination of the two (CAI). Bypassing the toxic conditions of the cervix, sperm is injected directly into the uterus. The overall success rate of artificial insemination is generally high (40 to 70 percent),[25] though several attempts are usually necessary to procure the results. From 1986 to 1987 an estimated 172,000 women in the United States were artificially inseminated. Approximately 35,000 births resulted from AIH and 30,000 from AID.[26]

For others, however, even more radical steps are necessary. For such persons, in vitro fertilization (IVF) holds some promise. Figures on success per retrieval cycle vary. Some suggest a low figure, claiming that about 14 percent of implantations result in pregnancies, only 10 percent of which come to term.[27] Others suggest a success rate between 20 and 25 percent.[28] Success depends on a variety of conditions, including methods of oocyte retrieval, techniques of fertilization used, potency of the sperm, freshness of the oocytes, and age. For example, a young woman aged twenty-five years or less has a 35 percent chance of achieving a clinical pregnancy and a 28 percent chance of delivering a live infant per IVF retrieval cycle, whereas a forty-year-old woman under similar circumstances has only a 12 percent chance of establishing clinical pregnancy and a 6 percent chance of giving birth to an infant.[29] Approximately 3,100 babies were born in each of the years 1990 and 1991 using this technique.[30]

As originally carried out, each attempt at IVF required that fresh eggs be extracted from the woman and fertilized. Acquisition of the eggs

25. Kamram S. Moghissi and Richard Leach, "Future Directions in Reproductive Medicine," *Archives of Pathology and Laboratory Medicine* 116 (April 1992): 439.

26. Office of Technology Assessment, Congress of the United States, *Artificial Insemination Practice in the United States* (Washington, DC: Government Printing Office, 1988).

27. Joe Leigh Simpson and Sandra Ann Carson, "Preimplantation Genetic Diagnosis," *New England Journal of Medicine* 327, no. 13 (24 Sept. 1992): 952.

28. David Levran et al., "Pregnancy Potential of Human Oocytes — the Effect of Cryopreservation," *The New England Journal of Medicine* 323, no. 17 (25 Oct. 1990): 1153.

29. Moghissi and Leach, p. 437.

30. Robert Pear, "Fertility Clinics Face Crackdown," *New York Times* 142 (26 Oct. 1992): A15.

necessitated multiple invasive surgical procedures. Technical developments now allow multiple eggs to be obtained nonsurgically through a needle inserted through the vaginal wall into a follicle.[31] The development of freezing techniques further reduces the need for the woman to undergo repeatedly even this uncomfortable procedure. Several mature eggs can be extracted and fertilized at the same time. The pre-embryos that are not implanted are frozen until needed. This means that should the initial implant fail, other pre-embryos are readily available for use.

Further developments have gone beyond implanting the fertilized egg in the uterus. Researchers have begun to implant zygotes (one-celled fertilized eggs) directly into the fallopian tubes (ZIFT — zygote intrafallopian transfer). A more promising procedure is GIFT (gamete intrafallopian transfer), where a mixture of sperm and eggs is introduced into the fallopian tubes. Here fertilization can take place within the natural environment. Whereas IVF-ET (embryo transfer) achieved a success rate of live births of 14 percent per attempt, GIFT reached 23 percent.[32]

A complicating factor is cost; each implant procedure costs between $3,000 and $8,000, averaging around $5,000. Since several attempts are usually necessary, the procedure can require a substantial outlay of funds, sometimes exceeding $20,000.[33] Since such procedures generally are not covered by medical insurance,[34] only the more affluent in society can afford to have children in this manner. However, this is not unique to assisted reproduction; in the field of medicine and in many other areas, often only the affluent can initially make use of advances in technology. Only later when procedures can be carried out or technology produced inexpensively and made widely available can the masses afford them.

For some women even this development is insufficient. They may suffer from uterine disease or may have had a hysterectomy, making it impossible for them to bear a child. In other cases, where for example the woman has heart trouble, pregnancy would be dangerous for the woman. In still other cases it might simply be an economic hardship or an inconvenience for a professional woman (e.g., a fashion model, physician, or

---

31. In 1989, "87% of all retrievals were done by ultrasound and only 8% by laparoscopy" (Moghissi and Leach, p. 436).

32. Moghissi and Leach, p. 437.

33. Pear, p. A15.

34. For a court case to the contrary, see Paul Reidinger, "The Cost of Conception," *ABA Journal* 76 (July 1990): 83-84.

corporate executive) to bear a child. For such women, there is the option of employing a surrogate mother to bear the child. Another woman is artificially inseminated with sperm from the husband, implanted with a pre-embryo produced by IVF from the egg and sperm of the prospective parents, or helped to conceive with GIFT. When the child comes to term, the child is turned over to the "adopting" parents.

Still other techniques such as cloning or the raising of children in artificial wombs are more futuristic and perhaps not even possible but nonetheless provocative of ethical reflection.

How shall we evaluate these endeavors to fulfill the desire to have children? What are researchers and health care providers required to do? What are they permitted to do? In short, what are the ethical implications of this desire to have children and of the activities that scientific researchers and medical personnel undertake to fulfill it?

The Ethic of Stewardship: Filling

In Chapter Three we constructed an ethical paradigm from the Judeo-Christian account of creation. We suggested that it contained three basic injunctions. Let us explore each to ascertain how it is relevant to questions about assisted reproduction, focusing in the remainder of this chapter on the injunction to fill, and turning to the injunctions to rule over and to care for in the following chapter.

There is a coherence between the divine command to fill the earth and both the natural desire to have children and the human psycho-physical structure that makes its fulfillment possible. Not only do we have sexual desire, but the very processes necessary to conceive children produce the psychological pleasure that motivates us to fulfill the desire.

But does having this desire and the requisite psycho-physical structures translate into the *right* to have children? Holding that we have a right to have children is a stronger contention than maintaining that we are structured psycho-physically to be able to procreate. The latter describes something about us; the former is a normative statement invoking considerations about what ought to be the case. Since rights entail obligations, to have the right to have children implies that others have some sort of obligation with respect to our procreative activity.

Before we proceed we must be clear on precisely what is meant by the claim that people have a right to have children, for the term *have* is ambiguous. One possible interpretation treats "having" as possessing something. The right is to *possess* a child, while the corresponding obligation on others is to provide a child for the claimants. Interpreted in this way, the claim that we have such a right is morally suspect, for it treats human beings as objects that people can own if they want and that some other person is obligated to provide for claimants, much as one might procure a furnace, stove, or television for people who legitimately demand one. Children are not an item of property to be owned.

A second understanding of the proposed right to have a child treats "having" as a process. The right to have children is the right to bear or father a child. Though the result is a child, the right has to do with the process rather than the product. Others have the obligation to make it possible for the claimant to bear or father a child. How this obligation might be understood varies, as we shall see shortly. We will focus on this process sense of "right to have a child" in our subsequent discussion.

Rights and Reproduction

Rights are frequently divided into three categories. *Natural rights,* a concept that arose in the Enlightenment, concern the inherent rights we have to such things as our life, liberty, property, and pursuit of happiness. Natural rights stem from what is necessary to maintain the integrity of the individual as an individual: the freedom to be, say, and do what we want (within limits imposed by our communal existence).

If we have a right to bear or father children and if it is a natural right, it would mean that others are obliged to leave us alone in our procreative activity. The right to have children is the right to have the liberty to procreate. The state, for example, should not dictate who can or who cannot have sexual intercourse or which women may or may not bear children. This right to bear or father children is frequently invoked in opposing state-sanctioned compulsory sterilization. This practice, it is argued, violates a fundamental natural right.

This is not to say that natural rights can never be overridden; there are times when depriving someone of their natural rights is justified and legitimate (e.g., in criminal incarceration). But where there are exceptions, the burden of proof rests on the person who denies another's rights to provide a clear justification for overriding the natural right of liberty.

Similarly with the natural right to procreate. Supposing that we have such a right, it does not follow that it can never be overridden by more significant concerns. Thus, for example, a parent might restrict a minor's liberty rights, on the grounds that he or she is not yet an autonomous decision maker and that his or her future autonomy must be protected. The restriction, however, is temporary until autonomy is gained.

Sterilization is a more radical choice, for it permanently removes the right to procreate. In the rare case where involuntary sterilization might be allowable, it will be on the grounds that the person restricted neither is rational or autonomous nor has a reasonable prospect of becoming so, and that sterilization will significantly benefit and protect the freedom of the nonautonomous. In such cases someone must be appointed as the guardian of that person's natural rights and interests.

The claim that the right to bear or father children is a natural right is important for addressing controversial cases in sexual ethics such as sterilization. However, our concern in this chapter is with the infertile and their potential claims for assisted reproductive technologies. Nothing follows from this natural right about whether infertile persons can demand from society, physicians, clinicians, or researchers that they assist them in bearing or fathering children. The procreative right as a natural right is the right to be left alone, whereas the infertile need interference with their infertility. Only if they are "interfered with" in ways that will bring about conception, only if some action is taken by those with technical knowledge and expertise, will the infertile be able to bear or father a child. By being left alone the infertility continues and the person's claim on their right is unmet. Hence, if the infertile are going to be able to claim that others are obliged to assist them to overcome their infertility, some notion other than natural rights must be invoked.

A second category is *rights based on special relationships*. By entering into a relationship with someone I obtain certain rights with respect to them. For example, if someone borrows money from me, I have the right to be repaid. This is an example of a right arising from a *consensual* relationship entered into voluntarily by both parties. We also enter into *nonconsensual* relationships that confer rights. For example, being born into a family gives the child the right to be nourished by the parents and the parents the right to be obeyed by the child. Though the child did not consent to being born to these parents nor the parents to have that particular child, the resulting relationship confers mutual rights and obligations.

If we have a right to bear or father children, there is reason to think that it is based on a special relationship. From the Christian perspective marriage provides the social and moral context in which procreation ought to take place. Marriage confers upon spouses conjugal rights that allow them to make claims on their spouse to perform sexual acts with them that can lead to procreation (1 Cor. 7:1-5). Interpreted in this way, the right to have children must be understood in a weak sense, for though entering into marriage confers the obligation to perform the natural acts that make it possible for someone to father or bear a child, the obligated cannot see to it that a child is procreated. They cannot bring about a pregnancy as a fulfillment of a claim.

Where does this leave the infertile? Though the right to have children creates obligations on another (i.e., the spouse) to engage in acts that might lead to pregnancy, the infertile need the assistance of someone other than their spouse to fulfill their desire to bear or father children. Do they have the right to demand this assistance on the ground that they have a right to have a child? Since the infertile are not naturally in a special relationship with a health care provider, the infertile cannot make a claim on the health care provider to take actions to assist them to bear or father a child on the ground of rights based on special relationships. Of course, they might enter into a consensual relationship with a physician or clinic to perform, for example, AIH. Once they enter into such a relationship, the infertile will have rights against those with whom they have contracted. But we are concerned with the question whether the right to have children might provide the basis for the infertile to demand that a physician, clinic, or researcher enter into a relationship to foster pregnancy. Understood in this way, a right based on special relationships does not apply.

A third category, termed *human rights,* includes a wide variety of rights: a right to vote, to have a national identity, to obtain employment, to be paid a fair living wage, to have a free public education, to social security, to equal access to society's goods, and so on.[35] With human rights, those who are obligated to fulfill the claims are generally asked to do more than simply to leave the claimant alone; they are to see to it that the claimant is provided for in some way. They are given a ballot, social security, a job, education, etc.

Some who claim that we have a right to bear or father children treat

35. These are spelled out most notably in the United Nations Declaration of Human Rights, adopted by the General Assembly in December 1948.

it as a human right, for human rights frequently are rights that demand that someone provide goods or services for claimants in a certain way.[36] Possession of a human right to procreate children would obligate someone to fulfill that right, that is, to bring about a pregnancy according to one's wishes. To claim that we have a human right to bear or father children raises at least two important issues.

First, is there such a right? If one holds that rights exist only where their fulfillment is possible, only recently could bearing or fathering a child be a general right at all. Only a few cases of artificial insemination were reported prior to this century, and human in vitro fertilization began in the 1970s. Is the presence or absence of human rights this dependent upon what we can provide by our technology? If so, this lays a heavy burden upon the creators of reproductive technology, for in creating the technology they simultaneously create human rights for individuals in a society.

Second, there is this even more problematic question: Against whom does someone have the alleged right to bear or father children? It is not a right against all individual members of society; I cannot approach just anybody and demand that he or she fulfill my right to bear or father a child. At the very least, this would be infringing on the natural right of persons to freely choose whom they want to marry and with whom they want to procreate. It cannot be a claim against the government, for governments cannot bring it about that somebody bears or fathers a child. Neither is it a claim against society at large, for society, like government, cannot cause a pregnancy.[37] The ground for excusing both the government and society at large is that there can be no obligation where the obligated cannot do anything about the alleged obligation. Consequently, the only persons the infertile could make a claim against would be the health care providers and the researchers who could lend them assistance, for these are the only persons who can do something about the infertility.

According to the American Medical Association, the "physician shall be dedicated to providing competent medical service with compassion and

36. John A. Robertson, "Surrogate Mothers: Not So Novel After All," *The Hastings Center Report* 13, no. 5 (Oct. 1983): 32-33.

37. Where the government or some particular segment of society was an identifiable cause of the infertility, it might be required to compensate persons for their infertility. For example, A. H. Robbins Company has a moral responsibility for the infertility caused by its negligence in manufacturing and marketing the Dalcon Shield contraceptive. But the right to be compensated for damages caused by gross negligence that resulted in infertility is not the same as the right to bear or father children.

respect for human dignity," though at the same time it affirms that a physician shall "be free to choose whom to serve."[38] The first clause leaves it open whether there is such a right as we have been speaking about, for it depends upon how one unpacks the concept of human dignity. If dignity implies this right, then physicians should satisfy claims made on the right, since thereby they are rendering a service to humanity consistent with this dignity. The second clause, however, gives physicians the prerogative to refuse to accommodate infertile persons, despite their demand. Though when a physician takes on a patient "he may not neglect him,"[39] this does not obligate the physician to fulfill this alleged right. Perusal of medical codes reveals that no such right is recognized by the health care profession. Physicians are obligated to serve and to "render service" in an emergency (referring undoubtedly to taking lifesaving measures), but the alleged obligation to aid the infertile does not fall under the category of emergency measures. That correcting infertility is not recognized by the health care profession as an obligation does not mean that having a child is not a right, but it does indicate that health care providers do not consciously intend to obligate themselves in this regard when they take up their profession.

Whether this is a burden that researchers take on when they study gynecological problems is a difficult question. The ethics of investigation certainly require people to use *well* what is discovered or created, but does it require them to *use* what is discovered? Research that has to do with lifesaving measures would be an obvious example where the obligation to use would follow from the research, though even here some restrictions having to do with on whom it is appropriate to use it would apply. One might appeal to the contention that as stewards we are to care for others, including the infertile, and that this obligation to care for others imposes obligations on researchers to use what they have discovered for the human good. More broadly still, perhaps the alleged right to have children creates a general obligation on science, but if so, it is a strange right, for there is no one in particular who is obligated to satisfy it. We will return to this general issue in Chapter Eight.

In sum, there seems to be no general human right to bear or father children that creates in others the obligation to fulfill that right. Bearing

38. *American Medical News,* 1/8 Aug. 1980, p. 9.

39. The 1957 ethical code of the American Medical Association, *Journal of the American Medical Association* 164, no. 10 (6 July 1957): 1119-20.

or fathering children is, from one standpoint, a privilege that is ours by virtue of our biological structure. It can also be an action that, in certain contexts, others might be obligated to facilitate. Spouses have conjugal rights regarding each other; that is, marriage confers consensual obligations to engage willingly in sexual activities that might result in procreation. Facilitation of pregnancy can also become an obligation of health care providers when they enter into special relationships with clients with the specific goal of making it possible for the couple to bear children. What becomes obligatory, however, is not responding to the general right to have a child, but rather responding to claims for assistance in conception set in the context of voluntarily assumed special relationships.

This coheres with our stewardship ethic, which in responding to the command to fill emphasizes the relationships that the steward establishes with God, with other stewards, and with creation. Procreation best grows out of relationships, not rights. We are to engage in sex with our spouse not because we have a right to it, but because it expresses our love for our spouse and accords with God's intent for creation. Engaging in sex is one way we can fulfill that intent. Health care providers also contribute to responding to the command to fill the earth when by virtue of their chosen occupation they respond to and assist those who desire but by themselves cannot have children. Again, however, this is best done not as responding to a right, but rather in the context of a vocation. One chooses to assist people in their legitimate reproductive concerns as a vocation in which one serves the Landlord as a good steward over the part of creation entrusted to one.

Who May Fill the Earth?

In addition to raising the question of rights, the injunction to fill the earth introduces another important question: To whom is this injunction addressed? Are all persons, married or unmarried, psychologically and emotionally qualified or unqualified, of any age, enjoined or permitted to fill the earth? Or are there some restrictions regarding who morally should undertake this filling role?

Traditionally, childbearing in most societies has been followed by childrearing within the family unit of husband and wife or more commonly within the extended family. Frequently both parents were in attendance to foster the offspring's development, though generally both parents did not contribute equally to the upbringing. The major burden fell upon the females (mother, older sisters, women in the extended family). Of course this did not prevent conception or rearing from taking place outside the family structure. Illegitimate children were not uncommon, nor was it unusual for fathers to be absent for extended periods of time at sea, work, or war. Yet society recognized that conception outside marriage and the family structure seriously breached morality and that raising children in such a state was not ideal. Such children unjustly carried a stigma, though their predicament was not of their own doing.[40]

In the final decades of the twentieth century, the traditional family structure is being altered. Some might say more strongly that it is under assault. For a significant sector of American society, single parenting is now the norm, usually forced, at times desired. One out of two marriages now end in divorce, and 61 percent of American children will live in a single-parent family sometime before the age of eighteen.[41] One of every four children is born out of wedlock; 17 percent of births to white mothers in 1990 were nonmarital, 57 percent of births to black mothers.[42] The number of households with children headed by single male parents increased 92 percent between 1980 and 1991; during the same period households with children headed by single female parents increased 25

40. As Rabbi Barry Freundel explains (in an unpublished response to a questionnaire generated by a subgroup for the Genetics, Religion and Ethics Conference, March 1992, Houston, TX, pp. 21, 25), this is very much the case within Judaism:

> On the father's side, there is a predisposition in Jewish law to maintain family ties and integrity even against circumstantial evidence to the contrary. In cases of extended absence (longer than nine months) of the husband, pregnancy was attributed to the action of demons delaying the husband's sperm. In cases of racial anomalies appearing in children, imaginative genetic theories are advanced to explain and maintain the integrity of the family. . . . Judaism has a concept of illegitimacy . . . that precludes an individual so stigmatized from ever marrying (or having any of his descendants marry) within the community. Because of the devastating nature of such a stigma, Jewish law works to preserve legitimacy, even in the face of circumstantial evidence to the contrary. For example, if one knows information about a family previously thought to be legitimate that would question its legitimacy, one is enjoined against coming forward with that information.

41. "Single Parents," *American Demographics Desk Reference Series*, no. 3 (July 1992): 14. According to John P. Dworetzky (*Psychology* [St. Paul: West Publishing Co., 1982], p. 357), 50 percent of American children will spend an average of six years living in single-parent households.

42. "Single Parents," p. 14.

percent to 6.8 million. In 1991, 28 percent of American children under eighteen did not live with both parents, up from 15 percent in 1970 and 23 percent in 1980. The situation for black children was more stark. In 1970, 41 percent lived with a single parent; in 1991 the figure was 64 percent.[43]

Single parenting occurs as a product of divorce (46 percent), unmarried pregnancies (26 percent), separation (21 percent), or death of a spouse (7 percent).[44] But increasingly singles, men as well as women, consciously choose to live singly while at the same time wanting to raise a family.[45] As many as 15 percent of the women coming to the Idant sperm bank for artificial insemination are single. This deviation from the two-parent family structure is pushed further by the desire of lesbians and gay men to raise children whom they have not begotten by intercourse with the opposite sex. So who, then, is to fill the earth? Does God's directive apply to singles as well as to married people, to homosexuals as well as to heterosexuals? Is it morally permissible for all to bear or father children, or are there moral limits to be adhered to in deciding who is to procreate?

The question has implications not only for those who desire children but likewise for the researcher and physician. To whom do they have an obligation to provide the services of assisted reproduction? To any who want and can afford it? And supposing physicians have no *obligation* to provide such services, are they even morally *permitted* to provide the services of assisted reproduction to any who want it? Are there no solid grounds, as some argue,[46] to deny AID to single women?

In this book we have chosen as our model the biblical creation injunctions. One thing that is clear from the Genesis context is that these injunctions are given to a pair of human beings, male and female. To them together are given the tasks to fill and rule the earth,[47] and they are to do it within the context of a family unit graced with heterosexual relations.

---

43. U.S. Bureau of the Census, *Statistical Abstract of the United States,* 112th ed. (Washington, DC, 1992), charts 56, 62, & 69.

44. "Single Parents," p. 15.

45. Jill Smolowe, "Last Call for Motherhood: More and More Single Women Are Choosing to Be Unmarried . . . with Children," *Time* 136, no. 19 (Fall 1990): 76.

46. Maureen McGuire and Nancy J. Alexander, "Artificial Insemination of Single Women," *Fertility and Sterility* 43, no. 2 (Feb. 1985): 184.

47. In Genesis 2, God gives the command to care for the garden only to the man, since the woman does not yet exist. It would be a mistake, however, to draw implications regarding application from this.

The very first story about human persons relates their aloneness without each other. God states — is it a decree or an observation? — that it is not good for the man he created to be alone. So God resolves to make for the man an appropriate companion, a complement to him, with whom he, having left his own family, will engage in sexual union and create a family. Oneness between complementary opposites is thus established as the ideal. (We will explore this in greater detail in Chapter Ten.) This does not mean that establishing a family is required of human beings or that being single is immoral. Indeed, the apostle Paul speaks of singleness as a gift (1 Cor. 7:1-7). What it does indicate, however, is that the two-complementary-parent family structure provides the norm from which the stewardly activities having to do with human procreation are to be carried out.[48] That is, *intentional* procreation in a single-parent context violates the norm.

In regard to filling the earth, not only the biblical model but also good secular reasons support maintaining the two-parent family structure. The literature is vast; we can only provide a summary of some of the more significant findings regarding the status of children in single-parent families. A 1988 National Center for Health Statistics survey "found that young people from single-parent families or stepfamilies were 2-3 times more likely to have had emotional or behavior problems than those who had both of their biological parents present in the home."[49] Such young people also tend to have less well developed social skills.[50] "According to research compiled by [Karl] Zinsmeister [a scholar at the American Enterprise Institute], more than 80 percent of the adolescents in psychiatric hospitals come from broken families. Approximately three out of four teenage suicides 'occur in households where a parent has been absent.'"[51]

48. The Christian can appeal to even stronger statements in the New Testament, where sexual activity outside of the marital bond is forbidden (e.g., Acts 15:28; 1 Cor. 5:1-5; 6:9, 13-20; Eph. 5:5). However, we make no appeal to this, since we claim that employing assisted reproductive techniques involves neither fornication nor adultery. These require sexual relations between persons, whereas we are speaking strictly about reproduction apart from having sexual relations with a member of the opposite sex.

49. David Popenoe, "The Controversial Truth: Two-Parent Families Are Better," *New York Times* 142 (26 Dec. 1992): 21.

50. P. S. Fry, "Father Absence and Deficits in Children's Social-Cognitive Development: Implications for Intervention and Training," *Journal of Psychiatric Treatment and Evaluation* 5 (1986): 113-20.

51. Joe Klein, "Whose Values?" *Newsweek* 119, no. 23 (8 June 1992): 21. It should be noted that the editors of *Newsweek* refused to share the source of these studies, which we were unable to verify independently.

Further, many argue that both male and female role models are needed for the mature rearing of a child. "Children may fare well in single-parent families, but the chances of problems increase. . . . Boys raised without fathers are likely to manifest gender-role deficiencies, especially if their father is absent during their infancy. Similarly, girls raised without fathers are more likely to show dissatisfaction and maladjustment in their female role and to have problems interacting with males."[52]

Academically, despite little difference in IQ scores, children from single-parent families have lower school performance evaluations than those from two-parent families.[53] This affects not only their grade point average but also their performance ratings by teachers in citizenship and behavior.[54] A study by the National Association of Elementary School Principals found that 30 percent of two-parent elementary students surveyed ranked as high achievers, compared to 17 percent of one-parent children. At the same time 23 percent of two-parent children and 38 percent of one-parent children were ranked as low achievers.[55] Furthermore, children from single-parent families are less likely to complete high school (42 percent for whites; 70 percent for blacks)[56] and college.[57]

The short-term economic prognosis for these children compares unfavorably with that for children from two-parent families. One study estimates that "45 percent of female-headed families with children at home live in poverty, as do 19 percent of male-headed families. But only 8 percent of married couples with children under age 18 live in poverty."[58] Similarly, "children in single-parent families are six times as likely to be poor. They are also likely to stay poor longer. Twenty-two percent of

---

52. Dworetzky, pp. 358, 357. Some dispute this claim; see McGuire and Alexander, p. 182.

53. Irwin Garfinkel and Sara S. McLanahan, *Single Mothers and Their Children* (Washington, DC: The Urban Press, 1986), p. 28.

54. Lisa Terre, William Ghiselli, Lina Taloney, and Eros DeSouza, "Demographics, Affect, and Adolescents' Health Behaviors," *Adolescence* 27, no. 105 (Spring 1992): 9.

55. Karl Zinsmeister, "Growing up Scared," *The Atlantic Monthly* 256, no. 6 (June 1990): 52.

56. Garfinkel and McLanahan, p. 29.

57. Barbara Dafoe Whitehead, "Dan Quayle Was Right," *The Atlantic Monthly* 271, no. 4 (April 1993): 74. "Among all the children who are over eighteen at the ten-year mark, *60 percent are on a downward educational course compared with their fathers and 45 percent are on a similarly downward course compared with their mothers*" (Judith S. Wallerstein and Sandra Blakeslee, *Second Chances* [New York: Ticknor & Fields, 1989], p. 157).

58. "Single Parents," p. 14.

children in one-parent families will experience poverty during childhood for seven years or more, as compared with only two percent of children in two-parent families."[59] Their economic setting in turn affects many other aspects of these children's lives.

Just as striking are the long-term prospects for children from single-parent families. "Among whites, daughters of single parents are 53 percent more likely to marry as teenagers, 111 percent more likely to have children as teenagers, 164 percent more likely to have a premarital birth, and 92 percent more likely to dissolve their own marriages."[60] When they enter the work force they have lower occupational status scores and lower earnings.[61]

Children from single-parent families have a greater likelihood of being victims or perpetrators of crime. Smith and Jarjoura, for example, argue that though poverty, race, and family structure are all factors in victimization, the significantly increased crime experienced by single-parent households with children between twelve and twenty does not result simply from poverty but also reflects family disorganization.[62] "A study that tracked every child born on the island of Kauai in 1955 for 30 years found that 'five out of six delinquents with an adult criminal record came from families where [a parent] was absent.'"[63] According to the Bureau of Justice, 70 percent of juveniles in state reform institutions come from single-parent or no-parent families.[64]

It might be responded that the data are merely correlative rather than causal. Being raised by a single parent will not cause one to take up a life of crime, attempt to commit suicide, or have identity problems. In each case it depends on the participants in the single-parent family and how they interact.

The point is well taken. Environment is one contributing factor; how people react to that environment is another. However, the amount of correlative data is striking, suggesting that in many cases the environment contributes significantly to the personal and social problems the child faces and that the resources for meeting these problems are not readily available. Children raised in this setting start out facing a battery of

59. Whitehead, p. 47.
60. Garfinkel and McLanahan, pp. 30-31.
61. Garfinkel and McLanahan, pp. 29-30.
62. Douglas A. Smith and G. Roger Jarjoura, "Social Structure and Criminal Victimization," *Journal of Research in Crime and Delinquency* 25, no. 1 (Feb. 1988): 47.
63. Klein, p. 21.
64. Zinsmeister, p. 52.

difficulties that many children in two-parent, functional, heterosexual families do not face. As the sociologist David Popenoe puts it, "Social science research is almost never conclusive. There are always methodological difficulties and stones left unturned. Yet in three decades of work as a social scientist, I know of few other bodies of data in which the weight of evidence is so decisively on one side of the issue: on the whole, for children, two-parent families are preferable to single-parent and stepfamilies."[65]

In light of this evidence, to *consciously* and *intentionally* undertake a pregnancy,[66] knowing that there will only be one parent there, only one half of the complement, is questionable morally, for one is putting the prospective child in a position that increases the likelihood of shortchanging his or her opportunities and development.[67]

Qualitative Concerns

Finally, we argued in Chapter Three that filling has not only a quantitative but also a qualitative aspect. This raises questions concerning what can and should be done with the fertilized egg before it is implanted in the woman. If we are under obligation to procreate well, then it is obligatory that the processes of assisted reproduction used not harm the potential child. This means that caution must be taken not only in implementing the procedures used in assisted reproduction but also in making as sure as possible through prior experiment that new reproductive procedures do not damage the fertilized egg that is to be implanted. But prior experiment

65. Popenoe, p. 21.

66. Lest we be misunderstood, we are not speaking about single parenting per se but about conceiving children to be raised in this state. It is important to distinguish intentionally rearing a child in a single-parent environment from having to do so because the marriage failed or one parent died. In the latter cases the child already exists, and one undertakes to provide the best one can for the child. In some circumstances where the intact family is dysfunctional (for example, where the wife or children are abused), maintaining the two-parent family might actually be more harmful than raising the child as a single parent. The choice in this case would be between the lesser of two evils.

67. It is distressing that out of fifteen select major committees reporting on reproductive technologies, only one-third counseled against restricting in vitro fertilization to couples, and only two indicated that the couples should be married. See "Ethics and New Reproductive Technologies: An International Review of Committee Statements," *The Hastings Center Report* 17, no. 3, supplement (June 1987): 6-7.

to assess these procedures will have to be conducted at some stage on viable human pre-embryos, a matter of serious moral concern.

Likewise controversial is the implication that if we are to procreate well, then it would be both permissible and obligatory to intervene, to the best of our ability, so that the pre-embryo will be as free as possible from genetic defects. The "best of our ability" refers here both to the limits of our knowledge and to the limits of what we can do safely to alter genetic structures. That is, a qualitative filling means that it is proper to do preimplantation diagnosis and take therapeutic measures to correct debilitating genes in the fertilized egg, and that where this can be done reasonably, safely, and within certain economies, we have a moral obligation to do it.

Though actual intervention of this kind remains futuristic, it is important to make the point at this stage, for such considerations imply that scientists are morally permitted to do something they have already embarked on, namely, to conduct relevant research to ascertain the genetic structures of pre-embryos and more significantly to determine how they can be altered when deficiency is detected. Both knowledge and technology depend upon research. Here again, that the research must be conducted on fertilized but not implanted (or implantable) eggs raises moral problems. We will return to this issue in the final section of our next chapter.

Still more troublesome for some, though no longer futuristic, is preimplantation diagnosis and selection from among pre-embryos of those that are most free from genetic defect, or better, free from a genetic defect or defects for which they are tested. Even now, fertilized eggs can be analyzed for whether they carry the defective genes that cause cystic fibrosis or whether they would be carriers of this disease. Those that have the disease or are carriers of it can be isolated and not implanted, so that the prospective child is free from this debilitating disease or from passing it on to his or her progeny. The first child conceived with this intent was born in 1992. The procedure appears relatively safe for the resulting child, though at present expensive ($2,000).[68]

What is problematic to some is not that we implant pre-embryos free from this disease, but rather *(a)* what happens to nonimplanted pre-embryos and *(b)* the slippery-slope concerns about what persons might do to pre-embryos with less serious deficiencies or with characteristics that are not deficiencies at all (such as gender or eye color). With regard to the

68. Simpson and Carson, pp. 952-53.

former, selective implantation seems to resemble abortion. With regard to the latter, if we select for certain characteristics, have we not affirmed thereby that not all humans are of equal value? Those with certain characteristics are more valuable than those with others.

With regard to the first problem, one has to inquire whether and to what extent causing the death of unimplanted pre-embryos, when some are chosen for selective implantation, resembles abortion. On the one hand, the process shares with abortion the termination of viable pre-embryos prior to birth. However, a significant difference remains. Whereas aborted pre-embryos, embryos, and fetuses could come to term if left alone in the womb, the same cannot be said for the fertilized egg outside the womb. That is, whereas the fetus, to be aborted, has to be killed directly, the killing of the unimplanted pre-embryos is a by-product of another action. Their death is indirectly caused. Thus, the morality of the process of implanting selected pre-embryos must be treated in the way appropriate to evaluating actions with negative by-products.

The ethical principle most often invoked to evaluate such actions is the *principle of double effect*. According to this principle, for an action that has negative by-products to be morally acceptable, four criteria must be met.

1. The action must be morally good or morally neutral.
2. The evil or bad effect must not be the means by which the good is achieved. It must be truly a by-product.
3. The evil or bad effect must not be intended, only foreseen. That is, the intentions must be good or moral.
4. The good effect must be equal to or preferably outweigh the bad or evil effect.

The principle of double effect does not apply to direct abortion, for the death of the fetus is not a by-product of the abortive act. Furthermore, even if one could apply the principle, direct abortion fails to meet the second condition above, because the abortion of the fetus is the direct means by which the good, whatever that might be, is attained.

The case of selective implantation, however, is different. Here the death of the unimplanted pre-embryos, though an evil, is a true by-product of the implantation process. Accordingly, we can evaluate whether the indirect killing resulting from selective implantation might satisfy the principle of double effect and hence be justified. First, the actions of fertilizing ova and implanting them in the womb can be held to be morally neutral actions; those who morally question these acts do so on the basis of their *consequences* for unimplanted pre-embryos and the *risk* imposed on the implanted pre-embryo, not on the grounds that in vitro fertilization and implantation are intrinsically wrong acts. Second, the death of the unimplanted pre-embryos is not a means to the good end of producing a healthy child for the couple. The other pre-embryos need not die in order for successful implantation to take place; their death is a true by-product of the process of helping the couple to conceive. Third, the intention is good, namely, to help the couple give birth to a healthy child. Finally, one might hold that the overall effect is good, in that a healthy child is born to a couple, whereas there was a reasonable risk that the child that might have been conceived by normal means would have suffered greatly after birth with its serious ailment. The death of the unused pre-embryos is lamentable; yet they neither suffered nor were aware of a lost existence.

This suggests that although initially the process of selecting some pre-embryos to be implanted and leaving others to die seems to parallel abortion, the two are different in a morally significant way. The one is a case of direct killing and hence must be evaluated on those grounds, whereas the other is a case of indirect killing that can be morally justified using the principle of double effect. Hence, though evil or bad results occur in both cases, the moral evaluation of the two acts will differ.

The second objection to selective implantation mentioned above is that to choose one pre-embryo over another violates the contention that all persons are of equal value. To select, one must determine that some people are less valuable than others and make decisions regarding their future based upon that determination.[69]

This objection is based on an inference that we want to deny. With the objector, we affirm that all human *beings* possess the same worth. By virtue of being human and being valued by God, all have equal value. But from this affirmation it does not follow that all human *characteristics* are of equal worth or value. Some characteristics, such as good health, the ability to walk or to use one's senses, or the ability to think, reason, or express emotions properly, are more desirable than others. Otherwise we would not employ physicians, therapists, transplant teams, or educators to facilitate them.

One might likewise draw attention to the contraposit inference of

---

69. I owe this objection to Heidi Wisner.

the above, which must also be rejected. We refuse to move from the contention that not all characteristics are of equal worth to the view that not all humans are of equal worth. When my children are ill, I strive to make them well. Qualities of wellness are superior to those of illness. But that one child is well and another ill does not make the children of unequal value. Rather, what we try to do is to change or remove the inferior characteristic so as to make the life of the ill child better.

But, the critic might respond, this only applies to cases where one replaces one characteristic with another. Where we make a selection among possible characteristics that can be added to a person who already exists, as when we move a child from sickness to wellness, we have not devalued the person but only replaced one undesirable characteristic with another. But in the case of selective implantation, the physician does not select possible characteristics that can be added to an organism that already exists or will be born (this is another, future scenario: prenatal genetic surgery). What is being selected is not the better characteristics per se but rather the being who possesses them. For example, where you have three pre-embryos, one with the defective gene for cystic fibrosis, one that is a carrier, and one that is free from that genetic defect, one implants the third pre-embryo, since being free from the disease is a better characteristic; the rest are left to die.

This objection is well taken. What must be compared is person P with desirable characteristic C, to person Q with less desirable or un-desirable characteristic D. But once it is admitted that C is more desirable than D, there is a basis for choosing between P and Q other than on the basis of their intrinsic value as human persons. And since the selection is not based on their human worth but on the possession of certain qualities, it need not follow that their human dignity is threatened.

Still, the critic might respond, since persons cannot be separated from their qualities and persons are therefore rejected, their human dignity suffers. Persons must be treated holistically.

To reply, one must distinguish between two types of cases. In some cases, we make value judgments about persons bearing characteristics and select some and not others in various contexts. Suppose two captains are choosing ball teams from a group of kids. The captains alternate in making selections until they reach Mike, the last child, whose selection will even out the teams. But Chris refuses to take Mike, because he is . . . the reasons might be many: unpopular, a poor batter, short, of a different race, and so on. Here Chris's refusal to accept Mike is an attack on his human

141

dignity. But this differs from cases where only a limited number can be selected. We make such selections when we choose a mate, select a six-footer over a four-footer for the last spot on the basketball team, or vote for one candidate over another. Selection based on qualities is acceptable and not degrading, because in the situations at hand, not all can occupy the desired post, whether as my mate, on the sports team, or in political office. Where there are restricted positions, having certain traits makes some more qualified than others to fill open spots, but it does not make them more human. Limited access to resources means that selection is necessary, and one appropriately chooses the political candidate or ball-player possessing the appropriate characteristics over another based on how their characteristics satisfy external demands. Both intrinsic and extrinsic worth must be measured. Where two beings have equal intrinsic worth and unequal extrinsic worth but only one can be selected, selection does not violate human dignity.

Similarly with selected implantation. Not every pre-embryo can be implanted. Hence, one properly selects out the most appropriate pre-embryo, in this case, the one lacking the defective gene. But human dignity is not thereby violated.[70]

The respondent might argue that, while in many cases selection for limited spots based on qualities is appropriate and not degrading, the situation changes when the selection is a matter of life and death. Where one is to live and the others are to die, selection on the basis of characteristics is contrary to dignity. Rather, random selection should be used to best preserve human dignity. A comparable example here might be the use of random selection in deciding among a large group of applicants who will have access to the kidney dialysis machine. Selection on other grounds affronts human dignity.

Yet, though randomness makes humans per se equal, it problematically also treats all their characteristics as equal, and we have denied that this is the case. Kidney dialysis well illustrates the point. Though randomness is employed, the selection process in the cases of kidney dialysis is

---

70. In fact, one might argue more strongly that in some cases not every person should be selected, on the grounds that it might not be good for themselves or for others. Suppose that little Mike wants to join the Big Bruisers football team and cannot be talked out of it. One might have to say no, in Mike's own interest. Or again, some candidates for office should not be elected, given their ineptitude, for the protection of the citizens. This raises the issue of paternalism (pure in the first case and impure in the second), which we will not evaluate in this discussion.

never based merely on randomness. Candidates are screened initially to determine which are most likely to benefit from the procedure and which are not. This initial screening recognizes not differing human worth but the importance of applying criteria to facilitate the best use of scarce resources and to evaluate the significance of quality of life.

So too here, in selection of pre-embryos there is an initial screening based upon determination of future quality of life. Such a screening will eliminate those that possess the defective gene for which we tested. Where multiple pre-embryos lack the genetic defect, random selection can be used to determine which is to be implanted, for now there are no relevant grounds for selecting one over the other; each is of equal worth.

The critic might advance one more argument. Suppose we developed an artificial womb in which to implant the remaining defective embryos; would we be obligated to do so in order to protect human dignity? If so, then we have a moral imperative to get on with the research to save lives. If not, then our argument based on limited options seems irrelevant.

The supposition suggests an important caveat. On the one hand, since artificial wombs have not been developed, the issue of limited options remains relevant in selected assisted reproduction. On the other hand, since the critic's objection has a point, and since we would answer that we would not be obligated, the ground for rejecting reimplantation in such cases shifts from limited options to two other grounds: (1) the future quality of life, and (2) the ambiguous status of the pre-embryo. That is, if we had an artificial womb that would take the pre-embryos possessing the defective gene to birth, we should not do so, both on the grounds that their future quality of life will be poor and on the grounds that the pre-embryos are potential persons to whom we have differing moral obligations than we have to fully actual persons. With respect to the former ground, we do not have the obligation to provide what is necessary for life no matter what. That is, we are not obligated to sustain life where there are serious doubts regarding future meaningful existence. Quantity of life considerations do not always take precedence over quality of life considerations. Of course, the difficult question arises of how one can determine when life is not worth preserving. What criteria must be met? Examples having to do with anencephalic fetuses and persons in persistent vegetative states are somewhat easier to decide, for here potency for meaningful life never was possible or else has departed. Other cases are more difficult to decide. For example, those afflicted with Huntington's disease can live into middle age; yet their decline is so painful and debilitating that the suicide rate is increased. Perhaps here the Golden Rule — do unto others as you would want to have done to you — provides the most help. Yet the application of the principle must be tempered with the concern that any decision be made from knowledge of what life is like for people suffering from these afflictions. To gain that knowledge, we must hear the stories of those who are currently afflicted.

The latter ground raises the question of the status of pre-embryos, to which we will return in the next chapter when we address the matter of abortion.

To summarize, selective implantation based upon knowledge of genetic structure creates the possibility that parents who have the likelihood of procreating children with serious genetic defects can have children free from those defects. Thereby the quality of life for the prospective parents is enhanced, and their offspring, too, can anticipate improved quality of life. Such selection, which involves indirect killing, can be justified where selection is grounded in genuine concerns about future quality of life. It is not parallel to abortion. Nor need it imply that people who already suffer from the genetic defect are any less valuable or have any less human dignity. At the same time, like any other procedure it is capable of being abused and hence should be carefully monitored.

In this chapter we have addressed questions about assisted reproduction from the dimension of our stewardship ethic that is concerned with filling. In the following chapter we will turn our attention to the issues that have to do with caring for (focusing on the means used to fulfill a couple's desire for children) and ruling over (focusing on abortion and the treatment of unused pre-embryos).

CHAPTER SIX

Stewardship and Assisted ——— *Reproduction: Methods* ———

Though William and Elizabeth Stern wished to have a child, they postponed the attempt until Dr. Stern completed her medical residency in 1981. In the meantime, Dr. Stern was diagnosed in 1979 as suffering from optic neuritis, a symptomatic indication that she probably had multiple sclerosis (MS). Some time after the diagnosis she conceived but did not bring the pregnancy to term. She was told that during the pregnancy she had suffered a temporary paralysis due to her MS, and that further pregnancies might result in temporary or permanent paralysis.

Eschewing further attempts as dangerous to her health, in 1985, through the Infertility Center of New York, the Sterns contracted with Mrs. Mary Beth Whitehead to bear a child for them. Mrs. Whitehead already had given birth to a son and a daughter by her husband. Mrs. Whitehead agreed to be artificially inseminated with sperm from Mr. Stern and that upon birth the baby would be turned over to the Sterns to be raised as their child. In return she was to be reimbursed $10,000 to cover her expenses. On March 27, 1986, Melissa (or Sara, as Mrs. Whitehead named her) was born.

Mrs. Whitehead gave the child to the Sterns on March 30, but the next day she went to their home and requested that she be given the baby for a week until she got herself emotionally in order. The following week she informed the Sterns that she had changed her mind and intended to keep Sara as her own daughter. The Sterns turned to the courts for assistance and received a court order to have the baby returned to them. When the order was served, Mrs. Whitehead handed the baby out her back window to her waiting husband, and the next day together they fled to her parents' home in Florida. It was not until July that authorities gained custody of the child and returned her to the Sterns.

The battle for "Baby M" now turned to the New Jersey courts, where Mrs. Whitehead and Mr. Stern fought to gain permanent custody of the baby girl. Mrs. Whitehead sought to have the surrogacy contract declared void on the ground that it violated a New Jersey law against selling babies. Mr. Stern endeavored to have the court hold her to the terms of the surrogacy contract she signed. Judge Harvey R. Sorkow ruled on March 31, 1987, that the surrogacy contract was legal and that Baby M belonged to the Sterns, who promptly legally adopted her in the judge's chambers. Mrs. Whitehead appealed the case to the New Jersey Supreme Court, which overturned the lower court ruling. The justices unanimously rejected the legality of a surrogacy contract involving payment, thereby holding that Mrs. Whitehead was Baby M's legal mother. However, in the meantime Mrs. Whitehead had divorced her husband and was pregnant with another man's child. The court decided that since the child's best interests should be uppermost, Baby M was to be given to the Sterns to raise, while Mrs. Whitehead was provided visitation rights.

The case tragically illustrates the desires and attempts to satisfy them, the decisions and potential problems created, and the possible conflicts and relational minefields that await human endeavors to assist reproduction. Such unusual, difficult cases draw our attention to the seemingly intractable moral dilemmas created. At the same time, these cases stand in contrast to the many instances where surrogacy has proceeded without incident to provide children for the infertile. Yet the moral dilemmas must be faced when we assist people to procreate.

The Ethic of Stewardship: Caring for the Infertile

We argued in the previous chapter that we have the obligation to fill the earth and that this justifies our efforts to help the childless to conceive, though it does not obligate us to make these efforts. The second injunction of our stewardship ethic — caring for — addresses the means we use to fulfill a couple's desire for children. Many questions might be raised; here we consider two facets: the use of genetic materials other than what comes from both parents and the use of means that replace marital intercourse to conceive children. We will consider the former in this section and a specific case of the latter in the next (recognizing, however, that often the two are not clearly separated in the discussion).

Genetic Exclusivity and the Use of Donor Material

Some Protestants and Catholics argue that the origin of the genetic material used in either artificial insemination or IVF makes a moral difference in the process. Humans should be procreated using neither donor sperm nor donor eggs but exclusively out of the genetic material of the prospective parents. Going outside the marital pair for sperm or eggs to artificially fertilize (heterologous artificial fertilization) "is contrary to the unity of marriage, to the dignity of the spouses, to the vocation proper to parents, and to the child's right to be conceived and brought into the world in marriage and from marriage."[1] It separates "parenting as an act of begetting . . . from parenting as a vocation to nurture."[2] The argument here combines concerns for both the means used to procreate (that it occurs apart from the conjugal sexual act and from the commitment to parent what one procreates) and the origin of the materials employed.

We shall say much more about the ethics of using artificial means in the next section, but a word about the issue is appropriate here. The phrase "unity of marriage" in the Catholic context seems, in part, to be a euphemism for the "inseparable connection . . . between the two meanings of the conjugal act: the unitive meaning and the procreative meaning." It should be granted that intercourse has at least these two functions and that prima facie these should be preserved within marriage. The one expresses the union of love, the other an end of marriage. However, it does not follow that it is always inappropriate or immoral to have acts from love without the possibility of procreation (though given the centrality of love in marriage, procreation should not occur without love). Even within the natural scheme of things there are natural methods (for example, the rhythm method) whereby love can be shown but procreation prevented, and there are times when these are appropriately employed. Use of these

---

1. Congregation for the Doctrine of the Faith, "Instruction on Respect for Human Life in Its Origin and on the Dignity of Procreation: Replies to Certain Questions of the Day," in *Intervention and Reflection,* ed. Ronald Munson, 3rd ed. (Belmont, CA: Wadsworth, 1988), pp. 458-59.

2. Hessel Bouma et al., *Christian Faith, Health, and Medical Practice* (Grand Rapids: William B. Eerdmans, 1989), p. 196.

(and other) methods cheapens neither love, procreation, nor the vocation of parenthood. Catholic theologians justify preserving the unitive and procreative meanings of marital intercourse on the ground that it promotes our "exalted vocation to parenthood."[3] But when sexual union does not promote that vocation and there are other means available that neither break the marriage bond and vows of fidelity nor destroy the unitive meaning of intercourse, the argument invoking the goal of parental vocation and furthering human dignity points to the legitimate and moral use of those means. The "goods and meanings of marriage" are preserved and furthered, for love continues to be expressed, and procreation and parenthood are made possible solely and wholly within the marriage covenant.[4]

Turning to the other issue, what can be said about the claim that the use of others' genetic material is immoral? The argument is that it "brings about and manifests a rupture between genetic parenthood, gestational parenthood and responsibility for up-bringing."

That there is a "rupture" cannot be denied; at least one person in the genetic chain will not play a role, either in the gestation or in the parenting of the child. But must this unity always be preserved? What is immoral about separating these connections in unusual cases, while preserving what might seem to be the more significant value of marital fidelity?

The Catholic response is that it violates the "right to become a father and a mother only through each other."[5] Whether there exists such a right is debatable, but even supposing there is, it in no way prohibits use of heterologous genetic materials. Rights are bases for making claims; they allow us to make legitimate claims on others. But they do not require that claims be made. That I have the right to go to the theater tonight does not imply that I must make a claim on that right and go. Whether I do or not depends on my choice. Similarly, married couples having this alleged right do not need to claim the right. By mutual consent they might decide to become a father and mother by use of heterologous material, without at the same time violating their vows of faithfulness to each other or their mutual dignity. The Catholic position stated here confuses rights with

3. Congregation for the Doctrine of the Faith, p. 460. Underlying this is a natural law ethic, about which we will have somewhat more to say in Chapter Ten.

4. A similar position is argued by Bouma, who favors "a model of sexuality that takes seriously the connection between sexual intercourse and the possibility of procreation" but at the same time allows for responsible use of contraceptives (p. 216).

5. Congregation for the Doctrine of the Faith, p. 458.

obligations; that they have a right to this does not mean that they have an obligation to procreate in this way, especially if they cannot become fathers or mothers in this fashion.

Some have claimed that the use of heterologous genetic material violates some rule of sexual morality. They liken use of artificial insemination using donor sperm (AID) or in vitro fertilization (IVF) using donor material to adultery. Yet adultery occurs only when married persons have sexual relations with someone other than their marriage partner. Since assisted reproduction occurs without the donor having any sexual contact with the prospective mother (or father), marital fidelity is preserved.

Perhaps what underlies the argument against using donor gametes is a claim that there is something significant about genetic exclusivity. This suspicion is strengthened by the claim that "it deprives [the child] of his filial relationship with his parental origins and can hinder the maturing of his personal identity."[6]

It is true that in our society most people want to have children genetically related to them. But there is no reason to think that one can draw any moral prescriptions or restrictions from this desire. Ancient cultures were not exclusivistic. In ancient Hebrew culture, artificial insemination was not available. In its place the custom of the levirate was instituted, whereby the brother of the deceased, through intercourse with the widow, whom he took as his wife, would procreate an heir to carry on his brother's line. The first son born to this union looked not to his biological father for his inheritance, but rather to the male in whose line he was born through his mother (Deut. 25:5-10).

In our own culture, adoption similarly illustrates the willingness to go outside genetic exclusivity to construct a family and fulfill the vocation of parenthood. Couples who cannot have children or who want additional children can legally and morally adopt others' biological children as their own. The adopted child is genetically distinct from both adopting parents, yet by entering their family the adopted becomes their child and they become his or her parents. In adoption there is no room for qualms over genetic exclusivity. Indeed, until recently it was impossible (and thought inappropriate) for the adopted to know the identity of his or her biological mother or father. The adopted's mother and father were truly those who fulfilled the parenting role.

There is nothing in the injunction to care for others that would restrict

6. Congregation for the Doctrine of the Faith, p. 459.

the material of procreation to that of the parents. We do have special obligations to care for those with whom we have a special relationship. For example, I have a greater obligation to support my spouse than I have to support my students, and I have a greater obligation to the students in my class than to other students at my institution. The relationship that creates the obligations might be genetic (I am especially obligated to care for my children), but often it is not genetic (as with my wife and students). I am obligated to nurture my children, whether they are biologically mine or mine by adoption, and failure to do so in either case is immoral.

Instead of appealing to a scenario of rupture to evaluate the morality of AID, one might suggest a parallel between sperm or egg donation and other types of donation, such as organ donation. The donation of an organ, whether from a cadaver or from a living human, can be the means to another's quantity of life (as with a donor heart, kidney, or liver) or quality of life (as with a cornea). Similarly, the donation of an egg or sperm can be the means to quantity of life for the conceived (for without this donation the egg under glass or descending down the fallopian tube will not be fertilized) and to quality of life for the prospective parents. In organ donation we require only that the donation be voluntary, that the organ be sound, and that the donor (if alive) or donor's representatives be honest in reporting what they know about the donor's health. Similarly, in the donation of genetic material we require only that the donation be voluntary, that the material be as free as possible from obvious genetic defects, and that the donor report anything that might jeopardize either the quality or the quantity of the life of the conceptus.

Hessel Bouma and his colleagues, arguing against the view that children are products to be made perfect or perfected in their making, note that children are a *gift* from God.[7] But this insight fits perfectly into the stewardship model, for though God can give his gifts directly to his stewards to care for, he most frequently gives them indirectly through others. In gamete donation, he uses someone who lies outside the marriage to give an important gift to the childless couple who wants a child.[8]

On this model, receiving assistance from outside the marital bond, whether it be a donor organ for survival or genetic material for facilitating procreation, is not an attack on the dignity of the couple. It accords with

7. Bouma et al., pp. 198-200.

8. This suggests that a noncommercial acquisition and deployment of donor gametes might be more consistent with our ethical paradigm.

the affirmation of our legitimate desire to procreate and to assume the parental vocation, both of which recognize and enhance the dignity of the spouses. It also furthers the realization that we are not completely self-sufficient, that there are ways in which others in the community can contribute to our quality and quantity of life.

The problem generated by genetic nonexclusivity often has more to do with the psychological than with the biological. It is argued that being conceived from donor material might create identity problems for the child. Similar claims have led to changing adoption procedures to allow children to discover and meet their birth parent(s). Procedural changes also might be made, where desired and desirable, with children born using assisted reproductive techniques.

Similarly, it is argued that using donor material might create psychological problems for those prospective parents who have the desire for genetic exclusivity in their offspring. They might have difficulty in accepting a child who is not strictly biologically their own, and this in turn would create difficulties for the child, who eventually will sense their rejection. With this possibility in mind, careful psychological screening should be done before using donor material for artificially inseminating the wife or for IVF. One way of addressing the psychological dimension is to use only the genetic material of the couple or to mix it with donor material so that there is at least the possibility that the child conceived is biologically continuous with those who will parent him or her. This process allows one to care for the sensibilities of the prospective parents while providing for the fulfillment of their natural desires to have children. At the same time this should not be considered a panacea, since the true parentage of the child can be determined genetically.

Not only should caution be used in using donor material, but study of the psychological impact should also be done on those who conceive and are conceived in this fashion. One should expect nothing less if assisted reproductive procedures are to be employed caringly. But though this caring approach recognizes the possible negative psychological consequences of using foreign genetic material, it does not proscribe its use.

Surrogacy

The stewardly injunction to care for the creation raises another issue: Do the means used to bear the child make a moral difference? We have already touched on this in the previous section, but let us consider this issue in more detail with respect to surrogate parenting. In most pregnancies the prospective female parent bears the child. But recently the maternal role has been expanded to include what are termed *surrogate mothers*. Actually, this term can be less than apt. When a woman, inseminated artificially with sperm from the prospective father who will parent the child, bears a child that is genetically hers, she is not a surrogate but the real birth mother. What makes this case different from AID is that prior to conception she has stated her willingness, usually expressed in a contract, to give the child to another couple for adoption, where another woman will take on the mother-parent role. The term *surrogate* applies more properly to a woman who, having had implanted in her womb a pre-embryo conceived by IVF from the egg and sperm of the prospective parents or by GIFT (gamete intrafallopian transfer), intends to give the child to its genetic parents. In this case she bears no genetic relation to the fetus but simply serves as its host while it develops. Nonetheless, a surrogate mother is still a mother, for it is she who originally nurtures and gives birth to the child, though she does so on behalf of another.

Surrogacy, it is claimed, falls prey to two moral failings. First, it degrades the woman, who now treats her body as a means to some other end. For example, the woman might trade her procreative life and its product (the child) for financial benefit. Second, it separates the decision to create children from the decision to raise them. That is, it separates procreation from parenting.[9] In effect, surrogacy is not the family building measure it is touted to be. "The dissolution of the marriage of the natural mother is not necessary to the success of the contract, but a severing of the mother-child bond is: the 'surrogate-mother' arrangement creates a family bond only by destroying a family bond."[10]

In reply to the first objection, it must be acknowledged that the typical commercial employment of surrogate mothers by for-profit firms smacks of the very thing that this objection claims. This is nowhere clearer than in the celebrated Baby M case, with which we began this chapter. The 1985 contract between Mary Beth Whitehead and William Stern provided that she undergo amniocentesis during the pregnancy and that

9. Herbert T. Krimmel, "The Case Against Surrogate Parenting," *The Hastings Center Report* 13, no. 5 (Oct. 1983): 35. See also Bouma et al., pp. 196, 203.

10. George Annas, "Death Without Dignity for Commercial Surrogacy: The Case of Baby M," *The Hastings Center Report* 18, no. 2 (Apr./May 1988): 21.

if the fetus were abnormal it was to be aborted. Should she refuse, she would receive no financial remuneration. A similar fate would befall her if she miscarried before the fifth month. If after the fifth month she miscarried or the baby was stillborn she would receive only a thousand dollars. If she successfully delivered a healthy baby, she would receive ten thousand dollars. The disparities in the fees and the fact that even in some circumstances beyond her control she could receive nothing for her work clearly indicate that the remuneration was not for her services (which were performed regardless of whether the baby was born or of its condition) but for the baby itself.[11]

In this regard, Judge Sorkow missed the point when he held that the money was used to compensate Mrs. Whitehead for her trouble, pain, and suffering. Judge Sorkow ruled that the money exchanged was not used to purchase the child, for the biological father, in virtue of the artificial insemination using his sperm, already "owned" the child and hence could not pay another to obtain it. You cannot, he said, purchase what already belongs to you. However, though Mr. Stern contributed to the procreation, so did Mrs. Whitehead; consequently she had an equal interest in it. This created the Solomonic problem of who gets the child when the woman changes her mind about following through on the contract.

The other feature of such commercial transactions that makes the procedure suspect is that it involves a third party whose primary concern might be to make a profit. Though they claim simply to be rendering a service, commercial surrogacy clinics often act primarily for their own benefit. This suggests, at least prima facie, that the child is being used as a medium in the transaction. However, one cannot always conclude that merely because a third, profit-making party was involved, other motives, such as that of benefiting the infertile, were not primary.

One must be careful not to lump all surrogacies together in the same pot. Though some surrogate mothers enter into a contract with the motive of mere financial gain, other women might have a sincere desire to be of service to the infertile couple. They might consent to an induced pregnancy out of truly altruistic motives. The model for reimbursement here would be parallel to the procedures followed for normal adoptions, where fees are paid to a nonprofit entity that assists the adoption and where the only money that goes to the birth mother is to cover childbirth expenses.

11. George Annas, "Baby M: Babies (and Justice) for Sale," *The Hastings Center Report* 17, no. 2 (June 1987): 13.

Though this removes the financial motive, it does raise the question of fairness. Would it not be morally proper to allow the woman to be reimbursed for carrying the child? After all, she has devoted considerable time and effort to the project and has experienced suffering, discomfort, and substantial risk. Where the financial considerations are clearly and strictly compensatory for the pain, suffering, inconvenience, and risk, so that the woman would be reimbursed no matter whether the child was born alive or not, it would be proper to reimburse her for bearing the child.

The critic might retort that this is precisely the moral problem, that the woman sells her body for another to use. It is immoral to traffic in human bodies. There is a real difference between renting a car and renting a womb.

But surely we cannot avoid using our bodies to achieve diverse ends. There are very few, if any, occupations in which one does not use one's body in the service of another for financial benefit. Whether one teaches, pumps gas, or works in a mine or a factory or behind a counter, one's body is being employed. That is, the person for whom the service is being rendered is "renting" the body and mind of the service-renderer.

There is nothing immoral about the mere fact that one puts one's body to use for one's own gain or for the benefit of another. This applies, whether it be one's feet, brain, hands, or reproductive system. What matters is how one treats the body in the occupation; that is, whether one uses or abuses it, honors or debases it. To abuse or debase one's body is failing to act in a stewardly manner, and this is immoral.

Even proper use of one's body can be accompanied by pain and discomfort and possibly some risk of function and life. Surrogate pregnancy is not unique is this. In our everyday employment, frequently some part of our body suffers: the feet of the clerk, the lungs of the miner, the voice of the teacher, the hands and arms of the factory worker, the tendons and joints of the musician or athlete. It is the extent and nature of the suffering that must be weighed, both in terms of the stewardly injunction to care for ourselves and in terms of the benefit that the action produces.

In introducing financial considerations, there must be no hint of the exchange of money for the child. What makes surrogacy questionable is not that the woman consents to use her reproductive system for another's benefit, nor that she is reimbursed for her labors, but rather that the product of the process, who is a human being, might be treated as a mere product to be traded away or bought and sold. If surrogacy is engaged in, it must be done in a way that avoids any such appearance. For example, it is preferable that the genetic material come from both the prospective parents through a pre-embryo transfer, not because we want to preserve genetic exclusivity, but because it further distances the process from mercantilism.

But what about the second objection, that it separates procreation from parenting? It is reasoned that we would have doubts about the moral character of a woman who conceives a child for the sole purpose of putting it up for adoption. She consciously destroys the mother-child bond. In such cases, we would argue, it would be better if she does not conceive at all. Does not the same hold true for the surrogate mother?

We agree, as the objection insists, that it is prima facie immoral to divorce procreation from parenting. This constitutes a primary objection against mere casual sex, for it is entered into with no thought of responsibility for the child that might result from it. One might wonder what the response would be if, during intercourse, the child appeared and queried, "Are you willing to be my parents?" The nine-month hiatus between conception and childbirth provides a convenient way for the male in particular to avoid making this connection. This break between procreation and parenting is the source of our moral qualms about a woman who conceives a child for the purpose of putting it up for adoption. As Bouma and his colleagues well put it, "Begetting is not merely biological and physical; it is essentially parental, entailing obligations for nurturing the child."[12]

But the case of surrogacy is different. It is not that the child will not be lovingly parented. Indeed, since it is so intentionally conceived, it might be more properly cared for than the child who results unwantedly from intercourse, within or outside of marriage. It is not the case that it will not be parented by someone biologically unrelated to it. The biological father remains the father; possibly the biological mother will be its parenting mother. It is rather that the woman surrogate consciously, with forethought, makes it possible for a specific, predetermined other to assume the parenting role. And she does so, not out of neglect, as does the woman who conceives for adoption or the woman who conceives out of casual sex, but out of self-giving love. The conception is not done with the thought of abandoning parenthood, but with the intention to hand that role over to another who desires that role *for the particular child*

12. Bouma et al., p. 196.

conceived who is related to them. The child, from the moment of conception, is a gift to another, and, if the process is conducted properly, finds itself in a family context of love and desire for its welfare.

A source of the difficulty comes with the term *mother.* An adopting woman will refer to herself as the child's parenting mother and to the woman who bore the child as the birth mother. Mothering can include both roles: the bearing of the child and the parenting of the child. Normally, these should not be separated. However, where parenting is impossible unless a child is provided by another, then a separation of the roles is necessary. But where the child is properly placed and loved by the parents, the mothering occurs in its most important role, providing for the full development of the child's potential. The giving is truly blessed.

As with any moral situation, there are always caveats. Could the payment for services be coercive for the woman, especially if she were poor or in dire financial straits? Obviously yes, and in such circumstances there is the potential for exploitation. It also clouds the self-giving factor. But then again, financial considerations are often a motivating factor for getting us to do things that we might not normally do. Financial reimbursement in and of itself is not necessarily coercive. The issues here are not all that different from those involved in obtaining consent for nontherapeutic experimentation. There, too, the question of the relation between reward and just payment and between coercion and free consent arise. But in both cases, the potential for exploitation does not imply that we should prohibit the procedure, only that it is necessary to establish guardians — both personal and impersonal in the form of laws — to protect the interests of those who could possibly be exploited by undue financial pressure. Poverty should never be a ground for commercial exploitation.

What if the surrogate mother changes her mind and wants to keep the child, as Mrs. Whitehead did in the Baby M case? On the one hand, one could follow a legal route and say that when a contract is broken, at most a fine is called for; the woman would pay a fine, return all monetary advances, and keep the child. As the surrogate brokers have often admitted, without proper legislation surrogacy contracts are legally unenforceable. On the other hand, the contract does serve the purpose of clearly setting forth the expectations and of guiding the conscientious in carrying forth the surrogate pregnancy. It provides the formal context of promise making, with the correlative obligation of promise keeping. However, should the promise be broken, some sort of creative possibilities that allow both the bearing mother and the genetic father to develop relations with the child might be necessary. In this sense the case parallels more recent arrangements between families of adopted children and their birth mother or between parents who have divorced. The arrangements are less than ideal because they apply to a less than ideal situation. Barring this, perhaps surrogacy contracts have to be written with the possibility in mind that the surrogate mother might indeed want to keep the child; the contract might specify that, because of the bonding that occurs during pregnancy, she would have the right to decide to keep the child until some specific time following the birth. The final decision would then parallel adoption procedures. The downside of such a provision, however, is that it significantly waters down the promise-making function of the contract and places a heavy and unwelcome psychological burden on both the birth mother (who now has an option at odds with the intention of the original act) and the prospective parents as they await the bearer's final decision.

In line with his covenant paradigm, Bouma and his colleagues have suggested that the notion of a covenant between parties might be more in line with a Christian ethic than a contract. In this way surrogacy resembles marriage, which works best as a covenant, not as a contract. And indeed, we think this is correct. Ideally, in surrogacy a woman should enter into a covenant with a married couple to bear their child, as they covenant with her to raise it properly and lovingly. Their covenant responsibilities are moral, taken before God. One problem, however, is that covenants have not fared very well in our society; 50 percent of marriage covenants sooner or later end up being broken. Without continuous commitment to the one covenanted with and to the one before whom the covenant was made, covenants, too, can collapse. A second problem is that covenants work best when those who covenant together have had and continue to maintain a close relationship, neither of which generally characterizes surrogacy. Where the surrogate mother is a stranger, it would seem that a contract similar to that enacted in business is appropriate. Furthermore, we live in a litigious society. This by itself might not be unusually troublesome, were it not for the fact that here the welfare of a third party is at stake. In order to protect the child's interests, contracts seem appropriate, though surely it would be advantageous to have them bolstered by covenants that take seriously the relationship into which the parties have entered.

There are always exceptional circumstances, not all of which can be anticipated by legal documents. When Patty Nowakowski discovered that she was carrying twins but that the inseminating father wanted only the girl, the question of the relation between the siblings was raised.[13] It is impossible to discern ahead of time the genuineness of the desire or willingness of the couple to take whatever children come (whether normal or physically or mentally affected, individual or twins). Yet one might again claim that the situation is not all that different from adoption, where siblings are separated or where the healthy normal child has an easier time finding adoption than the abnormal or unhealthy one.

What these cases show is not that surrogacy is immoral or that it ought to be prohibited outright, as has been done with commercial surrogacy in England. Rather, they show that, like all other human actions, it is fraught with dangers and complications. Human relations are not programmable. Assisted reproduction does not differ from normal reproduction in the possibility of presenting tough cases for ethicists. What is needed in dealing with individual cases is an approach that takes into account both the best interests of all the parties, especially the child's, and the obligations that the parties have assumed by entering into the specified relationships. Put another way, the stewardly injunction of caring for all persons, realizing that none will derive the full benefits because their interests conflict with those of others, must be followed in finding a solution to the moral dilemmas.

The Ethic of Stewardship: Ruling over Procreation

In vitro fertilization brings an additional dimension to the ethical consideration of assisted reproduction, one that makes it appropriate to consider the third stewardly injunction: the injunction to rule the earth on behalf of the Landlord for the good of others.

We have seen that in order to reduce trauma to the woman and increase the likelihood of success, more than one egg is extracted from the woman and exposed to fertilizing sperm.[14] Though multiple fertilizations occur in the petri dish, only two or so are transferred to the woman. Once

13. Patty Nowakowski, "How Could I Let Them Separate My Twins?" *Redbook* 175 (July 1990): 38, 40-41.

14. Some have questioned the value to women of freezing embryos. For a list of counter-reasons, see Andrea L. Bonnickson, "Embryo Freezing: Ethical Issues in the Clinical Setting," *The Hastings Center Report* 18, no. 6 (Dec. 1988): 26-28.

the transplanted egg has successfully attached to the uterus and begun to develop, the problem arises of what to do with the nonimplanted but fertilized eggs. Should the orphaned pre-embryos be respectfully destroyed, allowed to die, or provided to infertile couples to adopt? In some cases the nonimplanted pre-embryos are frozen, so that if the first transfer attempt fails others can be implanted in the woman without her having again to undergo egg extraction. It is estimated that in 1986 there were 824 stored frozen pre-embryos; in 1989 that figure increased to 9,000. Some of these await the success of pre-embryo implants fertilized from the same retrieval; others are orphaned — unclaimed, unwanted, or without traceable owners.

People who hold that fertilized eggs are human beings argue that IVF procedures that create multiple pre-embryos in order to reduce the hardship on the woman face an inescapable moral dilemma. On the one hand, the pre-embryos cannot be destroyed or left to die, for that would be taking a human life. Since, these critics maintain, morally protectable human life begins at conception and since conception has occurred, albeit outside the woman, the lives of these developing human beings is morally protectable. On the other hand, the only option to preserve the remaining embryos is to implant them. But not only does this violate genetic exclusivity; it is also practically impossible. It violates genetic exclusivity because the pre-embryos would be implanted in the wombs of genetic strangers; it is practically impossible because, since the eggs have to be implanted before they divide too far, it is unlikely that there would be enough women who would be prepared biologically at the appropriate time, let alone willing, to carry adoption beyond its normal scope and adopt a pre-embryo to bring it to term. It might be suggested that the pre-embryos be frozen until adopting women can be found, but even freezing pre-embryos is rejected as being contrary to their dignity. "The *freezing of embryos . . . constitutes an offense against the respect due to human beings* by exposing them to grave risks of death or harm to their physical integrity and depriving them, at least temporarily, of maternal shelter and gestation, thus placing them in a situation in which further offenses and manipulation are possible."[15] It is because of the unacceptable consequences of both horns of this dilemma that those who take this view of pre-embryos reject IVF.[16]

15. Congregation for the Doctrine of the Faith, p. 457.
16. Congregation for the Doctrine of the Faith, pp. 453-63.

We have already rejected part of the second horn of this dilemma, arguing that we see no moral grounds for genetic exclusivity. Indeed, it would seem that pre-embryo adoption would be on a par with adoption after birth and would be equally acceptable. The objection to freezing pre-embryos introduces a different set of problems. One concerns the factual matter of whether freezing increases risk to the pre-embryo. It is commonly reported that freezing does not harm pre-embryos,[17] though the weaker ones, which constitute one-fourth to one-half of all frozen pre-embryos, do not survive freezing. But then weaker embryos do not survive in the womb either, for only about 31 percent of all fertilized eggs result in birth.[18] Other studies, however, report that freezing weakens the pre-embryo, making it less likely that it can be successfully implanted.[19] If the latter is the case, then freezing lowers the chances that the pre-embryo will successfully develop into a child, and hence it might be understood as an attack on its human dignity. This introduces the more serious, moral issue raised by the objection. This problem, along with consideration of the first horn of the dilemma, forces us to consider the status of the pre-embryo and the morality of abortion.

Abortion

Detailed consideration of the morality of abortion would take us far afield from our intended subject matter. Yet we cannot completely avoid addressing the subject since so much having to do with assisted reproduction touches on it. The debate has to do not only with our ethical structures but also with our view of when and on what basis beings are morally protectable. We cannot touch on all the caveats but will only attend to the major arguments.

The predominant argument given in support of the pro-choice posi-

17. Kamram S. Moghissi and Richard Leach, "Future Directions in Reproductive Medicine," *Archives of Pathology and Laboratory Medicine* 116 (Apr. 1992): 437.

18. In normal procreation, out of 100 eggs exposed to sperm, 84 are fertilized, 69 are implanted in the uterus, 42 are alive a week later, 37 are alive to the sixth week, and 31 are alive at birth. Clifford Grobstein, "External Human Fertilization," *Scientific American* 240 (June 1979): 61.

19. David Levran et al., "Pregnancy Potential of Human Oocytes — The Effect of Cryopreservation," *New England Journal of Medicine* 323, no. 17 (25 Oct. 1990): 1153-56.

tion is founded on a woman's autonomy. Women should be able to exercise control over every aspect of their life, including their reproduction. In this regard they should not be compelled to become pregnant or to maintain a pregnancy once begun. We might put the argument as follows:

1. A woman morally may (has the right to) do what she wants to with her body.[20]
2. The fetus is a part of a woman's body.
3. Therefore, the woman morally may (has the right to) do what she wants to with the fetus.

The argument in this form is suspect, however. The first premise fails, for there are moral limits on what someone can do to and with their body. Not only are we stewards of others, but we are also to be stewards of our own body. The apostle Paul well writes that we are God's temple; hence we are to care for ourselves as if God were living in us (1 Cor. 6:19-20). For example, it would be immoral for me intentionally to maim myself by cutting off my fingers, as is sometimes done in order to obtain workers' compensation. The second premise is also false, for although the fetus depends upon the woman's body, it is not a part of it in the same way the woman's organs are a part of her. The fetus has its own unique genetic structure and exists in its own protected environment within the woman. It is a foreign body in her and eventually will be rejected by her body. Hence it would be more accurate to affirm

2a. The fetus is dependent on the woman for its existence.

Now, to derive conclusion 3, we must rewrite 1 to say,

1a. The woman may (has the right to) do what she wants to beings that are dependent upon her for their existence.

But although 2a is true, 1a remains suspect. For example, a woman morally cannot poison her family or throw herself in front of a door during a fire to prevent them from escaping. Even in matters of reproduction,

20. Judith Jarvis Thomson, "A Defense of Abortion," *Philosophy and Public Affairs* 1, no. 2 (Fall 1971): 54. This also seems to be the emphasis of those who interpret the Constitution as providing for privacy among the rights granted to citizens.

there are moral limits. For example, we should hold it immoral for a pregnant woman to abuse drugs, like cocaine or alcohol, that can seriously affect her baby. One cannot morally excuse a woman who has caused her child to suffer from fetal alcohol syndrome, on the grounds that she is an autonomous person and may do what she likes. The point here is that although autonomy is a value it is not the sole value. One must also take into account one's obligations to oneself and to others, at the very least not to do them unjustifiable harm but to do them good. The woman is a steward of what depends upon her for its survival and quality of life. This does not mean that she may never take the life of another, but it does mean that she may not do so unjustly. The question thus becomes this: When does justice allow the taking of another's life? What circumstances or conditions are relevant?

One response to our argument is to note that we have assumed that fetuses have the same moral standing as the mother, that they are human beings or human persons. But most pro-choice advocates deny that fetuses are members of the moral community such that they have full and equal standing or rights with the mother. The fetus is not yet a person in a morally protectable sense, is not a member of the moral community, because it has not yet achieved certain features, such as consciousness of one's world, the developed capacity to reason, self-motivated activity, capacity to communicate on a variety of topics, and self-awareness.[21] Hence the fetus is not something for which the woman is to be a steward in the same way that she is to be a steward for herself or for her already born children. There are no grounds for comparability. Thus, our example regarding causing fetal alcohol syndrome begs the question by assuming that harming a fetus is like harming a child. Fetal alcohol syndrome becomes a problem when it affects someone who will be born and will live with it.

But surely these features, such as the capacity to reason and communicate, place too great a stringency on the determination of moral protectability. They would be sufficient to characterize a morally protectable human being (that is, if something has these properties, it is

---

21. Mary Anne Warren, "On the Moral and Legal Status of Abortion," *The Monist* 57, no. 1 (Jan. 1973): 55. Michael Tooley argues that "an organism possesses a serious right to life only if it possesses the concept of a self as a continuing subject of experiences and other mental states, and believes that it is itself such a continuing entity" ("Abortion and Infanticide," *Philosophy and Public Affairs* 2, no. 1 [Fall 1972]: 44).

human), but they are hardly necessary (that is, a being that lacked these properties could still be human). To claim that these features are necessary would put a whole host of individuals in serious jeopardy, from neonatals and infants, to sufferers from severe forms of Down syndrome, to Alzheimer's patients, people in persistent vegetative states, and people nearing the end of their life. The claim that most people feel that we should not kill members of these groups and that our society can afford to provide for them, but do not feel this way about fetuses, hardly suffices to exclude fetuses from the category,[22] for not only should we not make our morality depend on how people feel about or view others, but there are many who would gladly adopt fetuses rather than see them aborted.

In short, there are serious grounds for doubting the strong form of the pro-choice position. An argument based primarily on autonomy ignores our additional moral obligations, and the admission of only "high achievers" to the moral community jeopardizes the status of "low achievers" whose participation in the community is not nor should be in doubt.

The major argument for the pro-life position takes the perspective of the fetus, arguing for its fundamental moral protectability or its right to life.

4. All human life is morally protectable.
5. Fetal life is human life.
6. Therefore, all fetal life is morally protectable.[23]

In her classic piece "A Defense of Abortion," Judith Thomson contends that, although the argument might be sound, the conclusion will not establish the pro-life position. Although the fetus might be morally protectable, it does not have the right to be given anything necessary to sustain it. "If I am sick unto death, and the only thing that will save my life is the touch of Henry Fonda's cool hand on my fevered brow, then all the same, I have no right to be given the touch of Henry Fonda's cool hand on my fevered brow."[24] It would be the charitable act of a Henry Fonda, as a good Samaritan, to do it, but his visitation is not morally

22. Mary Anne Warren, "Postscript on Infanticide," reprinted in Munson, ed., *Intervention and Reflection*, p. 92.
23. William E. May, "Abortion and Man's Moral Being," in *Abortion: Pro and Con*, ed. Robert L. Perkins (Cambridge, MA: Schenkman, 1974), pp. 14-15.
24. Thomson, p. 55.

required. Likewise the fetus does not have the right to be guaranteed the use of a woman's body, even though it needs it for survival. It is like the case of a woman who is kidnapped and attached to an ill violinist who needs the use of others' kidneys for nine months to save his life. Although there is a prima facie obligation not to kill another, and unattaching her would kill the violinist, she is not obligated to remain connected to him. Despite the circumstances surrounding her attachment, she might agree to remain connected, but her doing so is not an obligation or the violinist's right, but the act of a good Samaritan. Similarly, the woman who lends her body to the fetus is a minimally decent or perhaps even a good Samaritan, but no one can be compelled to be a minimally decent or good Samaritan.

What is lacking in Thomson's case, however, is a genuine appreciation of the relationship that holds between a prospective mother and what she carries. The relation is not the same as that holding between Henry Fonda and Judith Thomson, who neither know each other nor (presumably) have met; neither is it the relation holding between the good Samaritan of Jesus' parable and the beaten victim on the road, which only comes to be when the Samaritan voluntarily undertakes to assist the injured man; nor is it the relation between a violinist and someone unwillingly abducted and attached to him. Pregnancy creates special, nonconsensual relations that must be taken into consideration. The woman and man brought the fetus into existence, a foreseeable even if unwanted result of their action. Where taking the action is voluntary, they give tacit consent both to the action and by implication to its foreseen, reasonably possible consequences. When the voluntary action results in bringing into existence a being which is dependent upon them for its existence, their participation establishes a special obligation-creating relation to the fetus. This, in part, is what makes the example of a woman abusing drugs to the detriment of her fetus so telling, for in doing so she abdicates a responsibility that she has by virtue of her relationship to the fetus.

There is, of course, a caveat in this, for we have spoken of *voluntary* action. The case of rape presents a difficult exception. Here the woman, who is the exclusive caregiver for the first nine months, did not consent to the intercourse, so that the resulting nonconsensual relationship is forced on her and more closely resembles that between Thomson's violinist and the abducted woman. For such cases, Thomson's argument is relevant. Even if the fetus were to have a prima facie right to life, it would not follow that to take its life is unjustified, for the special relationship created was forced on the woman. At the same time, however, persons' origins should not be used as grounds for discriminating against them or for taking their life. Hence, in the case of rape one has a more straightforward conflict between autonomy and the affirmation of human dignity manifested in not having one's life threatened.

One way to address this conflict is to question the truth of premise 5. It is difficult to equate the life of the pre-embryo with the life of what we all consider to be a morally protectable human being. It is true that conception provides the only real genetic and physical discontinuity in the process of fetal development. Prior to conception neither the sperm by itself nor the ovum by itself will develop into a human being. Only when they conjoin can procreation occur. Further, not only is the biological break obvious, but the radical change in probabilities affecting on the one hand the sperm and egg and on the other what is conceived might be taken to signal that here something significant has occurred. The likelihood that an individual sperm in an ejaculation will fertilize an ovum is 1 out of 200 million. The likelihood that a particular ovum in a female will be ovulated is about 400 in up to a million. But once fertilization occurs, the probability that what is conceived will come to term dramatically increases, to 3 out of 10. It is true that this shift in probabilities does not establish humanity. Neither does it guarantee that the discontinuity has *moral* significance. But it does call attention to the unique significance of this point in human development.[25]

Yet despite this uniqueness, one is hard pressed to say that this being with human parentage is a human person. The pre-embryo is a long way from being a human person, for it lacks the features that we associate with human personality. Its most telling features are its preprogrammed genetic code and its continuity with the child that results, but at this point it has few person-generating capacities. What it does possess is the potential to develop and realize those capacities; it is, in effect, a potential person developing gradually but rapidly toward personhood.

People have sought other points of physical and social discontinuity that might bear on determining personhood. Paul Ramsey points to segmentation as a place of discontinuity. At about the seventh day after fertilization, the appearance of a streak in a blastocyst containing identical twins signals their separation into two unique individuals. This, he suggests, is the point

---

25. John T. Noonan, *The Morality of Abortion: Legal and Historical Perspectives* (Cambridge, MA: Harvard University Press, 1970), p. 56.

of significant discontinuity. The problem is that this occurs only in the case of identical twins, and most births are not of this sort. Hence there are no grounds for taking what applies to an unusual case and making it a standard for all cases, despite the advantage this might have for allowing intrauterine devices (IUDs) and morning-after pills to be used.

"Quickening," or the first movements of the fetus felt by the woman, might be a point of discontinuity for the woman, but not for the fetus, for it was moving long before the woman sensed it. Baruch Brody suggests that, as the termination of brain activity signals the end of life, so the commencement of neural activity at the eighth week signals life's beginning.[26] But intriguing and provocative as this symmetry is, the appearance of organs and the commencement of their functions is a process, not a punctiliar event. Furthermore, brain function is a *necessary* condition for personhood when determining the end of life, whereas here we are looking for a *sufficient* condition for personhood.[27]

The point chosen by the Supreme Court in *Roe v. Wade* was viability, but in many respects viability represents a change in the fetus's environment rather than in the fetus itself. Though it is true that currently there seems to be a point before which a fetus is not viable due to inadequate lung development (21 weeks), yet just as recent technology has pushed viability significantly earlier than ever before, this physical limit, too, someday might be overcome by advanced artificial wombs that can harbor someone from conception to birth outside the mother.

Despite the problems noted, locating discontinuity at fetal neural activity, viability, and birth has some merit. Though these are not punctiliar points of discontinuity, they do chronicle significant changes in the fetus, its abilities, and our relation to it. The fetus moves through stages quickly, however, and any attempt to delineate particular transitions, before which the fetus is not a human person but after which it is, is doomed to failure. This suggests that we should abandon the painful search for a clear point of discontinuity and instead adopt a developmental or gradualist view, according to which the respect due to the pre-embryo, embryo, or fetus is appropriate to its stage of development. Its stage of development will correlate with our relevant obligations toward it. This means that we will have differing obligations to pre-embryos than to ten-week-old fetuses or to viable ones.

26. Baruch Brody, "On the Humanity of the Foetus," in Perkins, ed., *Abortion: Pro and Con*, p. 72.
27. Bouma et al., p. 42.

That we have degrees of moral obligation dependent upon both the object itself and our relations to it is not a foreign notion. I have differing degrees of obligation to my family, my extended family, my students, people at my church, neighbors, and strangers, depending on their relationships to me. I ought to show up prepared to teach my students, but I need not share my paycheck with them or deeply love them; I ought to deeply love and share my paycheck with my family. Even in terms of preserving life I have different degrees of obligation depending on the person's proximity and relation to me. The obligations I have to feed my children differ from those I owe to Somali children.

At the same time, though, following from our stewardship ethic, I do not treat my students or Somali children lightly, especially when they are in need. Neither should one treat lightly the zygote, pre-embryo, previable fetus, or viable fetus. Even in its earlier stages, the fetus is a potential person, a being that will in time image God. Consequently, as a potential person, it deserves respect.[28] This need not mean that its life cannot be taken. What it does mean, however, is that there should be good justification for doing so, one that at the same time incorporates our stewardship obligations. The degree of justification required to take its life will increase as the fetus develops.[29] Early in its life, the pre-embryo has a potentiality that is physiologically and psychologically distant from the realization of its capacities; the realization of the potentiality is more near as the embryo and then fetus develops and becomes aware of its surroundings. The actual time gap between these points is rather narrow, however, requiring that if abortion is justified it be carried out sooner rather than later.

Hovering over this middle view should be two other concerns. One

28. Bouma et al., p. 45.
29. Bouma and his colleagues (pp. 47-48; chap. 7) wish to shift the object of discussion from the fetus as an imager of God to those who are considering taking its life. They use Jesus' parable of the good Samaritan to suggest that even were the fetus not made in God's image, it is a neighbor that should be treated as being in need of help. The shift is helpful but not entirely satisfactory. For one thing, we do treat different things differently. How I treat a cow differs from how I treat a child, and this does not depend simply on me as God's imager but also on the creature itself. Similarly, it is equally important to ascertain the status of the fetus, insofar as we deem appropriate responses to it. For another thing, our relation to the fetus goes beyond the relation the Samaritan had to the man attacked by thieves. Generally speaking, one parents because one procreates. As Bouma and his co-authors argue all along, procreation should not be disjoined from parenting without good reasons, some of which we have considered in this chapter.

is a realization of the ambiguous status of the fetus. The present debate is one sign of this ambiguity. At the same time, where there is ambiguity, one should act with caution and care. The other stems from the Christian emphasis on love. If love is properly characterized as self-giving, requiring sacrifice, then actions affecting others, including those who are developing toward personhood, even where ambiguity is present, should be taken with their good in mind as well as our own. Since we are to love our neighbor as ourselves, and since the fetus can lay claim to being a very near neighbor, our actions toward it should never stem from concerns for autonomy alone, for we are never strictly autonomous, but also should flow from the attitude of love that distinguishes us as Christians.

Research on Pre-embryos

Returning to the original subject, suppose that the process of IVF goes forward and research on pre-embryos continues. Is there a way in which the commands to care for and to rule over can be responsibly met? The lifespan of the pre-embryos will be brief, two weeks at most. Yet they contain an immense pool of information about the human genetic code. More importantly, they make it possible for us to learn about the very early stages of human development, knowledge that is valuable not only for its own sake but also for the benefits it makes possible to others.

For example, fetal research can help us better treat infertility and improve contraceptive techniques. It might also help us learn how to prevent genetic diseases such as Duchenne muscular dystrophy or cystic fibrosis. Experiments on pre-embryos might also make it possible for us to discover how certain drugs affect human pre-embryos. Drugs could be given to thirty-two-celled blastocysts without either causing them pain or suffering or affecting their life, since without being implanted they will die. This information can be useful in helping physicians know with what drugs to treat fetuses in utero.

At the same time this shows respect for the pre-embryos, for in this way they are contributing to the quality of life of others. It might be thought that this allowance runs counter to our above contention that it is not morally justifiable to treat potential persons only as means. But there is a significant difference between experimenting on pre-embryos that cannot and will not be implanted, and terminating to benefit another the life of a fetus that, if left alone, would come to term. If indeed the pre-embryos used for experimentation were potential humans in the sense that they could and would develop into full human beings if they were left alone and merely nurtured, this would hold true. However, pre-embryos that cannot and will not be implanted lack this potential, whatever their innate capacities. They are, as it were, doomed to die by not being implanted.

There is a parallel here with the use of aborted fetuses discussed in the previous chapter. We argued that use of aborted fetuses for transplants was morally permissible *provided that* they were not conceived or aborted for that purpose. Similarly here, adopting the more conservative limitations, it can be argued that experimental use of pre-embryos for human good is permissible, so long as they are not generated for that purpose but are the by-products of legitimate attempts at facilitating conception. In this way, they can still serve a function for humans, assisting those who will survive. As unconscious contributors to society, they still deserve respect. They are given a role — that of furthering our knowledge to enable others to have an improved quality of life — that is superior to mere death.[30]

Conclusion

Stewardly ruling, then, allows — indeed, encourages — us to rule even ourselves for our benefit. This ruling is to be done with care and respect; there is no allowance for the treating of humans as mere means. Hence in treating the infertile to allow them to fill the earth by their procreative activities, concern must be shown not merely for the infertile but also for the children who will result from assisted reproduction technologies.

Perhaps, in all the discussion, this is the most overlooked concern. In wanting to satisfy the desires of the infertile, serious attention is paid to the infertile and to the means used to satisfy their desires. But little is said about what will happen to the child once it is born. It is the process

---

30. For a thoughtful treatment of the variety of issues posed by freezing embryos, see John A. Robertson, "Resolving Disputes over Frozen Embryos," *The Hastings Center Report* 19, no. 6 (Nov./Dec. 1989): 7-12. See also *Final Report of the Embryo Research Panel* (Washington, DC: National Institutes of Health, 27 Sept. 1994), which takes a more liberal approach to producing preimplantation embryos specifically for research.

that commands attention, when what is of greatest significance is what happens to the product of assisted (and normal) reproductive activities. The surrogate mother is screened, as is the sperm donor. But to what extent are the prospective parents not only screened, not only counseled in their original decision to parent, but also assisted in their subsequent parenting? It is here that the primary work of mothering and fathering, of filling and caring for, are fulfilled. But once this point is reached, the problems are no longer unique to assisted reproduction, but in general concern the larger task of parenting.

CHAPTER SEVEN

Stewardship and the ───── *Human Genome* ─────

IN AUGUST 1990, Dr. W. French Anderson, then of the National Heart, Lung and Blood Institute, received preliminary approval from the National Institutes of Health (NIH) to use gene therapy to treat two children suffering from adenosine deaminase (ADA) deficiency. People with ADA deficiency lack an enzyme that is necessary to break down dangerous by-products that inhibit the growth of cells responsible for producing immunity. Without these cells they cannot fight the diseases or infections to which they are exposed. Previous drug therapy on the two children had proven unsuccessful; without this experimental intervention their prognosis was dim.

Anderson and his colleagues used recombinant DNA techniques to insert a copy of the gene that produces the missing enzyme into cells withdrawn from the children suffering from ADA deficiency. These altered cells were put back into the children through blood transfusions. The intent was that these new genes would become active and encode for the missing enzyme. Anderson reported to the Second National Conference on Genetics, Religion and Ethics in Houston in 1992 that dramatic effects were achieved after eight infusions over a ten and one-half month period, which would make this the first successful use of gene therapy on human beings.

The Human Genome Project

Humans normally have twenty-three pairs of chromosomes in the nuclei of each cell. These chromosomes are the "volumes" in the body's "encyclopedia" of information, which contains between 50,000 and 100,000 "articles" in the form of genes. Each chromosome is structured as a double helix with two strands of DNA held together by weak hydrogen bonds. Each strand is composed of four kinds of "letters," that is, molecular subunits called nucleotides, that are linked across the helix in pairs. Adenine (A) bases link with thymine (T), guanine (G) with cytosine (C). Although we know that these letters go together to make up articles in the encyclopedia's volumes, we have yet to determine where most of the genetic articles are located, what base-pair letter sequences compose these articles, and what the articles are about or say. Between the individual articles lie long strands of base pairs (strings of letters, if you like) that seem to play no informative function in the encyclopedia, or at least none of which we know. To carry out the project of deciphering the encyclopedia, researchers need to isolate the articles from the "background chatter" of the intervening segments of base pairs.

Scientists are already embarked on the human genome project, which is undoubtedly the most ambitious, if not the most expensive, scheme to date in the history of biology. The fifteen-year, $3 billion project will map (determine the location of) all the genes in the human chromosomal encyclopedia and sequence the base pairs (nucleotides) of each gene.

Just as encyclopedia articles need to be understood and interpreted, so must the information contained in the genes. Thus a further step, not officially part of the human genome project, will occur simultaneously with the sequencing in various laboratories around the world. Scientists will attempt to discover how the expression of the genes is regulated (how they are turned off and on) and how the genes function, that is, what information they encode to assemble amino acids into polypeptide chains, which then are linked together to form proteins. Only when biologists discover how the genes function or are expressed in the body will the project begin to attain its full intended usefulness.

Projected Uses of the Information

Why undertake a project that involves thousands of scientists in the international community at such a cost? It is not difficult to conceive of other uses for biologists' intellectual energy and for government resources. What do those involved in the human genome project hope to achieve from the research?

One of the more dramatic events in recent genetic research was the identification in 1993 of the gene responsible for Huntington's disease, which afflicts approximately 30,000 Americans.[1] This disease usually attacks people between their mid-thirties and mid-forties, though the onset can range from ages twenty to sixty. Destroying brain cells, it is degenerative and always fatal. It often begins with a slight slurring of speech and facial tics and then progresses into jerky movements, contorted facial grimaces, and bodily spasms. It eventually leads to loss of speech and of the ability to swallow, disorientation, unwarranted emotional outbursts, depression, and dementia. Children of those with Huntington's have a 50 percent chance of having the genetic disorder themselves. Though geneticists knew since 1983 that the defective gene was located on chromosome 4, isolating it proved difficult because of the more than one hundred genes present in the targeted area. Because the gene has now been isolated, a more accurate and less expensive test can replace the former diagnostic procedures, which, though 95 percent accurate, required taking blood samples from several members of a person's family to perform linkage tests to identify a genetic marker associated with the disease.

Huntington's is one of more than three thousand genetic diseases already identified in humans. "Genetic disorders are the second leading cause of death among 1- to 4-year-olds in the United States and the third leading cause of death in 15- to 17-year-olds. It is estimated that 25-30 percent of admissions to U.S. acute-care hospitals for persons under eighteen are for genetic conditions; about 1 percent of adult admissions are for genetically-related conditions."[2] Medicine to date has generally devoted its energy to treating the symptoms of illness. Discovery of the human

1. The Huntington's Disease Collaborative Research Group, "A Novel Gene Containing a Trinucleotide Repeat That Is Expanded and Unstable on Huntington's Disease Chromosomes," *Cell* 72 (26 March 1993): 971-83.

2. LeRoy Walters, "Genetics and Reproductive Technologies," in *Medical Ethics,* ed. Robert M. Veatch (Boston: Jones and Bartlett, 1989), p. 212.

genetic structure will allow physicians in some instances to discover, prevent the transmission of, and possibly treat the causes of genetically based deficiencies and illnesses.

Genetic Counseling

The human genome project ultimately will pay off by facilitating various ways of addressing genetic problems. First, it can assist in genetic counseling. Since defective genes can be passed on from one generation to the next, it is important that people have accurate information about their genetic constitution so that they can make informed choices about procreation. For example, about 1 in every 2,500 Caucasian newborns has cystic fibrosis, a genetically inherited disease that affects the epithelial cells lining the lungs, intestines, and pancreas. The affected cells are impermeable to chloride ions, so that patients cannot transport water into their airway passages. Thick, sticky mucus forms in the lungs and pancreas, blocking ducts and making it difficult to breathe and digest food. To sustain respiratory function, this mucus must be loosened and extracted regularly and frequently — a prolonged and painful process. The mucus provides a breeding ground for bacteria, so that patients often die of pneumonia. In 1989 the gene responsible for the most prevalent forms of this disease was isolated, and tests were devised to ascertain the presence or absence of the mutated gene in adults.[3] Before the isolation of the gene and the development of tests for the gene, physicians could only say that every Caucasian child had the same risk, about 1 in 2,500, of being born with cystic fibrosis. Though there are as many as 160 mutations of the gene, a single mutation, designated ΔF508, occurs in approximately 70 percent of all Northern European carriers. By testing for this and six or so other more prevalent mutations, one can assess more determinately the odds of passing on the disease. The risk is 1 in 1,000 (assuming a 90 percent carrier detection rate) where one partner has the gene and 1 in 250,000 where tests for the gene in both parents are negative.[4]

3. J. R. Riordan, J. M. Rommens, B-S. Kerem et al., "Identification of the Cystic Fibrosis Gene: Cloning and Characterization of Complementary DNA," *Science* 245 (1989): 1066-72.

4. "Statement of the American Society of Human Genetics on Cystic Fibrosis Carrier Screening," *American Journal of Human Genetics* 51 (1992): 1443-44. See also Nicholas Wald, "Couple Screening for Cystic Fibrosis," *The Lancet* 338, no. 8778 (23 Nov. 1991): 1318-19.

Less expensive and more accurate information about a client's genetic composition can increase the accuracy, informativeness, and relevance of genetic counseling. Prospective parents can ascertain who if any carry specifically-tested faulty genes and whether being a carrier poses serious risks for the potential child. This in turn will enable them to decide more knowledgeably and responsibly about their procreative activities in light of their obligations to future generations.

At the same time, however, genetic testing and accompanying counseling are not a sure-fire guarantee of more enlightened reproductive decisions. If the test is taken before the woman is pregnant, the specific test results and their meaning soon can be forgotten. Those counseled might not understand what the counselor is communicating at the time, might comprehend it at the time but after the counseling session become confused or forget the information, or simply ignore it in the passion of lovemaking. Further, the partner of the person tested might not be asked to take a test or might not consent to do so, so that the risk is more difficult to determine.[5] Where the test is taken after pregnancy has commenced, the decision whether to terminate the pregnancy has to be made quickly and under tension, which creates stress in a context where the stresses of pregnancy already are present. Here the role of the counselor is made more difficult, for he or she must avoid being directive in order to allow the couple to make their own, yet informed, decision.[6]

5. One study found that

[e]ven after counselling, a significant minority were unable to state correctly their risk of having an affected child, even to the extent that 13% believed that two carriers could not have a normal child. In the primary care context, only 57% [of those tested for cystic fibrosis] suggested a test to their partner, and only 87% of the partners underwent a test, so less than half (49%) the couples at higher risk had carrier status determined in both partners, in each case with only 85% efficacy. Thus only 35% of couples at risk of having a child with cystic fibrosis were detectable by this screening approach. ("Screening for Cystic Fibrosis," *The Lancet* 340 [25 July 1992]: 209)

6. Not being directive is not equivalent to being value neutral. Bouma and his colleagues suggest a covenantal model to govern genetic counseling. See Hessel Bouma et al., *Christian Faith, Health, and Medical Practice* (Grand Rapids: William B. Eerdmans, 1989), pp. 252-53.

Genetic Screening

The information gathered about human genetics will also provide the basis for more accurate genetic screening. Though currently many states screen newborn children for genetic conditions such as phenylketonuria (PKU), hypothyroidism, sickle-cell disease, and galactosemia, there are potentially hundreds of other genetic deficiencies or diseases for which health care providers could test in the future. Neonatal screening would allow for early diagnosis, whereas adult screening would provide information for genetic counseling.

Population screening by itself is of little value unless it is accompanied by some sort of action. In some instances, screening can lead to early (including prenatal) treatment of the symptoms of the disease. For example, neonatal screening for PKU allows physicians to detect this condition at an early stage; by putting the infant on a restricted diet low in the amino acid phenylalanine they can prevent serious mental retardation.

In other cases, screening of pre-embryos allows the implantation into the mother of pre-embryos free from certain known genetic defects. As we discussed in Chapter Five, pre-embryos conceived by in vitro fertilization can be screened for cystic fibrosis, so that only those pre-embryos whose test results indicate that they would not suffer from the disease, or perhaps would not even be carriers, can be implanted into the woman's uterus to grow and develop. In this way, children can be protected from certain serious genetic diseases that their parents are known to carry. Or put another way, couples who are carriers of seriously defective genes can have children free from those defects (though of course there is no assurance that they will be free from other genetically based diseases).

In still other cases the severity of the genetic defect might indicate the appropriateness of selective abortion to prevent a future child's extraordinary suffering due to severe genetic defects. Selective abortion for genetic defects presents one of the most difficult topics about which to generalize, for not only do genetically based diseases vary in severity, but even those suffering from the same disease can be affected and suffer to different degrees.

Some prenatally diagnosable illnesses seem clearly to warrant selective abortion. For example, in fragile X syndrome a repeating sequence of CCG bases, which may increase dramatically from one generation to the next, produces severe mental retardation. Infants with trisomy 13 (Patau's syndrome) have abnormal brains and defective hearts, kidneys, and gastrointestinal systems, so that they rarely survive longer than a few months.

Other cases are more difficult. Beta-thalassemia, in which the person loses one or both alleles of the beta globin gene, affects the oxygen-carrying capacity of the blood, leading to anemia. The disease is painful, and death can be postponed to the patient's twenties often only by expensive and frequent (as many as twenty per year) blood transfusions (which have their own negative side effects in the buildup of iron) and support therapy. The disease is especially prevalent in the Mediterranean area or among those of Mediterranean descent. In Sardinia the government currently employs a program of selective abortion to reduce births of afflicted children. The fetuses of pregnant women are routinely screened for the disease, and where tests yield a positive result, abortion is counseled. With this program, the incidence of thalassemia in live births went from 1 in 250 to 1 in 800. It is claimed that "parents who were at risk of having an affected child had virtually stopped reproducing before antenatal diagnosis, but are now having a normal number of children, as it can be guaranteed that all of these will be healthy [free from thalassemia]."[7] Thus, where parents once faced the frightening prospect of having children suffering from the disease who would not have access to life-prolonging transfusions, now parents can have children free from this genetic illness.

Yet such peace of mind is not without cost, for the tragedy of aborting those who would have suffered from the disease must be factored in as well. The decision regarding terminating pregnancy involves a painful weighing of potential prospects. The lives of these children are more than likely to be brief and painful. Yet brevity is a relative notion, extending from infancy to over twenty years. Thus parents must face the difficult question of whether it would be better to bring the child to birth knowing that his or her life may be very short and painful or to abort the child and spare him or her the pain but also deprive him or her of future existence. Along with such painful decisions, however, the testing has brought the positive effects of freeing parents from the fear of the disease that may have kept them from conceiving children. It has also delivered women known to be carriers of the disease from discrimination in the search for marriage partners. With the testing, parents have the freedom to conceive, knowing that their children can lead a life free from beta-thalassemia's painful and fatal effects.

In cases such as these, mere quantity of life for the prospective child

7. Bob Williamson, "Thalassemia: From Theory to Practice," *Nature* 292 (30 July 1981): 406.

cannot be the sole governing factor; stewards must also consider the quality of life, difficult though that may be to calculate. Indeed, it is the difficulty of determining what that quality will be that creates the moral ambiguity and the difficulty of making decisions in such cases.

Hessel Bouma and his colleagues suggest that selective abortion would be warranted when the child would not be able to become a person or have God-imaging capacities.[8] They understand this not in terms of having a soul, but in terms of what we are and what we are able to do. Bearing God's image means specifically representing God — in the model we are using in this book, being his stewards. This requires "the capacity for reflective choice-making," the ability to "not only make choices but also reflect on them and make choices about our choices, not only have desires but also have desires about our desires, not only evaluate but also evaluate our evaluations, not only think but think about our thinking."[9] To have God's image requires self-consciousness and the ability to enter into relationships with God and the community. This approach makes it easier to decide regarding those genetic diseases that bring death within the first year of life, when these dimensions of personhood would never be realized. But the more difficult cases occur where the person could live past this stage, to achieve the abilities Bouma delineates, but whose suffering and dysfunction are so very great.

One worry of those who counsel against terminating pregnancy based on prenatal genetic testing is that abortion can become a welcome tool for us as we become seekers of genetically perfect children. "We must not reduce individuals to their diseases,"[10] for to do so can mislead us to surmise that since diseases can be, indeed where possible ought to be, eliminated, those suffering from them ought to be eliminated as well. As stewards we must reject any such reductionism of those whose lives are in our hands and in whose interests we act on God's behalf.

At the same time, we must avoid the cruelty of giving and sustaining life no matter what the cost to the sufferer. There is the possibility in this of a selfishness that will not let the other person go but hangs on in the name of humanity and caring. The heroism of sacrifice for the child might be a good only for the hero.[11]

8. Bouma et al., pp. 248-49.
9. Bouma et al., p. 32.
10. Bouma et al., p. 245.
11. Bouma et al., p. 251.

Beta-thalassemia illustrates the problem, the promise, and the challenge. We have delineated the problem of suffering and premature death. The promise is that the disease can be treated. Currently, bone marrow transplants make it possible for the patient to be disease free. As of 1990, the longest-surviving patient had been free from the disease for six years.[12] And the success rate of transplantation — 75 percent after one year — provides additional hope, though at the same time it provokes caution, for the procedure must be done early, before the disease weakens the child and hence lessens the transplant's chances for success. The challenge is twofold. One challenge is to find the resources to treat children so that treatment can replace abortion. The other is to find ways to treat the disease more effectively, either by drug or genetic therapy or in the future by altering the gene in the germ cell. Unless the underlying genetic defect is addressed, by preserving children into their procreative years one is increasing the genetic load for the disease, that is, making it more likely that there will be more children needing the expensive transplant treatment.

In the interim, in those cases where selective abortion is justified, time matters. Given the view of abortion we developed in Chapter Six, selective abortion should be done as early in the pregnancy as possible, when the fetus is still only a potential person. Hence, science that enables earlier screening is to be encouraged.

Postnatal screening is also problematic, though in a different way. Given the current state of research, ethical and decisional difficulties arise when a gap exists between the time when a genetic defect can be determined and when treatment will be available for the disease. For example, tests for specific DNA markers on chromosome 16 can now give a pre-symptomatic indication whether someone will suffer from adult polycystic kidney disease, which affects 1 out of every 500 to 1,000 persons and accounts for 5 to 10 percent of end-stage kidney disease. Persons normally will experience the first symptoms around age forty, and 50 percent will have chronic renal failure by age seventy.[13] Death comes usually of uremia or hypertensive cardiovascular disease. However, to date nothing can be done about the disease. Should physicians who test for and find the

---

12. Guido Lucarelli et al., "Bone Marrow Transplantation in Patients with Thalassemia," *The New England Journal of Medicine* 322, no. 7 (15 Feb. 1990): 417-22.

13. "Autosomal Dominant Polycystic Kidney Disease," *The Lancet* 339 (9 May 1992): 1146-50.

condition in an infant inform the afflicted person when he or she gets older, the afflicted person's parents, or (later) the spouse?

It would seem that mandatory screening should be done only for genetic defects that can be treated or cured. Where treatment is unavailable, there is little point in screening, unless persons wish to be screened, for unless one overrides their autonomy and invokes mandatory sterilization to prevent transmission, nothing can be done with the information that will benefit the sufferers or their descendants. Furthermore, screening can lead to stigmatization and discrimination, as occurred in the past, for example, with sickle-cell screening. Screening in the early 1970s led to discrimination in getting life insurance. Despite the fact that there was no evidence that those with sickle-cell trait (as over against sickle-cell disease) had any shorter life expectancy, 50 percent of the insurance companies increased the premium for those diagnosed with the trait. The Department of Defense refused to allow those with the trait to attend the Air Force Academy or have aviation or flight crew training, though paradoxically they allowed them to advance in rank through the university ROTC programs. Even businesses like the airlines screened for the trait and made it a factor in hiring, believing that those with the trait could faint at high altitudes.[14]

Genetic Therapy

In addition to genetic counseling and screening, genetic therapy will gradually become available to people suffering from serious genetic defects. In genetic therapy, not the symptoms but the underlying genetic defect will be treated. In what is termed *somatic cell therapy*, the body cells of a patient are treated so that the individual (and that individual alone) is changed and hopefully benefited. In one method, normal DNA sequences are attached to modified retroviruses and inserted into cells extracted from the patient. These altered cells are grown in culture and then infused into the patient's body, where it is hoped they will not only divide and grow but also have their added genes turned on and made to function properly.

Prior to 1990, attempts at genetic therapy had not yet successfully

14. James E. Bowman, "Genetic Screening Programs and Public Policy," *Phylon* 38 (June 1977): 128-29. For an excellent discussion regarding the issues concerned with genetic testing, see John Rennie, "Grading the Gene Tests," *Scientific American* 270, no. 6 (June 1994): 88-97.

treated genetic disease. However, the case with which we opened this chapter signals the beginning of what will become increasingly possible. ADA deficiency made an ideal first target for gene therapy because it is a single-gene disorder and because the drugs that have been developed to control the symptoms are unable to eliminate the defect and pose risks of serious side effects. Following the initial NIH approval, research in gene transfer and genetic therapy has been extended to other diseases, such as malignant melanoma, acute myeloid leukemia, neuroblastoma, brain tumors, and cystic fibrosis.[15] In May 1993 physicians at the Children's Hospital of Los Angeles used blood drawn from his umbilical cord to infuse new genes into an infant named Andrew Gobea, who was born with severe combined immune deficiency.[16]

Safety remains a primary consideration for somatic gene therapy. Getting the gene into the body poses certain risks, including the possibility of overdoses of the virus, spread of the virus in the body, or that it could be contagious to others. Many potential difficulties have been lessened, though not eliminated, by the use of weakened retroviruses.[17] The attempt to get the corrected gene into the proper place on the chromosome poses a more difficult problem. The substantial risk is that the gene will insert at an inauspicious place — for example, where it might turn on an oncogene or shut down genes that suppress tumors, thus creating a cancerous condition. "That's why researchers are embracing the less-than-ideal: a temporary fix in which genes are transferred. In this case, the nuclear machinery will read the gene's instruction and make the protein, but the gene won't be copied or passed on to future generations of that cell."[18] The downside of this procedure, however, is that gene therapy must be repeated regularly, particularly for areas of the body like the lungs where cells are frequently replaced.

Treating each individual for a genetic defect is not as efficient as being able to stop the transmission of the defect to future generations. The human genome project has the potentiality to cast its lines into the

15. For a list of protocol descriptions and locations of research, numbers of patients, and starting dates, see "Human Gene Marker/Therapy Clinical Protocols," *Human Gene Therapy* 4 (1993): 847-56.

16. Sheryl Stolberg, " 'Bubble Baby' Is First to Get Gene Therapy," *St. Paul Pioneer Press* 145, no. 20 (17 May 1993): 1A.

17. Kurt C. Gunter, Arifa S. Khan, and Philip D. Noguchi, "The Safety of Retroviviral Vectors," *Human Gene Therapy* 4 (1993): 643.

18. Joe Palca, "The Promise of a Cure," *Discover* 15, no. 6 (June 1994): 84.

future by providing information that might make *germ line therapy* possible. Here the fertilized eggs themselves would be treated, so that the children produced would possess a "normal" copy of the particular gene.

> In studies involving mice, for example, genes have been added to one-cell mouse embryos after the sperm had penetrated the egg but before the genetic material[s] from the sperm and egg are joined within the same nucleus. If the experiment is successful, these added genes are then adopted by the embryo. As the embryo grows and the number of embryonic cells increases, the added genes become part of every new embryonic cell. Later, when the sperm or egg cells of the mouse develop, the added genes are included in approximately half of these reproductive cells. Thus, when the mouse reproduces, some of its progeny receive the added genes, and so on through the generations.[19]

The application of germ line therapy to humans, however, is far from straightforward. If germ line therapy is to succeed, it must be done when the organism is extremely simple, that is, composed of undifferentiated cells, so that the necessary changes can be encoded in all the cells. This would require that gene transfer be done prior to the embryonic stage. Only a small window — a day at most — is available to carry out the required early screening to determine whether genetic intervention is required and to conduct the desired genetic therapy.[20] This means that the procedure could probably be used feasibly only on pre-embryos conceived by in vitro fertilization.

Though it sounds simple in concept, germ line therapy in humans faces substantial technical difficulties. W. French Anderson notes three of them. First, germ line experiments with mouse pre-embryos have resulted in very high mortality, either because tampering with the cells damaged them, or because the new genes are lethal to the cells into which they are introduced, or because the cells do not reproduce, or because the genes fail to express properly. "In one recent experiment involving microinjection of an immunoglobulin gene into mouse eggs, 300 eggs were injected, 192

---

19. Walters, p. 221.

20. Some gene detection in mice can now be done in less than seven hours. In addition, "59% of the biopsied embryos established pregnancy by day 6.5, compared to 88% of unmanipulated controls." Christopher M. Gomez et al., "Rapid Preimplantation Detection of Mutant (Shiverer) and Normal Alleles of the Mouse Myelin Basic Protein Gene Allowing Selective Implantation and Birth of Live Young," *Proceedings of the National Academy of Science USA* 87 (June 1990): 4481.

(64 percent) were judged sufficiently healthy to be transferred to surrogate mothers, only 11 (3.7 percent) proceeded to live birth, and just 6 (2 percent) carried the gene."[21] Such a pattern of early clinical trials of germ line therapy in animal experiments portends an unacceptably high risk for human pre-embryos, raising serious questions as to when and under what conditions this type of intervention will be warranted. Before such procedures could be implemented on humans, the rate of failure in animal experiments must be reduced significantly.

Second, the randomness of the recombinant process poses problems for germ line therapy. The researcher cannot fully control whether or not genes incorporate, how many do so, whether they do so at the desired location on the chromosome, and whether they express properly. Failure in any of these areas can produce pathological conditions that might not only affect the initial individual but also pose potential problems for subsequent generations. For example, if different pre-embryos incorporate the genes at diverse locations and these persons then interbreed, their offspring might have a serious pathological condition that is both transmissible and eventually fatal. Before researchers proceed down this road there must be some assurance of control over the process of insertion and expression.

Third, substantial questions arise regarding the current usefulness of the procedure. Not every pre-embryo produced by a carrier of a genetic defect possesses the genetic defect. Even when both parents carry the recessive gene, there is still a 25 percent chance that the pre-embryo is neither a carrier nor affected by the disease. Other procedures, such as pre-embryo selection, pose much less risk and a higher percentage of success.

Should the procedures be perfected, the substantial potential technical problems circumvented, and adequate safeguards provided (no mean tasks, we must emphasize), germ line intervention might begin to provide the basis for making helpful genetic changes in future generations. In this way we might be able not only to cure genetic diseases but also to enhance or develop human capacities. In what follows, when we speak about germ line therapy we will assume its technical feasibility in order to focus on the relevant moral issues.

---

21. W. French Anderson, "Human Gene Therapy: Scientific and Ethical Considerations," *The Journal of Medicine and Philosophy* 10 (1985): 284.

The Ethic of Stewardship: Filling

The questions that now face the human genome project no longer concern whether the task is feasible, but rather how it will be accomplished and how the technology can be developed to facilitate it, when it will be accomplished, what its true costs will be and whether these will be matched by proportionate benefits, and what the ethical implications of carrying it out might be. The last question, which is our immediate concern, requires us to consider the human genome project and its implications in light of the paradigm of Christian ethical reflection we adopted.

Turning first to the command to fill, we have already noted our concern with the qualitative dimension of filling. When we fill the earth we bring about changes in it, and in doing so we assume moral responsibility for the quality of those changes. The human genome project promises a more direct and controlled application of the command than hitherto attainable. What is learned about the human genetic structure potentially will enable us to medically treat those suffering from genetic diseases. In rarer cases, it will enable us to alter the human genetic structure, to correct deleterious genes or substitute qualitatively better ones.

The human genome project by itself cannot achieve these changes. It must be coupled with other significant developments in genetic technology that will make it possible to treat the individual person. With respect to genetic therapy, treatment will proceed by successfully introducing recombinant DNA into a person's cells and then by getting the cells both to reproduce with the new genes and to express them in a regulated manner.

Both somatic cell and germ line therapy make possible the qualitative alteration of the human person, either a single individual or future generations. Some have sought to draw a line between these two types of therapy, giving their blessing to the former but not the latter.[22] But how are they different, and is that difference morally significant?

One difference concerns their scope — that is, how many are affected by the procedure. Somatic cell therapy, since it is limited to the patient, has at best short-term results. The changes, if successful, will alter that person's genetic structure, but that person's descendants will be left untouched. Thus

22. For a recent articulation, see "Position Paper on Human Germ Line Manipulation, Presented by Council for Responsible Genetics, Human Genetics Committee, Fall, 1992," *Human Gene Therapy* 4 (1993): 35-37.

the good to be considered in evaluating whether gene therapy should be applied is solely the good of that person (the patient).[23] With germ line therapy, however, the scope extends to future generations. The addition of a good gene and the improvement or enhancement of genetic capacities in the reproductive cells is intended to eliminate the disease caused by that gene or enhance capacities for many of the descendants. Thus, in doing germ line therapy one is engineering for future persons. It is not so much that something different is done in germ line therapy; rather, the difference is that it affects or is intended to affect people other than the individual whose genes are altered.

But the mere fact that more people are affected, all else being equal, is not a difference that is morally significant. What matters is not the number affected but how they are affected and whether they are affected for good or ill. One might affect many for good and few for evil, or vice versa. The amount of good or evil resulting from the genetic changes in germ line therapy might be much greater because it affects more individuals than somatic cell therapy, but the mere fact that it affects more than one person does not by itself alter the moral significance of the action. Unless the actions of genetic engineering are intrinsically immoral, the total possible good or evil achieved must be weighed. In each case one must evaluate how much good or evil is produced by the intended therapy.

Another difference might concern the risk involved. To alter the human genome for future generations is riskier than altering it in a way that only affects one person. Many argue that it is too dangerous to alter the human genome; we simply do not know what the final outcomes might be.[24] Now, it must be granted, as we have granted above, that danger, perhaps even significant danger in some cases, might lie behind germ line therapy. In this sense somatic cell therapy is safer, though it, too, faces the risks of unknown and possibly deleterious consequences for the patient.

But by itself, the possibility of danger does not rule out doing germ line therapy. In other words, the possibility of danger does not mean that

23. This is, of course, a bit simplified, since other issues — such as using scarce resources for genetic engineering that might be employed to treat people suffering from other diseases — impinge on the decision.

24. Ruth Macklin, "Moral Issues in Human Genetics: Counselling or Control?" *Dialogue* 16, no. 3 (1977); reprinted in *Intervention and Reflection,* ed. Ronald Munson, 4th ed. (Belmont, CA: Wadsworth Publishing Co., 1992), p. 446. See also "Position Paper on Human Germ Line Manipulation," p. 35.

germ line therapy should not be done; rather, it means that, because germ line therapy affects more than the solitary patient, much greater care and safeguards should be taken both in setting up the treatment and in carrying it out. Previous testing on animals should clearly establish the effectiveness and safety of the procedure.[25] The procedure should be done under strict controls and guidelines, with the seriousness that matches the seriousness of the possible consequences. With reference to our ethical model, it means treating future, possible persons with the respect due to them as individuals for whom we have a stewardship responsibility. What we do must be to benefit them and not do them greater harm, while at the same time realizing that doing them good might involve some harm. But a stewardship ethic is not risk aversive. To the contrary, in any medical procedure, from drug therapy to surgery, the assumed risks for harm must be weighed carefully against the possible benefits.

The more general point here is that although there clearly are specifiable differences between somatic cell and germ line therapy, it is less clear that these differences are morally significant ones. This does not entail that we should do germ line therapy. As pointed out above, there are at present significant technical difficulties and little reason to use the technique, given other, safer options. What it does mean is that, *all else being equal,* if in a particular case somatic cell therapy is moral, so is germ line therapy. Similarly, if in a given case, *all else being equal,* one is immoral, so is the other. That still leaves open for rational discussion whether, in any case, all else is equal. Given the present state of our technology, there are readily apparent differences in the outcomes of somatic cell and germ line therapy, though this might change at some time in the future.

But some balk at germ line therapy for other reasons. (1) Some suggest that it entails making the value of the person dependent on an ideal of biological perfection, that it promotes human hubris by suggesting that "all limitations imposed by nature can and should be overcome by technology," and that this striving for perfection would be available only to the socially privileged.[26]

But these objections apply no less to somatic cell therapy than to germ line therapy. The critical issue in all this is whether we can be adequate

25. W. French Anderson also stresses the importance of testing procedures with an openness that allows for public awareness and approval (p. 286). The ethical implications of experimentation on animals will be addressed in Chapter Eight.

26. "Position Paper on Human Germ Line Manipulation," p. 35.

stewards of the technical resources and abilities given to us, so that we can use them appropriately to assist others who are in need. As we have argued all along, a stewardship ethic sees technology as a gift to be used to benefit some while not degrading or devaluing others. To recognize that someone is in need biologically and to develop ways to meet those needs is not to demean their personhood; it is to recognize that they are persons for whom God has given us stewardship responsibility. When Jesus healed the crippled woman (Luke 13:10-17), for example, he had a biological norm or ideal in mind, an ideal that the woman herself recognized. His loving act helped her to come up to that norm so that she could live her life more fully. It does not matter whether this is done by miraculous healing, medicine, or genetic therapy. We are to act on behalf of God, not out of human hubris. At the same time, as with all technology, justice requires that we make that which meets basic human needs available to persons regardless of their social position, consonant with the restrictions imposed by our limited resources.

There are, of course, legitimate worries about gaining power over our genetic makeup. Such power, for example, could be used in ways suggested in Huxley's *Brave New World*. Knowledge and instruments of change can be used for good or ill. Tending the garden does not entitle us to ignore the snake that also inhabits it; rather, we are to be vigilantly on guard. We will return to the matter of snakes shortly.

(2) It is claimed that to eliminate undesirable genes in germ line therapy might be dysgenic.[27] One objector gives the example of Abraham Lincoln, who may have suffered from Marfan's syndrome. If we eliminate such genetic diseases, she argues, we might prevent the existence of another Lincoln. But this type of ad hominem argument is irrelevant; Lincoln's contribution to our society was not because of his disease but in spite of it. Hence to eliminate this disease from future populations by counseling or genetic engineering would not make it impossible for there to be great human beings who contribute significantly to society. Those who defend this objection must also be committed to the absurd position of advocating unlimited population growth, lest some individual "destined" for greatness be denied existence.

The objection has a point, however, that must be kept in mind no matter what kind of therapy is employed — namely, the recognition that many traits might have a function beyond what we now know. Thus, particularly with germ line therapy, a conservative approach, which em-

27. Macklin, p. 447.

phasizes safety, prior animal testing, and the requirement that there be substantial justification for the application of the technique, is called for.

(3) It is contended that in somatic cell therapy the autonomy of persons is respected, for their consent must be obtained before therapy morally can be undertaken. But in germ line therapy consent by future generations cannot be obtained. Since future generations have no opportunity to consent to or reject the therapy or to specify what characteristics they would want, we should not undertake such therapy, for it violates human autonomy.[28]

This objection launches us into consideration of the validity of paternalism (sometimes called parentalism), in which one makes decisions for others on their behalf, for their benefit. Where persons are autonomous, it is generally best to involve them in the decision-making process. There are situations, however, in which intervention is justified in order to put persons in a position where they can make a decision and to allow them to do so. That is, where the situation might produce irreversible changes which would severely restrict their later freedom (John Stuart Mill gives the example of selling oneself into slavery)[29] or where there are unusual or undue pressures that temporarily affect their judgment (which might lead to suicide, for example), intervention is justified so as to restore persons to a position where they can, with a sound mind, make a decision for themselves (which need not necessarily agree with that of the paternalist).

With nonautonomous persons, paternalism is justified in those cases where the persons, if and when they became autonomous, would probably see the wisdom of and sanction the actions we took.[30] But future persons are neither autonomous nor nonautonomous, since they do not yet exist. In order to respect their autonomy one could delay the choice until such time as they do exist, but then of course the matter of genetic composition is already decided; they cannot make the decision for themselves since they already exist with their unique genetic structures. It would seem proper, then, to treat future persons as we treat nonautonomous persons, for whom we make decisions that we feel will be in their best interests, that we believe they will approve of, and that will provide them significant freedom (that will not unduly restrict the scope of their choices). The latter implies substantial limits on the kinds of alterations that would be permitted under germ line therapy.

28. This objection was raised in correspondence by Hessel Bouma.
29. J. S. Mill, *On Liberty* (Indianapolis: Bobbs-Merrill, 1976), p. 125.
30. Gerald Dworkin, "Paternalism," *The Monist* 56, no. 1 (Jan. 1972): 76-84.

(4) Finally, it is argued that at this stage we generally are ignorant of the precise correlations between specific genes and particular phenotypic characteristics. For example, we do not know what effect altering certain genes will have on such physical characteristics as height, eye color, or baldness or such psychological traits as patience, kindness, lovingness, and the like. Neither do we know what genes, individually or in groups, to alter in order to bring about these traits, or what other consequences altering these genes might have. In fact, most traits appear to be polygenic in influence or origin.

This is true, and it constitutes a substantial general worry that calls for two things. First, it calls both for further research on the genetic basis of physical and psychological traits and behavior and for extreme care in making genetic changes. It does not, however, call for abandonment of the project. It might be the complexity of the project that makes it ultimately impossible to carry out, but this should be decided after, not before, conscientiously conducted investigation. Second, it calls us to restrain our optimism about genetic research and its claims. Probably most human characteristics cannot be linked with specific genes; and even where they are linked with groups of functioning genes, they are joint products of the environment, a person's complex genetic composition, and the person's choices and actions. Hence, germ line therapy has at best limited usefulness.

In sum, these objections do not rule out genetic therapy, either somatic cell or germ line. But they do suggest extreme care and caution in using these instruments to change humans, and perhaps a healthy dose of scepticism about what can be accomplished thereby. Three principles should be followed in the use of both somatic cell therapy and germ line therapy: (1) Is the procedure necessary? Are there no other effective means of treatment available? (2) Is the procedure effective? Can altered DNA be inserted into the chromosome where it is needed? Will it replicate there? Will it turn on to function properly in the cell? (3) Is the procedure safe?

The Ethic of Stewardship: Caring For

This leads to the second command, to care for the creation. The Christian ethicist will recognize that the garden over which we are stewards is not only a delight to the eyes but also contains snakes. Snakes come in different varieties.

Confidentiality

One snake is the possible misuse of the genetic information obtained. As we noted above, one purpose of the human genome project is to make it possible to ascertain the genetic structure of individuals, either through voluntary screening of adults or through pre- and postnatal screening. But once the genetic information is obtained, to whom should this information be made available and to what use should it be put?

The question of access is important. At the very least, individuals whose genetic structure is analyzed should have free access to the information so that they can make informed choices. This information can be important for them in many ways. For example, persons diagnosed as having the gene for Huntington's disease can know for certain, before the onset of any symptoms, that they face the prospect of an early death. Knowing this might assist them in making wise decisions about marriage, what life-style to adopt, what jobs to take, and what insurance to purchase, as well as matters concerning their death (for example, whether to sign a living will). This information might also have a bearing on decisions regarding whether or not to procreate. Since there is a 50 percent chance that each offspring will suffer from the same disease, as responsible stewards of future generations they should seriously consider their obligations to those who will result from their procreative activity. At the same time, conveying the information to the person tested also has risks, for it might increase the likelihood of suicide and social stigmatization. This means that the information must be conveyed carefully, in a context of compassion and assistance.

But who else should be privy to the genetic information discovered by testing or screening? Should the family or dependents of the person afflicted by seriously defective genes be informed? Here the snake raises its head. Suppose that someone agrees to genetic testing for Huntington's disease and tests positive. It would seem that the responsible thing for them to do would be to volunteer this information to their family, should their family wish to know. But suppose the sufferer chooses not to reveal the information, perhaps out of the belief (mistaken or not) that such information would cause undue marital stress. Should the physician or counselor inform the spouse and children anyway, the spouse because of the additional burdens and responsibilities that he or she will bear when the afflicted begins to exhibit symptoms of the disease, and the children because there is a 50 percent chance that they, too, have the gene? Or should the physician and counselor respect the autonomy of the patient by maintaining confidentiality?

The situation is not easy to resolve in a caring manner, for one of the conditions of testing might be that the testee's wishes for confidentiality be respected. Unless these wishes are honored, those who need or want the testing might not seek it. A policy of revealing confidential materials threatens the very process of genetic testing. At the same time, the interests of others must also be protected. Perhaps the most caring thing to do would be to make sharing the information with the relevant family members part of the initial testing agreement, prior to any received results. Should the testee not agree to this, the only option that remains is to attempt to persuade the testee that he or she has a moral obligation to share this information with the family, based on his or her responsibilities to others.[31] Practically, if Huntington's sufferers feel that they cannot handle the information, the counselor might seek to find ways to assist such persons in coming to grips with the information, so that their family or dependents can be provided for more adequately.

Another way of putting this invokes more explicitly our stewardship responsibilities. Where genetic information bears on others for whom someone with a serious genetic disease is a steward, there are prima facie grounds for believing that the person with the disease should share this information with those who might be affected seriously by the onset of the disease. The privacy of genetic information does not imply that it need not be shared with others. Stewardship provides the moral grounds for developing obligations for the sufferer to know and then share the information.

At the same time, however, the responsibility for sharing the information belongs to the affected person, not to the counselor or health care provider. Since the spouse's life is not threatened, there is no compelling interest overriding the confidentiality under which the testing was done to share the information. And since nothing at this point can be done medically to assist the children, the same holds true for them. If, however, there comes a point when the disease can be treated if caught early, this would provide an overriding justification, for the stewardship obligation to care for others at this point would require sharing the information so that any who are afflicted could be treated for the disease.

---

31. Bouma and his colleagues stress the covenantal aspect of testing, which both encourages and put limits on confidentiality (pp. 253-54).

Others beyond the family might also claim an interest in the information. Suppose that the testing is done in conjunction with an application for insurance. Currently, insurance companies generally require physical examinations for those who apply for insurance. If the human genome project is successful, insurance companies might want to include genetic testing among their requirements to obtain life or health insurance. But should the insurance company be privy to this information? It would be virtually certain that, should the insurance company discover that applicants have a genetic predisposition that could cause early death or protracted illness, the company would treat them as having a preexisting condition and either refuse to insure them or charge them costly premiums. Instead of divulgence, should there be a veil of ignorance drawn over one's genetic structures, so that the lottery of life is more evenly spread among the population?

But what are the grounds for specifying that genetic information is unique, that although the insurance company has the right to know the present and past states of the prospective insuree's health, it cannot have access to his or her genetic information? The relevancy of genetic information does not differ from the relevancy of information about chronic asthma or heart problems. Both concern the applicant's very personal matters, but both might be essential in determining the various risk factors involved in insuring a particular applicant and indirectly involved in protecting the economic interests both of the insurer and of other insurees. It is only that genetic tests give information about the causes or potential causes of problems, rather than information weighted toward the symptoms.

One response is that the information shows nothing about any present symptoms; it only specifies future or possibly future conditions, and so it is irrelevant to an assessment of the applicant's current health. Yet, in an inaccurate way, genetic information is already gleaned through the insurance application. Questions are asked concerning the health of the applicant's parents, whether they are dead, and if so at what age and of what causes they died. This family information is correlated with the health information of the prospective insuree to create an overall health profile. Genetic testing would surely provide a more accurate assessment than that currently obtained through the inferential method.

Should one's employer or potential employer also be privy to this information? One might think that this information is not relevant for the employer but only for the employee. Yet the present and future health of employees is a major concern of the employer, for both directly through work performance and indirectly through increased health care costs, employee health affects the company's profit margins and consequently its fiscal health. Recent moves by companies to reduce medical coverage for employees suffering from AIDS or for retirees illustrate their concerns. And there is anecdotal evidence about discrimination based on genetic testing.[32] In a sense, everyone in the company has a stake in the health of the other employees.

Further, the company might be working with materials or chemicals that, when combined with certain genetic predispositions, could deleteriously affect the employee's health. In such cases genetic information would be important to both the employer and the employee. It would not be good business to hire someone whose genetic predisposition indicates possibly significant costs for health care. In short, information about the genetic makeup of employees is relevant to the fiscal stability of the employer, particularly in the case of small businesses.

Not only does access to genetic information raise moral problems, but even the testing itself of actual or potential employees is laced with moral ambiguity. If the health care provider or tester was hired by the employer to test prospective or actual employees, the tester's primary obligations are to the employer, who established a special relationship by hiring the tester. Yet in the testing the most private domain of the employees is being invaded. Who then protects the interests of prospective or actual employees in the testing process? They could, it is said, go elsewhere for employment. But this merely shifts the burden onto another business or group of individuals (assuming, of course, that these others were not wise enough to do genetic testing for their own companies).

Given the relevance of genetic information to both insurance companies and prospective employers, the question remains regarding who will care for and protect the social and economic interests of the genetically impaired. Should the lottery of life determine that they cannot get insurance or a job? Should the balance sheets of companies deprive actual or prospective employees of health protection and employment opportunities? This question leads directly into public policy concerns. If the individual insurance company has no responsibility for providing insurance for those with deleterious genes and if the individual employer

32. David Stipp, "Genetic Testing May Mark Some People as Undesirable to Employers, Insurers," *The Wall Street Journal,* 9 July 1990, p. 1B.

is not responsible for protecting the interests of those who are only prospective and not actual employees, it becomes the responsibility of the society, through its government, to see to it that this sector of society is provided for or protected. That is, if genetic testing or screening is mandated by employers or insurers for getting or keeping a job or insurance, and if significant social decisions are based upon the knowledge acquired, then some sort of social structure must be put in place to prevent discrimination against individuals on the basis of genetic tests and to provide future protection in society for those who have significantly deleterious genes.

This might be accomplished in a number of ways. With regard to insurance, the government might institute national health insurance. This has the advantage of bringing all persons under the health safety umbrella, regardless of their economic and medical conditions. The risks are spread nationwide. However, this solution has the disadvantage of getting the government into a business for which it may be ill equipped. Barring this, the government might outlaw insurance discrimination by private companies based upon genetic deficiencies (or, more broadly, based on having certain medical or presymptomatic conditions). This, perhaps accompanied by some subsidy for the medically or occupationally disadvantaged, would provide equal access to all regardless of their medical condition or genetic predisposition.

Handling genetic discrimination in the marketplace might prove more difficult. A governmental guarantee of jobs to all is economically unwise, given the results of such experiments in the socialist countries of eastern Europe in the twentieth century. A more feasible approach would encourage the government to outlaw job discrimination based upon genetic deficiencies. Despite the relevance of genetic dispositions to the employer's bottom line, such legislation might be the best and most effective way meaningfully to care for persons with genetic deficiencies, for in this way the lottery of life belongs not just to the individual but is shared by other members of the society. Precedent for this can be found in the Americans with Disabilities Act of 1990, which prohibits discrimination in job application, hiring, promotion, training, and compensation on the basis of disabilities, where the person can perform the essential functions of the job.[33] This act could be interpreted broadly to cover genetic as well as phenotypic disabilities. To protect financially vulnerable businesses, such legislation should be coupled

33. *Congressional Quarterly* 48, no. 30 (28 July 1990): 2237-38.

either with some form of government-assisted health insurance or with legislation, similar to that proposed by the Clinton administration in 1993, that prohibits insurance companies from charging differential rates for preexisting conditions or genetic disparities.

To return to the issue at hand, our discussion has to do not only with financial and health outcomes but also with confidentiality. On the one hand, it is important that confidentiality be a priority in caring for individuals. To keep information confidential respects persons and their autonomy. It shows concern for them as individuals by giving them the dignity of making their own choices and allowing them to reveal what they want to others. This is a clear deontological value, for we allow people to keep or reveal this information regardless of the use others might make of the information.

On the other hand, personal autonomy is not the sole or supreme value. We are more than individuals; we live in communities, characterized by special relationships, in which as stewards we are obligated to care for others. At times others who might be affected negatively by someone's genetic disease need to be a party to the information. Such persons might include family members and perhaps even employers. But if the information becomes known to others who have power over the afflicted, then social policies must be in force to protect and care for that person. The steward will see to it that those who are losers in life's genetic lottery are given the dignity all stewards should possess as beings valued by God.

Discrimination

A second snake in the garden is the failure to pay attention to the human as well as the financial costs of genetic engineering. One human cost to be addressed is how the revelation that someone has a genetic defect will affect our view of the individual, whether that individual is another or ourself. The human genome project might encourage stigmatization of those with genetic abnormalities and their exclusion from normal community interactions.

We have already alluded to the problems of discrimination that arose with sickle-cell anemia testing in the 1970s. Sickle-cell anemia is a debilitating disease that results when genetically defective blood cells impair circulation and damage internal organs. It particularly affects black persons. The ability to ascertain relatively inexpensively who possesses

195

the gene responsible for sickle-cell anemia led legislatures in at least a dozen states to pass laws in 1971 and 1972 mandating sickle-cell screening for blacks.[34] Candidates for marriage were screened, as were prospective employees. On the basis of the test some employers refused to hire individuals who manifested the trait. Realizing the potential abuse of the screening, Congress in 1972 enacted the National Sickle-Cell Anemia Control Act, which required states to make screening voluntary rather than mandatory and to provide genetic counseling for those tested.[35] This was an attempt to rectify the failure on the part of many, including lawmakers, to differentiate between sickle-cell trait (where a person has inherited the gene from only one parent and experiences no symptoms) and sickle-cell anemia (where the person has inherited the gene from both parents and can experience severe symptoms), to prevent possible discrimination based on the confusion, and also to acknowledge that even within the class of those who had anemia the severity of the symptoms varied greatly.[36]

The disastrous results of the initial sickle-cell screening are sometimes contrasted with the positive results of the program developed for screening for Tay-Sachs disease, which particularly affects Ashkenazi Jews. In the Tay-Sachs program, education preceded the screening, so that people knew what was involved and what the tests meant. Instead of the population-wide screening for sickle cell zealously recommended for getting jobs and marriage licenses and also done in schools or prisons, screening for Tay-Sachs was conducted within the community in synagogues, storefronts, and community centers. It was also targeted to young people, who could both comprehend the significance of the test and use the information to make procreative decisions. Hence it was not perceived as an outside

34. Munson, ed., *Intervention and Reflection*, pp. 412-13.

35. Public Law 92-294, May 16, 1972. It was amended in Public Law 93-82, August 2, 1973.

36. In 1974 in New York State, for example, every baby born was to be tested for sickle cell; yet often follow-up was inadequate because of limited funds. R. Grover et al., "Current Sickle Cell Screening Program for Newborns in New York City, 1979-1980," *American Journal of Public Health* 73 (March 1983): 249. On April 27, 1993, the Agency for Health Care Policy and Research, a federally funded group of physicians, scientists, and patient representatives, recommended that all newborns be screened for sickle-cell disease and, where they test positive, be treated with penicillin. Universal screening is currently done in thirty-four states. "New Sickle Cell Guidelines Issued," *St. Paul Pioneer Press*, 28 April 1993, p. 5A.

(racial) threat against the community, as was the case with sickle-cell screening.[37]

Another example of a case in which the concern for the possible human costs of discovery of genetic makeup played a significant role involved the study of the possible link between an extra Y chromosome in males and aggressive behavior and mental illness. In 1961 the British medical journal *The Lancet* carried the report of a man who had a history of barroom brawling and who was discovered, when he was tested because he fathered a Down syndrome child, to have a unique genetic structure. Instead of the normal XY genotype, he had an XYY genotype.[38] Four years later *Nature* carried a study by Patricia Jacobs and several colleagues of 197 men from a high-security Scottish mental hospital (Carstairs).[39] Of these, what appeared to be an abnormally high proportion (3.5 percent) had the XYY genotype. On the basis of this, Jacobs and her colleagues theorized that male aggressiveness and mental subnormality was connected with this genotype. The extra Y chromosome somehow predisposed the male to a life of mental illness and violence. By the early 1970s some hospitals in the United States, England, Canada, and Denmark were regularly screening newborn males to see whether they were XYY, and in some instances this information was communicated to the parents. It was argued that "even if the XYY genotype could not yet be conclusively linked to criminal aggression, it seemed that parents of an XYY baby at least had the right to know that such a possibility, however remote, existed."[40]

The numerous subsequent studies often began with a biased population, selected because they seemed to exhibit the traits that were thought to be associated with the XYY genotype: violent, antisocial behavior, mental deficiency, and tallness. The studies found a higher than expected proportion of XYY males in the population studied, but they neither gave any information about the populace at large (for example, what proportion of XYY males actually exhibit such behavior or other behaviors; what was

---

37. Leslie Roberts, "One Worked; the Other Didn't," *Science* 246, no. 4938 (5 Jan. 1990): 18.

38. A. A. Sandberg, G. F. Koeph, T. Ishihara, T. S. Hauschka, "An XYY Human Male," *The Lancet* 2 (1961): 488-89.

39. Patricia A. Jacobs, Muriel Brunton, Marie M. Melville, R. P. Brittain, W. F. McClemont, "Aggressive Behavior, Mental Subnormality and the XYY Male," *Nature* 208 (25 Dec. 1965): 1351-52.

40. David Suzuki and Peter Knudtson, *Genethics* (Cambridge: Harvard University Press, 1989), p. 153.

the percentage of XYY males in the general population) nor helped to determine whether there was in fact any relationship between the additional Y chromosome and the behavior studied (that is, did unincarcerated XYY males display similar behaviors). The reports were often "anecdotal, impressionistic, and subjective."[41] Meanwhile, "in one institution in Maryland, XYY inmates were treated with female sex hormones in an attempt to restore 'normal' behaviour."[42]

A proposed study in 1968 at the Boston Hospital for Women brought the matter to a head. Child psychiatrist Stanley Walzer and pediatrician Park Gerald intended to screen newborns for genetic defects, including XYY, and then to trace their subsequent life history in order to determine whether there is any connection between the genetic defect (here, the extra Y chromosome) and certain behaviors (here, violence and mental illness). Objectors to the study (the Boston Science for the People group) argued that the very study, which involved informing the parents, would so stigmatize the testees that they might develop the behaviors sought simply as a reaction to how they were treated by others or to how people expected them to react. Parental prejudice or expectations might create a self-fulfilling prophecy. Furthermore, since there was no known therapy for the XYY condition, there seemed to be little use for the study. Eventually enough pressure was brought to bear on the researchers by both the public and other researchers that they abandoned the study in 1975.[43] In effect, the fear that serious discrimination would result, not only from what the tests would discover but from the testing process itself, terminated the testing.

Lessons can be learned from this case. On the negative side, public concern about the potential misuse of test results or even of the test itself brought to an end what some deemed a legitimate scientific study. On the other hand, this case shows that genetic research must be conducted within the bounds of ethical imperatives, that potentially discriminatory use of the results of the experiments and testing are grounds for serious concern. It forces the scientific community to pay serious attention to the research design, in terms of both experimental method and ethical implications.

41. Stanley Walzer, Park S. Gerald, Saleem A. Shah, "The XYY Genotype," *Annual Review of Medicine* 29 (1978): 564.

42. Jon Beckwith and Jonathan King, "The XYY Syndrome: A Dangerous Myth," *New Scientist* 64 (14 Nov. 1974): 474.

43. Beckwith and King, pp. 474-76.

This duality suggests an ambiguity for scientific research. It must be free from public pressure so that it can arrive at the truth; yet at the same time it must be responsive to the legitimate moral concerns of the community that the human dignity of the subjects not be imperiled or compromised.[44]

Not only must we be concerned with the possible stigmatization of others who have serious genetic defects; the potential discovery of our own genetic heritage also might affect how we understand ourselves. It is estimated that every person carries five to ten seriously defective genes. If this is the case, how will we deal with ourselves when we realize that we are not perfect, that every cell in our body is flawed in some way? This can become psychologically important when we consider procreation. Some might feel undue anxiety about what genes they pass on, worrying about whether those genes are defective or not. Am I, it might be wondered, worthy of procreating? Others might take on unwarranted guilt for the genetic diseases or disabilities of their children. They may see a genetic defect in their child as an occasion to blame either themselves or their ancestors. In short, the revelation of our own genetic fallibility can be the occasion either for acceptance of the human condition of finitude or for guilt and self-incrimination.

Biological Ideals and Norms

The third snake lurking behind this discussion is our obsession with normalcy. In the past we consigned people who looked abnormal to circus sideshows as freaks of nature to be excluded from everyday society. Though fortunately we no longer do that and have slowly begun to welcome them into the mainstream of society, we still have a fetish about the normal or the beautiful. The power to peer inside the womb enhances that appeal, for now with amniocentesis and ultrasound we can evaluate the fetus before it is born, and given the present legality of abortion in society, we can decide whether the conceived is worthy or perfect enough to be born. In a sense, there is the danger that the child can assume the character of a

44. In what seems almost a case of déjà vu, in April 1993 the NIH terminated a $78,000 grant to the University of Maryland that had been designated for a conference to explore a possible genetic basis for crime. The director of NIH, Dr. Bernadine Healy, explained the termination: "We as scientists and physicians believe it was an inappropriately ill-conceived, dangerously inflammatory conference in the way it was promoted." "Genetics-Crime Study Canceled," *St. Paul Pioneer Press*, 25 April 1993, p. A2.

commodity; we may feel that the child ought to be perfect since we are paying so much for it, and if it fails to meet the guarantee, we should be able to return it.

One of the potential, paradoxical results of the human genome project might be to show that there is no genetic ideal to be approximated. Each human has a diverse set of genes, most of which are good (properly functioning), some of which are seriously defective,[45] and a host of others that work well or poorly in combination with other genes. We are all flawed in some way. Some genetic flaws become phenotypically manifested; others do not. Yet in our imperfection we are equal in our humanity with each other.

Further, the human genome project may show that genes which might not convey any benefit in some circumstances might be very beneficial in others. Outside of the tropics, the sickle-cell gene, which can cause painful sickle-cell anemia when it occurs in pairs, conveys no apparent benefit. Yet in Africa, when it occurs singly, it enables people to resist a particularly prevalent strain of malaria, so that those who have this trait have a better chance of survival.[46]

Even if there were a human ideal it would be unattainable, for reproduction brings constant genetic variation. Our offspring result from a genetic mechanism that involves random (unpredictable but not uncaused) combinations. Hence, curing one genetic disease does not mean that another cannot begin, or even that the cured is not a latent carrier of another genetic disease that might manifest itself in future generations. To look for a genetically perfect human ideal is not only to treat humans as unchanging, which certainly is not our lot, but also to ignore our human creatureliness.

Once we abandon the concept of a genetically perfect human specimen, we can also abandon the thesis that the goal of genetic therapy is to create persons satisfying some specified general ideal or norm. At the same time, however, we cannot abandon norms or ideals altogether. Norms play an important role in our medical practice. We know what it is like not to feel "normal," not to have the normal range of motion in a limb, or to have an abnormal or defective heart. Physiological and psychological norms provide standards against which we can measure our bodily and mental performance. They are also standards that assist us in helping people — in dispensing medicine, giving physical therapy, conducting surgery, providing educational opportunities, and the like. Thus, though

45. Suzuki and Knudtson, p. 205.
46. Suzuki and Knudtson, pp. 200-202.

we want to reject the idea of a general human ideal by which all humans should be measured, we want to employ norms to assist us in developing individual human potential and in understanding the unique roles that genetics might play in that development. (We shall return to the matter of norms in the next section.)

The Ethic of Stewardship: Ruling

Consideration of the first two ethical injunctions — to fill and to care for — leads us to the third: to rule over and subdue the earth. The command to subdue gives authority to the stewards to rule over the creation, to change it for the benefit of the Landlord, the stewards themselves, and the creation. That God gave us this authority raises critical issues about the extent of our obligations, to whom we are obligated, and the limits of our obligations with respect to genetic therapy. Let us consider these in turn.

Therapy vs. Enhancement

According to our stewardship ethic, we have obligations to care for those whom we encounter and can benefit. Generally it can be said that we have obligations both to do no harm and also to benefit others. We are to do the best by them that we are able, given our other commitments and total resources. It is true that in ranking benefit and harm, not doing harm should be uppermost. At the same time, concentrating on simply not harming, which is the buzzword morality of our society, is nothing more than moral minimalism. Stewards are obligated to go beyond the minimal, negative duty to seek ways to benefit others.

When the duties of health care practitioners are listed, both not doing harm and benefiting appear. However, by virtue of the oaths taken, health care providers have a stronger obligation to benefit. Indeed, in the Hippocratic Oath, the obligation to do good precedes that of doing no harm: "I will use treatment to help the sick according to my ability and judgment, but I will never use it to injure or wrong them."[47] It is often

47. Nigel M. de S. Cameron, *Life and Death after Hippocrates: The New Medicine* (Wheaton, IL: Crossway Books, 1991), p. 25.

the case that harm must be done (as in surgery) in order to bring about a greater good, but the harm is a means to the patient's greater good.[48]

When we turn from our contemporaries to future generations, it would seem that our obligations to them coincide with those to present generations, though undoubtedly with decreasing urgency and degrees of obligation as generations become increasingly distant from us. My degree of obligation not to harm my potential children is much greater than my obligation not to harm the fortieth generation of my descendants, though I have some obligation not to harm them as well. I am not obliged to provide for their welfare in the same way or to the same degree that I am with regard to my own children, though I should not bestow on them a world worse than I inherited.

Some have questioned what kinds of *positive* obligations we have to future generations. Hans Jonas has written that "our descendants have a right to be left an unplundered planet, but they do not have a right to new miracle cures."[49] Progress, he goes on to say, is more attributable to grace than to obligation. Jonas is correct about miracle cures; at the same time, however, where we can benefit and not harm future generations with minimal cost or dysfunction to ourselves or others, there does seem to be some obligation to do so.

But how are we to carry out our obligations to present and future generations? As we have noted above, in our ordinary life we see to it that our children are benefited in a variety of ways. We work not only to correct their ills but also to improve them by providing preventive medicine, good education, a stimulating environment, recreational opportunities for physical and emotional development, worship experiences for spiritual growth, and so on. But if we carry out our stewardship roles toward our actual children in such ways, it would be appropriate similarly to seek good for our prospective children. We owe it to our Lord to do them good. This would mean that, where feasible, scientific researchers have obligations not only to try to find cures for those currently ill but also to try to improve the human genetic heritage.

Improvement of genetic heritage is a broad notion. If one is going to seek to improve future persons, certain characteristics must be chosen

48. It might be wondered how, given this, physicians can engage in abortion. The answer is that those who do so consider the woman, not the fetus, to be the patient. Hence they are taking action to benefit the patient.

49. Hans Jonas, "Philosophical Reflections on Experimenting with Human Subjects," *Daedalus* (1969); reprinted in Munson, ed., *Intervention and Reflection,* p. 349.

as end points. It is true that it is easier to reach agreement about what constitutes a defect to be remedied than to decide on changes that do not remedy defects but might constitute improvements or enhancements. Classically this has led to a distinction between what is termed *negative eugenics* and *positive eugenics,* or between what might be termed *gene therapy* and *enhancement genetic engineering.*[50] The former has wider approval; the latter is widely held to be suspect. Yet it is difficult to discern a morally significant difference between them.

One way to differentiate them relates to an established norm. Some suggest that gene therapy or negative eugenics seeks to improve persons by bringing them *up to* a norm. One adds to or repairs abnormally functioning genes where these cause seriously debilitating physiological and psychological symptoms. Genetic enhancement or positive eugenics seeks to improve persons *beyond* a norm, to add to or enhance features they already possess. But granted this difference in terms of departure from a norm, is the difference morally significant? Departure from a norm would be morally significant only if the norm were treated as an absolute standard, the departure from which would be morally unacceptable. On this view we are permitted to find ways to correct the genome of those who fall below the norm, whereas it would be immoral to improve someone beyond the norm. However, human norms tend not to be absolute standards; rather, they are statistical averages arrived at by analyzing the population. By improving those below the norm one is in effect raising the norm. This leads to the unacceptable paradox that what would have been immoral under the old norm would now become moral, for whereas improvement was forbidden when the person was slightly above the old norm, now that very same action is permitted because the person falls below the newly adjusted norm. Thus, although the use of norms is important, it is not by itself indicative of the moral unacceptability of one type of eugenics over the other.

More helpful might be a discussion of *ranges* of characteristics clustered around statistical norms. We might alter the human genetic structure within broadly conceived ranges to both correct and improve; however, there are points beyond which alteration becomes morally suspect. We encounter contextualism here; the location of those points depends on the characteristics under discussion. For example, it might be

50. W. French Anderson, "Genetics and Human Malleability," *The Hastings Center Report* 20, no. 1 (Jan./Feb. 1990): 21-24.

morally acceptable to increase a prospective child's height to six and one half feet from a projected six feet, but it would be morally unacceptable in our present context to increase it to twelve feet, despite the economic potential the child might have in a professional basketball career. One might reply that this leaves a lot of room in between, and so it does. The facts of the case — how children of certain sizes will be treated, why the change is being made, whose benefit is being sought, what the side effects of the change will be — will all be relevant factors in making genetically engineered changes.

Of course, all this talk about genetic enhancement makes it sound quite feasible, very easy, without danger or risk. To the contrary, not only are the techniques exceedingly difficult and as yet quite unperfected, but there is also the additional problem of correlating genetic changes with specific physical, psychological, and behavioral traits and characteristics. With few exceptions, where genes play a role at all, traits result from many genes cooperating together.[51] Thus enhancement could be medically hazardous should proper antecedent testing and procedures not be employed.[52] What we have argued is not that genetic enhancement is possible, easily accomplished, or risk free; rather, we have argued that if it becomes possible and can be done safely, genetic enhancement may morally become one of the instruments used by stewards to assist future generations, when it is done within the context of genuine stewardship concerns and obligations.

The most frequent objection to genetic enhancement is that we are too ignorant to do it right.[53] To change our environment is one thing; to change future persons is another. Not only do we not know what characteristics to choose or emphasize now; we are also ignorant of what the future results of our changes will be. For example, as the environment changes, genetically enhanced individuals might be deprived of the genetic resources to adapt to the change. In effect, we do not have enough wisdom to know for what characteristics to engineer. Do we want our children smarter, taller, darker or fairer, brown- or blue-eyed, with curly or straight hair?

In reply it may be said that the choices stated in this common objection are often trivial. Characteristics such as tallness or fairness may

---

51. Anderson, "Human Gene Therapy," p. 289.
52. Anderson, "Genetics and Human Malleability," p. 23; Macklin, p. 384.
53. Macklin, p. 381; Anderson, "Human Gene Therapy," 289.

or may not be desirable, and perhaps these are not worth doing research on or engineering for. But surely intelligence is another matter. Why not make our offspring smarter? Indeed, we continually make environmental changes to this end when we design and redesign our school curricula, seriously seek ways to improve the quality of education for our children, and recommend certain diets and the avoidance of exposure to lead. We even make changes in the children themselves when, for example, we administer drugs that have serious side effects to control hyperactivity so that they can learn and concentrate better. Even if there is a debate over making our children more intelligent (since most of us do not exhaust the intelligence we already possess), there is no debate over whether we want them to be healthier, more disease free, and better able to cope with and adapt to the world. We do have a fairly good idea of many types of characteristics that really would be desirable and worth promoting. Some, such as hair color and height, are matters of individual choice and are not likely to be top-priority demands in genetic engineering projects. Others, such as intelligence, ability to control one's emotions and behavior, and freedom from serious, genetically based diseases, are worthy of research and technological introduction.

What is frequently overlooked in the discussion of genetic enhancement is that all endeavors directed toward improving the world make significant choices about what will be desirable or good for future generations. Trying to improve education by changing its methods; developing solar power as an alternative to depleting oil and coal resources; installing asbestos to make schools more fire resistant; turning forests, wetlands, and deserts into farmland; eliminating small pox; installing clean water systems; building expensive flood control systems; conducting an expensive war to bring peace to a region — all of these endeavors are intended to improve the quality of life for ourselves, for others, and for successive generations. Sometimes the results are good and sometimes not; but we are justified in trying to change the world for the better as we perceive it, provided that there are no other, more desirable means available, that the method seems to be safe, and that there is good reason to think it is effective (that in any given case the benefits outweigh the harms). We can never be assured of the outcome. Yet if we are justified in selecting what is good when we try to improve our world, the same applies when we make changes in humans themselves. Genetic enhancement would simply be one of the possible instruments for effecting those changes.

It might be replied that these other cases affect humans indirectly, whereas altering human genes affects humans directly. True enough; but that by itself does not necessarily mean that one is more or less dangerous than the other. The ways in which we have dramatically altered the environment in many cases have more long-term negative consequences for humans than carefully controlled genetic therapy might have. To mention just one example, removing carcinogens and toxins like PCBs, lead, and mercury from our ground water supply will take many generations.

An even greater issue concerns what genetic enhancement might lead to. The specter of Nazism frequently arises out of the discussion. If we begin genetic enhancement, who will be enhanced, and what will happen to those who do not receive enhancement? Will they become the object of discrimination, both social and perhaps eventually lethal?[54] The worries are legitimate enough for us to insist that any research toward genetic enhancement be subject to the closest public scrutiny, and that any implementation be judged first and foremost under the rules of love and justice for those whom we are attempting to serve.

Yet doubts remain. In an articulate piece, Alasdair MacIntyre lays out seven traits or virtues that people might want their descendants to have: "an ability to live with the uncertainties of an unpredictable environment; a capacity for finding a particular and local way of being at home in the world; a commitment to nonmanipulative modes of relationship with people and nature; an ability to find a work that is peculiarly our own to do in the world; a recognition that there will come a time when life is complete, when it is time for us to die; and accompanying each of these a spirit of hope for which there cannot be adequate empirical grounds"; finally, "a willingness to go to war and to acquire and use the skill necessary to win a war." But, he continues, persons who have these traits "would be unable, by virtue of those very traits, to adopt manipulative, bureaucratic modes of planning. . . . They would be quite unwilling in turn to design *their* descendants" — and by implication, so should we. "It would clearly be better never to embark on our project at all."[55]

But is MacIntyre's inference, that persons with these virtues would not design their own descendants, correct? The answer is both yes and no.

54. "Position Paper on Human Germ Line Manipulation," p. 36; Anderson, "Genetics and Human Malleability," p. 24.

55. Alasdair MacIntyre, "Seven Traits for the Future," *Hastings Center Report* 9, no. 1 (Feb. 1979): 7.

MacIntyre is correct that such persons would not design in any manipulative or deterministic fashion their own descendants with these virtues; rather, they would want them to develop and nurture these virtues for themselves. In fact, one can make the more general point that virtues do not seem to be the sorts of things one can design future persons to have. People develop virtues as responses to what happens to them. But if this is the case, these (and other) virtues are not the sorts of things to which we refer when we speak about "designing future persons." Rather, what we design relates to the mental or physical abilities with which we provide our offspring to enable them to cope better with what they encounter.

MacIntyre believes that we cannot predict the future and hence future environments, and again he is both correct and incorrect. True, we cannot predict specific environments, the tools and equipment we will use, or the dominant paradigms or technology. As he notes, to predict what these will be specifically is to make that very innovation. At the same time, we have general ideas — as MacIntyre's own discussion of the virtues suggests — of the kinds of abilities and skills that enable persons to cope more effectively with their environments. Insofar as these have genetic roots, it is proper to look toward providing future generations with the bases for developing these skills and abilities. We are not thereby bureaucratically manipulating them; rather, we are providing the possibilities under which they can enrich and enhance their own lives. In this sense, genetically providing for future generations does not differ from the way in which we currently provide for our children through education and experience.

It is not morally right to manipulate our children into becoming certain persons. In this sense, MacIntyre is right: it would be immoral for us to blueprint or mold future persons. Yet emphatically it *is* morally proper, if not obligatory, to train and educate them and provide appropriate environments so that, as they mature and become autonomous, they have enriched opportunities for developing the values, virtues, and skills needed to adapt to their changing environments. We cannot and should not raise our children in a moral vacuum, in a value-free or value-neutral environment. Not every claim is true; not every idea is equally worthy; not every claimed good is really a good. Each of us believes in certain ideals and truths around which we have shaped our lives. We would be foolish to abandon those values, to refrain from attempting to inculcate them in our children. That would be tantamount to believing that they are not ideals or not really true. We properly want our children to share and adopt these ideals and truths, but as matters of their free choice. Hence we instruct, educate, and train our children and attempt to model the values we believe in. Thus we do — and should — seek to design our descendants in that we provide opportunities for their development and advancement, ideals to adopt as their own, and the encouragement and opportunity to become their own persons. Genetic enhancement can be one tool to facilitate the development of abilities that can provide the possibility of human growth.

Obligations to Future Generations

To whom do we have obligations? With regard to somatic cell therapy, it is clear that our obligations are to the patients who are being treated. When we consider germ line therapy, however, we must speak about our obligations to future generations. But does it make sense to claim that we have obligations to future generations, especially since they do not yet exist? How can one be obligated to beings that do not exist?

To answer this question we must determine to whom we have obligations. Judgments of obligation involve three elements. First, a judgment of obligation claims that the person is obligated to *do* or *become something*. I am obligated to care for and nurture my children or to keep a promise I made to a friend. Second, a judgment of obligation claims that the person is obligated to do something *for* or *to somebody*. My children are to be benefited or my friend is to receive the fulfillment of my promise. To distinguish this from the third feature, sometimes we can speak of obligation *toward* rather than *to* another. Third, a judgment of obligation claims that the person is obligated *to someone* to do or become what he or she ought to do or become. Having obligations is a referential notion. One never has obligations per se; rather, one always has obligations *to* someone or something. This someone might be the same as the person we are to do something for or toward (I am obligated to my children to provide for them), or it might be someone different (I am obligated to the person to whom I made a promise to distribute their goods to their relatives). But in either case, this person or persons to whom we have an obligation can hold us accountable for fulfilling our obligation.

It is the referential aspect that is problematic when speaking about obligations to future persons, for future generations do not yet exist for us to have obligations to. What meaning, then, can there be to the claim that by engaging in germ line therapy we are satisfying our obligations to future generations or that by failing to do experiments on the human genome we are failing in our obligations to them?

One response is to tie obligations to rights and rights to interests. Whatever can have interests can possess rights. For example, I have an interest in the actions taken by my government, for its actions can benefit or harm me. Accordingly, I have certain rights with respect to that government, including probably the right to some form of participation in determining who governs. Further, as we noted in Chapter Five, if a person has a right, someone else has an obligation to the possessor of the right. If I have the right to dispose of my property by selling it, you have the obligation not to interfere with that right. With respect to many rights, the obligation of others is negative — that is, they are obligated not to interfere with persons exercising their rights, but to respect their freedom of action. Regarding human rights, the obligations of others are more proactive. If I have a right to a free education, someone has an obligation to do more than leave me alone; they ought to assist me in claiming that right where possible and feasible. Simply put, rights incur duties or obligations on the part of others, either to leave me alone or to provide for me when I claim my right.

Although they do not yet exist, future people will have interests. Actions can be taken now that will benefit or harm them when they exist in the future. Whether future generations have rights is thus contingent on the future existence of persons who will have those rights. Should they exist, they have the right not to be harmed. We can be reasonably sure that there will be future persons who will make claims and who will have interests that can be affected by what we do now.[56] Hence it is reasonable to claim that, in this conditional sense, we have obligations to future persons who will be touched by our actions.[57]

Obligations to future persons are like obligations to distant persons. We can speak legitimately about our obligations to people in other parts of the world even if we do not know who they are. If we have good reason

56. Robert Elliot, "The Rights of Future People," *Journal of Applied Philosophy* 6, no. 2 (1989): 161.

57. For a more detailed presentation of the arguments that are only briefly sketched here, see Bruce R. Reichenbach, "On Obligations to Future Generations," *Public Affairs Quarterly* 6, no. 2 (Apr. 1992): 207-25.

to think that someone lives in a certain location, even if we do not know who is there, we have an obligation to them not to air-drop toxins on that location. Knowledge of who the affected persons might be in this case is irrelevant to determining obligations to them. What is relevant is that there is good reason to believe that someone lives there who will be affected by our actions. It is those who will be affected by what we do, whether in distant lands or in the future, to whom we have obligations.

Costs and Limited Resources

With regard to genetic therapy, the question of what proportion of our resources should be devoted to this comes under the general obligations we have toward others. It is our obligation to rule in such a way that we do others good and not ill. Where people suffer and where we can alleviate that suffering and significantly improve their existence at a reasonable cost to ourselves and risk to them, we are obligated to do them good.[58]

But here the proviso about reasonable cost and risk assumes significance. The increasing demands placed upon our limited resources, particularly those allocated for medical care, make it more difficult to justify the great expenses incurred by exotic procedures, which genetic therapy might be thought to be. Yet with some of the more common serious genetic defects, such as cystic fibrosis, the cost of developing a genetic cure might ultimately be less than the costs incurred in caring for those afflicted. Currently, according to the Cystic Fibrosis Foundation, to treat mild cases costs approximately $10,500 per year, moderate cases $26,000, and severe cases $48,500. Since approximately 30,000 Americans suffer from the disease (not to mention those affected worldwide), the annual expenditure of more than three-quarters of a billion dollars on treatment makes developing genetic therapy fiscally feasible and morally responsible. Similar cost/benefit analyses might justify genetic research into and the application of genetic technology to adult polycystic kidney disease, which affects 300,000 to 400,000 Americans, and to Alzheimer's disease, which affects more than two million Americans.

In other cases, where the genetic defect is rare, the costs might far exceed the benefits. It is true that in many cases the health care costs for

58. Where people go beyond that reasonable cost, we say that their action was not obligatory but supererogatory, that is, especially deserving of praise.

those afflicted are very high. Yet these are offset by the low incidence of the disease. For example, in the U.S. less than 100 people each year are born with ADA deficiency, which is the object of one of the initial trials of genetic therapy. If there were unlimited resources, there would be no question about developing genetic therapy to treat rare diseases, for it is human beings who suffer. The suffering of a few from a rare disease is as bad as the suffering of many from a more common disorder. But with limited resources, one must always ask what other goods are being sacrificed in order to make this therapy available to a select few.

And there are hidden costs to be considered as well. If we succeed in prolonging the life of ADA deficient persons to an age when they can procreate, we potentially increase the number of persons who are afflicted by or carry the defective gene. This in turn will require the use of a greater and possibly disparate share of health care resources to treat increasing numbers of ADA deficient persons. In effect, long-term implications must also be considered in making choices about the wise use of scarce resources.

It might sound crass or unfeeling to say that we have to consider the costs involved in saving a life or improving the quality of life. Our common response to a disaster such as a mine cave-in or a child dying of liver failure follows the dictum that life should be saved no matter what the cost.[59] But where resources are limited, the saving of the few must be weighed against the effects that applying those funds elsewhere will have. One cannot justify saving a life simply because it is a human life when the resources used to save that one life might seriously jeopardize the lives of others.

This is not to suggest that one sufferer is more valuable than another. Such a position would be incompatible with our Christian ethic; God values all persons equally. Rather, the contention is that ethics involves, at least in part, a calculation involving costs and benefits. Where resources have to be taken from some in order to treat others, calculation of the greater good must include not only the prospective good achieved by providing treatment to the few but also the harm done by denying use of the resources to others. This calculation is never easy to make, for it involves not merely quantitative considerations, that is, considerations

59. The more than one million dollars used to rescue two whales stranded in the Arctic ice in 1988 provides an interesting example, especially when the efforts to rescue these whales are compared with the contemporaneous killing of whales elsewhere.

of the numbers of persons involved, but also qualitative considerations, that is, considerations about the potential quality of life for all those concerned. It might even necessitate comparing quality of life for some with quantity of life for the few. But behind the calculations stands the deontological contention that persons are valued by God. Hence resources must be distributed in ways that are consistent with the view that all — those who are immanently threatened as well as those whose lives will be worsened or even threatened by denying them use of the resources — are valuable.

This problem is not unique to genetic therapy. Wherever the resources needed to maximize the good are in short supply, the problem of calculation arises.[60] This is not a reason for abandoning the utilitarian dimensions of ethics, as some would do. Rather, this simply indicates that we are stewards over a finite world. Indeed, the finitude of the world is what necessitates our stewardly role; we need to apportion wisely and justly the individual finite goods to achieve the greatest overall good. Thus, matters of justice (the way we rule over the creation to satisfy human needs) meet concerns for utilitarian goods on the battlefield of moral and social action.

We have already alluded to the primacy of justice, which lies behind almost every stewardly decision. Considerations of justice are especially important in allocating scarce resources. Not everyone can or ought to be treated equally. One must distinguish between cases in which persons are equal and cases in which they stand unequal. Justice requires that those who are equal be treated equally. But where persons are unequal in some relevant way, they need to be treated unequally. One inequality has to do with physical or mental health. Those who are ill justly demand more resources than those who are well. Life-saving surgery takes precedence over cosmetic surgery.

60. Genetic therapy is not the major locus of the problem of allocation of scarce resources. For example, the amount of resources we use after the age of sixty-five contrasts greatly with the amounts used earlier in life, both medically and socially (including education), to provide a higher quality of life. J. Lubita and R. Prihoda, "Use and Costs of Medicare Services in the Last Two Years of Life," *Health Care Financing Review* 5 (1984): 117-31; Steven A. Garfinkel et al., "High-cost Users of Medical Care," *Health Care Financing Review* 9, no. 4 (Summer 1988): 41-52. This, however, should not be taken to mean that the expenditures are inappropriate or wasted. See Anne A. Scitovsky, "Medical Care in the Last Twelve Months of Life: The Relation between Age, Functional Status, and Medical Care Expenditures," *Milbank Quarterly* 66, no. 4 (1988): 640-60.

Philosophers have suggested different criteria for determining how to distribute resources justly in cases involving inequality. In some cases *need* determines how we allocate resources. To those who are most needy and able to benefit we give first call on treatment. For example, where two people are injured, one seriously and the other not, despite the pain of the one not seriously injured, we treat the one whose life is in serious danger.[61]

In other cases *merit* plays a determining role in distributing justice. For example, justice demands that we give the top grades not to students who need them to pass or graduate but to those whose work merits them. In medicine, however, this criterion for the distribution of resources rarely applies, although perhaps one might allocate them based on how well the person takes care of himself or herself or follows instructions, placing those who continue to abuse their bodies lower on the priority scale for treatment. It would be unjust to continue giving resources that could be profitably used elsewhere to those who consciously waste them. (We shall return to this issue in Chapter Ten.)

In still other cases, the allocation of scarce resources depends on *randomizing* the applicants. Those who receive assistance are either selected in a lottery or treated on a first-come first-served basis. For example, in deciding who should get a scarce organ, consideration of need and possible benefit come into play in determining the initial selection; but after this, random selection might be the appropriate method used to select the final recipient. In short, justice, like concern for beneficence, demands calculation of whether the persons involved are equal or not and of which principle should be properly employed in deciding the basis for unequal distribution of the scarce resources.

God has given his stewards a mind to reason how these aspects are to be weighed and a heart to show compassion to all concerned. But God has not given us any easy formula for calculating what to do in each and every case, if indeed any such easy formula exists. Only if we or God knew every result of every action and choice could a precise calculus apply. But this is not available to us; all we can do is to make the best choices we can in each situation that arises. Neither is it available to God; God's

---

61. Even in this case there are exceptions. During the Second World War, medics often treated those with slight wounds before those who were more seriously wounded, and hospitals gave the scarce penicillin to those suffering from sexually transmitted diseases rather than to severely wounded soldiers, on the grounds that these could be returned to combat more quickly than the others.

foreknowledge applies to what will happen, not to what would have happened had other things been done or other choices been made. We return again to a theme developed earlier: being moral agents involves risk taking. God has given to his stewards, besides the responsibilities of stewardship, the tools for making situational moral calculations, based upon moral principles.

Conclusion

A realistic appraisal of the human genome project reveals several misperceptions. One is that genetic therapy is a panacea for the human predicament, that it will resolve all human medical and behavioral problems. Yet even in the most controlled genetic experiments, there is an element of unpredictability. Moreover, human beings are more than mere products of their complex genetic heritage. Their environment and how they respond to it will greatly affect their medical and mental conditions. Genetic therapy is only one tool in the complicated practice of medicine, and a minor (though often spectacular) one at that.

A second misperception is that mapping the human genome will remove individual responsibility for human behavior. To the contrary, however, though our genes in part determine what we will be like, how we respond to that condition will affect the quality of our existence just as much — if not more than — our genetic makeup. As we shall argue in subsequent chapters, though the characteristics we have are determined in part by our genes, moral questions remain about how we act, given our genetic predispositions. Hence, euphoria about the benefits that genetics may be able to bring about must be constrained by a parallel concern for stewarding the environment and with encouraging individuals to be good stewards over that for which God has given them responsibility.

CHAPTER EIGHT

———— *Knowing and Doing* ————

O UR CONSIDERATIONS to this point have focused primarily on issues that arise from applying science to human affairs. We have used a biblical paradigm derived from the Genesis creation stories to guide our consideration of three major ethical issues relating to the biological sciences: the environment, reproductive technology, and genetic engineering.

But the applicative side of science — what we shall refer to as *doing* — presupposes a prior stage, namely *knowing*. To be able to screen embryos conceived by in vitro fertilization for the gene responsible for cystic fibrosis or to find ways to tap the pharmaceutical potential of the Costa Rican tropical forests — to take only two examples — requires that we know the relevant features of human genetics or animal and plant ecology. Doing intimately relates to knowing.

This relationship raises three major, broad issues for us to address in this chapter. First, is it proper to try to know everything, or are there limits of some sort, moral rather than technical, on knowing? Is not some scientific knowledge intrinsically dangerous and hence better left unpursued? Second, is knowing justified only because it leads to doing (practical application)? What other justifications might be given for engaging in science in order to know? Are some justifications better than others? Third, what obligations do knowers assume by virtue of their creative speculations, data gathering, interpreting and theorizing, and testing and applying? Do they thereby incur obligations to society in general, including social groups, and if so what are they? What obligations do they incur to particular individuals who might be benefited or harmed by the research, either directly through the research process or indirectly? And

215

what obligations do they have to other members of their profession or the scientific community? Let us begin with the first issue concerning the potential dangers of knowing.

The Danger of Knowing

An ancient Greek myth warns about indulging the urge to know. The god Prometheus, defying Zeus's orders, stole the divine fire for human use and comfort. Furious at this affront to his supremacy, Zeus ordered Prometheus to be captured and eternally bound to a stone on a mountain in the Caucasus as punishment for his disobedience. But his victory did not assuage Zeus's anger. To ensure that humans would not threaten his rule by usurping his throne, as he had usurped his father's, he concocted a devious plan to debilitate them.

Zeus commissioned Hephaestus, the craftsman god, to make a beautiful woman named Pandora. He sent her to Epimetheus, Prometheus's brother, under the pretense of showing that he harbored Epimetheus no ill will following his brother's revolt. Though Prometheus, whose strength was foresight, had warned his brother not to accept any gifts from Zeus, Epimetheus was enthralled by Pandora's beauty, quickly forgot the stern admonition, and married her. But Pandora did not come alone; she brought with her a gift from Zeus, a beautiful amber chest decorated with flowers and pomegranates, with two golden handles in the shape of snakes.

Only after he took Pandora into his house did Epimetheus, whose strength was hindsight, recall Prometheus's warning. But it was too late to return Pandora. Nonetheless, he suspected Zeus's other gift and so charged Pandora not to open the amber chest, lest whatever mischief it held be loosed on the world. The beautiful box pleased Pandora, who initially was satisfied that at least she could keep the box, even if she could not open it. But the more she looked at the box, the more curious she became about its contents. Eventually she yielded to her desire to know and slowly lifted the lid. In an instant a flock of sprites flew out of the chest into the air and through the open door. Startled, she closed the box, but it was too late. She had loosed on the earth the evil devices of Zeus: diseases, disasters, woes, and cares of all kinds.

Distraught and crying, she told Epimetheus what had happened. But what was done was done. He asked Pandora whether all had escaped from the box, and she replied that she thought so. Again, however, doubt and curiosity overcame her; longing to know whether any more lurked within, she opened the box once again. Under the chest's gold rim clung a last sprite with drooping rainbow wings. She put it to her breast to warm it back to life, and from there it crept into Pandora's heart. It was Hope.

This ancient myth captures the suspicion some have about science and its task of acquiring knowledge. As Epimetheus distrusted the amber box, so for many there lurks the not always quiet distrust of what may lie inside nature's gilded box. They feel that nature is better left sealed. Unlike Pandora's box, the very difficulty of discovering its secrets suggests the forbidding nature of its mysteries. Were science readily accessible to the public, we might feel comfortable with it. But it is penetrable only by a select few, the modern-day successors of the curious, not to mention the wizards and alchemists, who undergo years of specialized training. They speak unique languages, filled with esoteric mathematical formulas and multisyl-labic words, and work in secret, sterile laboratories with complicated machines of enormous power. Their exotic chemicals and devices rend apart what nature has labored to put together. One need only envision the cyclotron, where one atom is smashed with a force greater than a speeding freight train against another, to blast it to smithereens, with the hope of catching a glimpse of some elementary particle released by the collision.

For the uninitiated, the practitioners of science represent a closed society. The initiation rites — passing difficult courses and exams, memorizing mathematical formulas, and spending long hours in laboratories — exact far too dear a price even to be tempting to many. The stereotypical character of the "evil scientist" often seen in the cinema does nothing to dispel popular mistrust of science. Dressed in an oversized white coat, hair flying Einstein-like, the mad scientist is bent on knowing something he should not, regardless of what others might suffer as a result. In the end he is often either made mad by his single-minded pursuit of knowledge or unwittingly becomes a victim of the evil unleashed by his experiments.

This uneasiness with knowledge is not found only in the mythology of Greece, the culture that paradoxically forms the seedbed for the Western pursuit of scientific knowledge. One can hear echoes of the same theme in the very tradition we have taken to supply the model for our ethical considerations. The Genesis account of the human fall from innocence contains intriguing parallels to the Greek myth, though it is derived from very different cultural roots. Like Pandora, the first woman expressly created by the Greek gods, Eve faces a situation that evokes her curiosity. God has permitted Adam and Eve to eat the fruit of all the trees of the garden except one. Why that one, the tree of the *knowledge* of good and evil? An answer is suggested by one of the story's characters: "God prohibits it," the tempter assures her, "because God is jealous. If you eat of it, your eyes will be opened and you will become like God, knowing good and evil." The attractiveness of the fruit, indicative of the attractiveness of knowledge, overcomes any reticence Eve may have to disobey the commands of her maker. She eats the fruit, she shares it with her companion, and ignorance falls from their eyes. They attain knowledge of — they experience — their true condition.

Eve's disobedient eating of the fruit, like Pandora's disobedient opening of the amber chest, results in dire consequences. Pandora's act releases the plagues of Zeus on hapless humanity. Eve's action brings God's punishment in the form of pain in childbearing and human banishment from the luxurious divine park to the dry deserts, where farming is rewarded with thistles and disappointment, and death waits around every rock.

The message of both stories, it seems, is that the heady pursuit of knowledge is dangerous and yields only disaster for humankind. Satisfying the desire to know is illicit, fraught with human hubris, and destined to release upon humanity ills now bottled up. Humans can and should live without knowing too much: knowledge can be dangerous.

If science is so suspect, why engage in it? Should we not leave nature alone — leave Pandora's box closed and Eve's fruit hanging on its tree — and humanity safe? We can avoid the dangers if only we do not seek to know, or at the very least, if we do not seek to know too much.

Two responses lie immediately at hand. One is to deny the presupposed claim that knowledge is dangerous. There is, it might be argued, no Pandora's box full of evil-causing sprites, no out-of-bounds tree harboring forbidden fruit. Knowledge is good; hence there can be no reason not to acquire it. Indeed, contrary to the pessimists, we should be about the business of acquiring all the knowledge we can; one can never get enough of a good thing. If anything, we are unique because of our rationality, which fits us for this destiny of knowing. Not to know contradicts our human nature and leads to the denial of human fulfillment. The *imago Dei*, the affinity we have with God, consists in this divinely appointed ability.

This reply, however, suffers from the same defect as the position it seeks to refute. Both suggest that knowledge is something that can be isolated from its possessors. For those who are suspicious of knowledge,

218

knowledge is intrinsically evil, functioning quite apart from the characters of Pandora or Eve; its very release into society brings evil and destruction. For critics of this view, knowledge is good in itself and hence is to be valued regardless of the uses to which people might put it. It is a boon to be sought with all one's energies. Both positions place intrinsic value on knowledge; they differ only on what that value is.

But both views suffer from the same misunderstanding of knowledge. As we argued in Chapter Two, knowing is a human endeavor. Knowledge belongs to some person or persons. Accordingly, it is neither intrinsically evil (as the Greek myth suggests) nor intrinsically good (as the critic envisions). It is not "out there," something that exists in the world waiting to be found. Knowledge is an epistemic property of particular persons. It is something we possess that plays a role in forming our attitudes, beliefs, desires, and actions.

The view that knowledge belongs to or is a property of persons leads to our second response: the possibility that we can have knowledge without necessarily using it. Our claim is not that having knowledge requires its use, only that knowledge is not something existing apart from knowers.

If knowledge is personal — or, when belonging to a community of persons, communal — it cannot be separated from those who have it. Knowledge thus stands to be either used well or misused. Using our knowledge, we can develop good desires that issue in good action. But just as easily we can develop evil desires that issue in evil acts. One cannot have lived in the twentieth century without realizing that acquired knowledge can be employed to the detriment of humankind. The bombing of London or Dresden or Baghdad cannot be separated from knowledge of aeronautics and sophisticated guidance and propulsion systems. The use of Agent Orange to defoliate the forests of Southeast Asia cannot be separated from knowledge of chemicals, their properties, and what results from their use. Knowledge of the structure of the atom led almost inexorably to the crafting of the atomic bomb, almost as if the discovery and development of the destructive power unleashed by the splitting of the atom was part of the original intent of the atomic scientists' program.[1] Yet it was scientists, not knowledge per se, that turned theory into superheated mushroom clouds. In these cases and countless others, it was not the knowledge that was at fault, but the persons who used that knowledge for destructive and arguably immoral purposes.

1. Robert Jungk, *Brighter than a Thousand Suns* (New York: Harcourt Brace, 1958).

In focusing on the possible destructive or immoral uses of knowledge, it is easy to overlook its tremendous creative and beneficial applications. While our century contains numerous instances where people used scientific knowledge for evil ends, it also abounds in the use of knowledge for good. Knowledge of viruses and antibodies has brought about immunizations for polio, measles, and whooping cough; knowledge of chemicals has resulted in powerful drugs to treat any number of illnesses and in consumer products that make our life easier and safer. Knowledge of radiation has resulted in treatment for cancer. Knowledge of the human genome portends future cures for genetic illnesses. We can use what has the potentiality for evil (the knowledge of viruses and chemicals can be used to construct potent biochemical weapons) for good.

This suggests that knowledge involves risk, not in the sense that attaining it is risky (though this might be true, as in the case of Pierre and Marie Curie and their discovery of and death from the effects of radium), but in the sense that knowers can do a variety of things with the attained knowledge, and we cannot predict whether they will use that knowledge for good or for evil. But unless we are willing to risk bad consequences, we cannot accomplish the good that someone possessing that knowledge might produce. Put another way, although it may be true that a particular evil might not have occurred if we had not explored such and such an area, it may also be true that a particular evil would not have been averted if we had refrained from exploring that particular area.

Controlling the Abuse of Knowledge

If, as we have stressed from the outset, knowing is a human act, and if knowledge is a human possession that can be seriously abused, then to control the possible misuse of knowledge we must control those who have it. One way of asserting this control is to prevent persons from having the knowledge in the first place. To deny access, to keep the knowledge private or secret, accessible only to a select few, might prevent it from getting into the wrong hands and being abused. The fewer who know, it might be claimed, the easier it is to control possible abuses.

Though in principle censorship often sounds promising, in fact knowledge control is neither desirable nor successful. First, who are to be the censors? In order to act as the censor, someone has to know what is to be forbidden for others to know. Someone, then, will have access to the relevant information that we are afraid to reveal to others. What qualities must the censors possess to raise them above others who allegedly might abuse the knowledge? They must possess honesty, integrity, courage to pursue truth, farsightedness, wisdom to know what to do with the knowledge, care for others, the ability to discern who can be trusted, and so on.

Whatever the qualities listed, one cannot escape the fact that the censors are themselves human, caught in the same human predicament as we all are. They are no more free from human hubris and fallenness, no less subject to temptations to power and personal gain, than any others. Consequently, their access to the information also needs to be monitored or controlled. Who then will monitor or control the censors? Who prevents the censors from being the abusers of the knowledge to which they have privileged access? And who in turn will monitor these monitors? Soon we will be invoking an infinite series of monitors, and even then we cannot be sure that the knowledge will not be abused.

Second, one might argue that knowledge so censored is more dangerous than knowledge widely accessible, for even if only a little truth lies in the old saw that knowledge is power, the censors by means of their control over what can be known possess power over others. And in the case of the censors it is an uncheckable power, since purportedly only they have access to the knowledge. If power can be dangerous, unchecked power is even more so. Thus, only when information is widely accessible can the abuses we abhor be checked and prevented. Were this not true, free media would be much less of a threat to the powerful.

Third, rarely does censorship work, at least in the long run. Often the more rigid the controls on knowing, the more attractive the forbidden fruit becomes and the harder those who really want to know try to find out. Neither Eve nor Pandora could resist what was forbidden to them, even by a divine command. From the breaching of the intricate network of controls invoked to maintain our atomic secrets to the revelations of Watergate, the history of attempts to maintain secrecy records failure.

Rather than invoking censorship, we should work to establish controls that regulate our ability to use knowledge to carry out our beliefs and desires in immoral ways. The establishment of relevant, workable safeguards is the responsibility of society and its institutions. The burden primarily falls upon governments to take action through their legislative, executive (regulatory), and judicial systems. However, other societal organizations play a definite role. Schools should educate societies' members about both empirical and moral facts, so that they can make wise judgments about the activities of those in positions of knowledge, power, or control, including people in government and science. Civic watchdog groups, though themselves rarely without bias, should be encouraged to monitor government officials, for as independent agencies they can provide the needed expertise to assess and criticize the excesses and abuses of those who allegedly protect society's interests. Professional organizations should set standards for and assess the performance of the members of their profession in the interest of the public good. The church has the divine mandate to speak with a prophetic voice to society about the proper uses and abuses of knowledge.

Christianity has several distinctive contributions to make with regard to the misuse of knowledge and how to prevent it. First, Christianity claims to understand both the basis and the cure for the human tendency to misuse knowledge. Traditionally, orthodox Christians have held that all humans are infected by what is termed original sin. Through Adam, the apostle Paul writes, we all were made sinners (Rom. 5:19). It is true that Christians disagree about precisely what the doctrine means and about its ontology (whether there is some condition that is actually transmitted and if so exactly how it is transmitted or affects us).[2] One interpretation is that original sin corrupts our nature by affecting human dispositions. We have a disposition, either innate or mediated through the environment, to do what is selfish and immoral, and this disposition is actualized by each of us, so that Paul can write assuredly that all sin (Rom. 3:23). It is a disposition that centuries of wrongful, selfish, disobedient actions reinforce. Some claim further that human desires over the centuries have entered into a "collective unconscious" in which we all participate that helps to form our dispositions and character. However original sin is understood, the prevalence of moral failure means that possible human abuses of scientific knowledge pose a real concern. The concern is grounded in the very character of human beings, a character that is disposed to do evil and difficult to control and change.

Set against the doctrine of sin is the Christian doctrine of salvation. Through spiritual regeneration, persons can be changed, and their dispositions can be altered. Old desires are gradually replaced with those that

---

2. For a summary discussion, see Millard J. Erickson, *Christian Theology* (Grand Rapids: Baker Book House, 1985), pp. 31-39.

accord with God's will; through the Spirit we become empowered to do what God wants us to do. It is not that Christians do not sin, so that they can be trusted with knowledge while non-Christians cannot. It is not that by virtue of salvation they deserve to be the censors. Rather, the grace that works in believers begins the process whereby, following the model set by Jesus, knowledge can be used by a regenerated self with love and justice to bring about good.

This process is far from complete, however. Regeneration does not bring moral perfection. St. Paul himself notes the struggle that even Christians face, a struggle between wanting to do the right and actually doing it (Rom. 7:15-25). Weakness of will to do the right persists. Thus the mere fact that someone is a Christian, though it should give us expectations and even hope that he or she will do the right, provides no guarantee. Society's controls must apply to Christian and non-Christian alike.

Second, Christianity also has something to say about how to control evil in society. Controlling evil, it contends, is a legitimate function of government. St. Paul in Romans 13 writes that all authority has been instituted by God for the encouragement of rightdoing and the discouragement of wrongdoing. For the Christian this means that the government is legitimately empowered to create institutions, agencies, and just laws to regulate human conduct in social settings. Applied to our issue, it means that governments and their administrators have the responsibility to see to it that scientific (and other) knowledge is not used for immoral or unlawful purposes.

That governments are ordained by God to foster the good provides no guarantee that the state will intervene in the practice of science in the right way, at the right time, and for the right reasons. Indeed, libertarians, following John Stuart Mill, have always been suspicious of governmental intervention, and at times justifiably so. Not only does government have the tendency to interfere ineptly, Mill argues, but giving it the authority to interfere adds inordinately to its power.[3] Yet, although Christians might applaud Mill's emphasis on liberty, they cannot adopt the complete libertarian package. Biblically, government is given a legitimate controlling role that extends beyond self-defense. On the one hand, if human tendencies and dispositions are to do evil, the government must intervene to mitigate evil-creating actions that follow from those tendencies. On the other hand, the state has a positive role as well, namely, to promote justice in the

3. John Stuart Mill, *On Liberty* (Indianapolis: Bobbs-Merrill, 1956), pp. 33-36.

society. The exploration of how it is to do this, who controls the government, and what principles will guide its action must be the subject of another treatise. Our point here is simply to emphasize that regulation of the use of scientific knowledge and encouragement of ways to find just applications of this knowledge are consistent with the Christian conception of the person and the state.

At the same time as they accord this role to the government, Christians must also be wary, for the government is operated by human beings whose condition of fallenness does not differ from that of the rest of the populace. Thus Christians cannot turn this role over to the government and then forget about it; vigilance over the government is necessary to make sure that it works for the common good, and not for the interests of its own members or those special interest groups or persons with whom it is in collusion. Put another way, Christians are to be both the strongest supporters and the severest critics of the government, insofar as it fulfills its role of bringing about good and repressing evil, making sure that knowledge is both widely distributed and properly used for the common good.

Third, Christianity also has a prophetic role for the church with regard to the pursuit and use of knowledge, a role that unfortunately has not always been prominent. This role embodies at least two dimensions.

(1) The church is to uphold the primacy of truth. It should not be afraid of the truth. Indeed, classically it has affirmed that truth is one, that there is no difference between God's truth and human truth. This accords with our biblical, ethical paradigm, for all truth is grounded in God's creation, which we attempt to know and understand. This is not to claim that we now have access to all truth or that we ever will; surely we are finite and truth is infinite. Neither does it mean that everything we or the church now proclaim is true. That all truth is God's truth is a claim about truth, not about human truth claims — that is, about what we take to be true. Put another way, it is a claim about God's truth: God knows every true proposition, and insofar as we know that same proposition, there is no disparity between the truth that we know and the truth that God knows, though undoubtedly there is a disparity in the richness with which the truth is known. Consequently, Christians need not fear scientific or any other truth, for insofar as it is true it concurs with the truth that God knows.[4]

This prophetic emphasis on the truth runs counter to any attempt

4. See Arthur F. Holmes, *All Truth Is God's Truth* (Grand Rapids: William B. Eerdmans, 1977).

to censor science, to prohibit the search for and teaching of what is thought to be true. This is one of the lessons of the oft-recounted early-seventeenth-century clash between Galileo and the Roman Catholic Church. It will be recalled that serious opposition on the part of the clergy and the Catholic Church to Galileo began after the publication of his *Letters on Sunspots*. Galileo, the clergy argued, had contradicted the Bible by asserting that the earth moved while the sun stood still. Galileo was denounced by the Dominicans to the Inquisition in Rome, which eventually condemned Galileo's writings, demanded that he recant his theories, and sentenced him to house arrest. At various times he was instructed that he could not hold or defend the Copernican system as describing reality, though he could discuss it as a mathematical supposition. Prior to his final trial the pope told him that he could discuss the respective merits of the competing cosmologies, provided he did not conclude that the Copernican one was true but instead agreed ahead of time with the papal view that it was impossible to know which system is correct, for by his omnipotence God can bring about the effects in the universe in various ways. In all of this the church attempted to protect a tradition that averred that Aristotelian science provided *the* biblical model.

The church in this instance misused its power and role by seeking to restrict the search for and promulgation of knowledge.[5] Even if the Copernican theories were contrary to Scripture, censorship was not to be preferred to reasoned argument and investigation of the evidence for and against the claims. Christians hold that truth is one, and that if this is the case it should stand the tests of human reason and experience — not that all will agree with any given position or that certainty will be achieved. But to establish or preserve a scientific thesis by dogmatic, authoritative imposition is never ultimately successful. To short-circuit reasoned inquiry by dogma is not prophetic but repressive, denying the proper functioning of God's gift of reason.

(2) The church in its prophetic role is to proclaim concern for the objects of knowledge. It affirms that knowledge should be both obtained

---

5. The case of Galileo is more complicated than merely the church against science. Both politics and personalities (including those of Galileo and Pope Urban VIII, whom Galileo satirized) were involved. Further, censorship was carried out largely on Galileo's manuscripts that were circulated in Italy, for it was considered an "Italian affair." See Owen Gingerich, *The Great Copernicus Chase* (Cambridge: Sky Publishing Corporation, 1992), chap. 14.

and used justly. In this regard Christians in and through the church are to be the conscience of their society (Jesus called it being salt), the watchdogs for the right and for justice. They are to speak with a voice that can be heard in all sectors of society, from the political to the scientific, calling for employment of knowledge in ways that treat the objects of knowledge with respect and dignity. In this we reaffirm the stewardship model of Christian ethics that we have invoked throughout this book.

At the very least, for a Christian biological ethic this means that in trying to gain knowledge about human beings or other creatures or aspects of nature, we are to treat them as more than merely means to expand our horizons of knowledge. The church has rarely been in the forefront of those who call prophetic attention to the treatment of research subjects, whether they are humans or animals. Yet as we have argued in Chapter Four, animals as well as humans are beings for whom we have a steward's obligation, both as individual creatures and as members of the larger biological community. While animals do not have intrinsic value, they have value derived from God who created them and should be treated accordingly.

Experimentation on Animals

We can apply this concern for prophetic stewardship to the use of animals in research and testing. The history of experimentation on animals is checkered. While many experiments on animals made scientific advances possible, especially in regard to human health and well-being, others were of dubious value. As one author puts it regarding the value of experiments in his field, "Most experiments conducted by neurobiologists, like scientific experiments generally, may be seen in retrospect to have been a waste of time, in the sense that they did not prove or yield any new insight."[6]

---

6. C. R. Gallistel, "Bell, Magendie, and the Proposal to Restrict the Use of Animals in Neurobehavioral Research," *American Psychologist* 36, no. 4 (April 1981): 358. Roger Caras states: "Perhaps none of us knows quite as much as we should. But I do know this from long association with the scientific community (not as an adversary but as a friend): about 80 percent of what goes on in the laboratory has nothing whatsoever to do with the good of mankind. Only 20 percent can be exalted to that level. That remaining 80 percent is for the fun, profit, reputation, or other benefit of the experimenters." Roger Caras, "Are We Right in Demanding an End to Animal Cruelty?" in *On the Fifth Day*, ed. R. K. Morris and M. W. Fox (Washington, DC: Acropolis Books, 1978), p. 130.

It is to the credit of animal-rights activists that they have drawn persistent attention to the misuse of animals in experiments. Many experiments have been ill conceived and useless, merely confirming long-accepted, obvious, or previously established truths. They have involved significant pain and suffering for their subjects. Furthermore, the acquisition of animals has often led to the deaths of many animals in their capture and transportation,[7] and the conditions under which the animals have been kept or housed have often been deplorable.

Death is a common end for experimental animals. In the United States in 1986 alone, 49,000 primates, 54,000 cats, 180,000 dogs, and 12-15 million rodents were killed as parts of experiments studying everything from vision to heart surgery techniques to cancer, as well as testing new drugs for safety and effectiveness.[8] It is true that the number of dogs and cats used in experiments in a single year equals the number destroyed by animal shelters and pounds in a single week.[9] Yet even though this is a drop in the bucket compared to the 5 billion animals slaughtered yearly for food, and though it represents a 40 percent decline compared with the number of animals used for research in 1968,[10] what gives pause is the enormous numbers of animals used as surrogates for humans in experimentation and the sometimes unnecessary suffering and death animals undergo.

This is not to deny that animals have played an important, even indispensable, role in some very significant research that benefits both animals and humans.[11] Of the seventy-six Nobel Prizes awarded in physiology or medicine between 1901 and 1988, fifty-four

---

7. With chimpanzees, "capture involves very heavy mortality. Almost the only way to obtain a chimpanzee is to kill its mother, so that for every chimp caught at least one other . . . dies. Furthermore, there is very heavy mortality, generally as high as 50 per cent, in route from the country of origin to the end user." Jeremy Cherfas, "Chimps in the Laboratory: An Endangered Species," *New Scientist* 109, no. 1501 (27 March 1986): 38.

8. "Winning through Intimidation," *U.S. News and World Report* (31 Aug. 1987): 48-49.

9. "Winning through Intimidation," p. 49. In 1988, whereas 150,000 dogs were used in experiments, "11-16 million dogs and cats were destroyed by public pounds, municipal animal shelters, and humane societies." *Animal Research and Human Health: Understanding the Use of Animals in Biomedical Research* (Washington, DC: Foundation for Biomedical Research, 1992), p. 11.

10. *Animal Research and Human Health: Understanding the Use of Animals in Biomedical Research*, p. 10.

11. For examples from psychology, see Neal E. Miller, "The Value of Behavioral Research on Animals," *American Psychologist* 40, no. 4 (April 1985): 423-40.

have been for discoveries and advances made through the use of experimental animals. Among these have been the Prize awarded in 1985 for the studies (using dogs) that documented the relationship between cholesterol and heart disease; the 1966 Prize for the studies (using chickens) that linked viruses and cancer; and the 1960 Prize for studies (using cattle, mice, and chicken embryos) that established that a body can be taught to accept tissue from different donors.[12]

When Joseph Murray and E. Donnall Thomas received the 1990 Nobel Prize for medicine, they "insisted that their lifesaving work on bone marrow transplants to fight leukemia and other blood diseases would not have been possible without testing on live animals."[13]

It also is true that most animal experiments are not overly painful for animals. A report by the U.S. Secretary of Agriculture for 1990 stated that 58 percent of the non-rodent animals used were not exposed to procedures involving pain or distress, while of the remaining 42 percent, only 6 percent received no pain relief by anesthetics, analgesics, or tranquilizers.[14]

Despite this record of achievement, attention still must be focused on the use and treatment of animals by the scientific research community. Patrick Cooke writes,

> Not every experiment is destined for a Nobel prize. Rats have been placed in jars of water and made to swim until they drowned. Why? To confirm a suspicion that female rats don't drown as quickly as males. Thousands of dogs, rats, and other animals have been poisoned by biological chemicals that the Department of Defense acknowledges are not warfare threats anywhere in the world. Various animals have been methodically zapped with electrical currents to see how much it would take to kill them. To fully exhaust cats so that they would remain asleep during sleep experiments, researchers have forced them to stay awake for days perched on tiny stools lest they tumble into surrounding pools

---

12. "Use of Animals in Biomedical Research: The Challenge and Response" (American Medical Association White Paper, 1988); published in *Animal Rights: Opposing Viewpoints,* ed. Janelle Rohr (San Diego: Greenhaven Press, 1989), p. 62.

13. Patrick Cooke, "A Rat Is a Pig Is a Dog Is a Boy," *In Health* 5, no. 4 (July/Aug. 1991): 59-60. The 1989 Nobel Prize winners in medicine, Varmus and Bishop, used chickens to study retroviral oncogenes; the 1991 winners, Neher and Sakmann, used frogs to study the chemical communication between cells.

14. *Animal Research and Human Health: Understanding the Use of Animals in Biomedical Research,* p. 13.

of water. Baboons have been strapped to impact sleds and slammed into walls in crash studies; other researchers later pointed out that the animals' anatomical shape was too different to reliably predict human injuries. More than 700 cats were shot through the head in a gunshot study that replicated the results of a nearly identical experiment carried out in 1894.[15]

Given the history of opposition by the scientific community to legislation protecting animals, concern is warranted. Attempts by Senator Joseph S. Clark in the early 1960s to enact legislation to require the humane treatment of animals during research met stiff opposition from the medical and scientific communities. They argued that any attempt to regulate the conditions of animals in research would limit the freedom of researchers and lead to the requirement that they obtain federal approval before they conduct certain types of experiments.[16] It was not until the 1965-66 session that Congress enacted legislation to cover the acquisition of animals for research and the humane handling and treatment of animals by research facilities. The impetus was a case involving the alleged theft of a family dog, which was subsequently sold to Montefiore Hospital in New York for research. Again, however, the National Institutes of Health (NIH) and the National Society for Medical Research (NSMR) argued that the legislation should only apply to animal dealers, that it should regulate the treatment of only dogs and cats, and that it should not impose any regulations on research facilities. A Congressional compromise was finally reached. The legislation, known as the Animal Welfare Act (AWA) of 1966, set standards governing humane housing and treatment of dogs and cats by research facilities as well as animal dealers, but it exempted research facilities from meeting those standards while the animals were the subject of research. Furthermore, the institutions themselves were to determine when the animals were being used in research and thus were not subject to regulation.[17] Much of the impact

---

15. Cooke, p. 60. A note of caution is warranted here, for Cooke provides no citations that would enable the reader to make an independent judgment about the legitimacy or illegitimacy, meaningfulness or meaninglessness, of the experiments. A better documented discussion can be found in Andrew N. Rowan, *Of Mice and Men: A Critical Evaluation of Animal Research* (Albany, NY: State University of New York Press, 1984).

16. "'Dognapping' Bill Enacted after Heavy Lobbying," *Congressional Quarterly Almanac* 22 (1966): 366.

17. Public Law 89-544, sec. 18, Aug. 23, 1966, known as the Animal Welfare

of the bill itself was lost when only $300,000 of the proposed $1.65 million was appropriated for administering the bill's provisions, including licensing and inspection.

The self-policing of institutions was more a hope than a reality. "Though the 1970 amendments to the AWA [Animal Welfare Act] were meant to stimulate more institutions to form committees, the 1978 ILAR [Institute of Laboratory Animal Resources] survey found that only 37 percent had their own IACUCs [institutional and animal care and use committee], the same proportion as in 1968. And NIH site visits to ten institutions in 1984 found that a significant number of committees existed in name only."[18]

It was not until 1985 that Congress amended the Animal Welfare Act by enacting requirements for minimum standards regarding "animal care, treatment, and practices in experimental procedures to ensure that animal pain and distress are minimized, including adequate veterinary care with the appropriate use of anesthetics, analgesics, tranquilizing drugs, or euthanasia."[19] Congress recommended that alternatives to experimental procedures that cause pain or distress be sought and that veterinarians be consulted about pain-causing procedures. It also mandated that the research facilities establish committees to oversee, by semiannual inspection of the facilities, the care of the animals used, to review the research and testing protocols, and to engage in education of those involved in the research at the center. Though recognizing the need to use animals for certain research, it recommended that measures be sought that would "eliminate or minimize the unnecessary duplication of experiments on animals."

Since the enactment of the federal legislation governing experimentation on animals and the NIH revision of its animal care guidelines, conditions in the laboratory have improved. At the same time, however, doubts remain concerning the kinds of treatment that animals receive within some laboratories. In response to a letter from a researcher who contended that researchers do care for their animal subjects, Peter Singer

---

Act. This provision was repealed in Public Law 91-579, Dec. 24, 1970. The 1970 amendment also removed the restriction of the act to only dealers and research facilities handling cats and dogs, defining "animal" more broadly to cover all warm-blooded animals that were intended for use in research, testing, experimentation, exhibition, or as pets.

18. Andrew N. Rowan, "Ethical Review and the Animal Care and Use Committee," *The Hastings Center Report*, Special Supplement, 20, no. 3 (May/June 1990): 20.

19. Public Law 99-198, Dec. 23, 1985, amendment to section 13 of the 1966 Animal Welfare Act.

records the details of a recent experiment with monkeys, published in *Radiation Research.*

> At the US Armed Forces Radiobiology Research Institute, in Bethesda, Maryland, researcher Carol Franz spent nine weeks forcing 39 monkeys to run on a cylindrical treadmill called an "activity wheel." If they did not run for the prescribed periods, they got an electric shock. The monkeys were subsequently irradiated; those receiving higher doses vomited repeatedly. All the monkeys were then put back into the activity wheel. The shock intensity was increased to 10 milliamps — an extremely intense electric shock. Some monkeys continued to vomit while in the activity wheel; they took between one-and-a-half and five days to die.[20]

It is difficult to imagine that the researcher in this case had a steward's care for her subjects; they were no more than means to an end. Even using utilitarian calculations, one must wonder whether the suffering and ultimately the death of the thirty-nine monkeys were justified by the knowledge derived from the experiment.

Animals play a role in science education, from rats and worms in elementary school to rats and primates in graduate school. Yet even this use of animals deserves careful scrutiny. Sigma Xi, the Scientific Research Society, encourages the use of alternatives (such as computer programs or biochemical procedures) where appropriate but defends hands-on animal experimentation as essential to professional training (such as in teaching surgery to veterinary or medical students or morphology to biology students).[21] The British, however, have banned use of animals for educational purposes (with the exception of their use under anesthesia to teach the techniques of microsurgery).

The role of animals in furthering human knowledge raises important and relevant questions. We will consider five of these in turn.

(1) Do we really need to know the information? This question concerning the necessity and centrality of the research is a legitimate one in the case noted by Singer. But even this question is complex. There are certain cases in which we might need to know the information if we want

20. Peter Singer, "New Attitudes Needed on Animal Testing," *New Scientist* 127, no. 1729 (11 Aug. 1990): 16.
21. "Sigma Xi Statement on the Use of Animals in Research," *American Scientist* 80 (Jan.-Feb. 1992): 75.

something or want to do something, but in which the moral question is whether we *should* want that thing. For example, until recent public pressure forced changes in the practice, live rabbits were used to test the toxicity levels of cosmetics and personal hygiene agents for human safety. In the Draize test, for example, chemicals were dripped directly into live rabbits' eyes until they were permanently blinded. The concentration at which the blindness occurred determined the safety threshold or toxicity level of that particular chemical agent for humans.

But was the development of new cosmetics or personal grooming products necessary, given the animal suffering that it occasioned? Even if one accepts the claim of the cosmetic industry that such tests were necessary to determine the safety of their products, and that every change in the product required testing the new formula in rabbits' eyes, a claim often disputed, the mere fact that the public might want the cosmetics or that the manufacturer needs to bring out a new line of products is not by itself sufficient justification for bringing extensive suffering to other members of the biological community. Our stewardly obligations are not to produce more goods for human consumption but to provide a quality of life that brings meaning and significance to human existence, while at the same time doing it in a morally responsible manner, that is, a manner that is consistent with our role as stewards of God's creation.

The point here is that our wants and desires, often created or stimulated by the economic desires of others, drive the wheels of science, technology, industry, and commerce. The result can be the misuse or abuse of the environment for what might not increase our quality of life at all, or for what increases it in ways highly disproportionate to the overall animal suffering or environmental dislocation (e.g., in obtaining the animals for experiments) caused. A Christian biological ethic must speak not only to the question of the need for animals to serve in the experimental capacity but also to the larger social and personal issues related to our wants and desires. Is the alleged need genuine, or is it merely another cog in the producer-consumer wheel of consumption? Unless the larger moral issues about human consumption are faced, our progress in bringing individuals to realize their environmental responsibility to other members of the biological community will be slight indeed.

Suppose, however, that our wants do express legitimate needs, the question still remains whether the information obtained from any particular experiment will significantly contribute to our knowledge. It is true that often the researcher does not really know what the outcome might show; indeed this constitutes the very point of the research. The proposal might look promising but in the end yield little of significance. Here one must be satisfied with anticipated results, that is, that the projected or anticipated results are worth whatever inconvenience, pain, suffering, or dislocation is imposed on the subjects. The crucial elements in judging the moral adequacy of the proposal will be that the prospective research program is carefully designed, is justified in the sense that the potential benefit of the research significantly outweighs the harm done to the animals, the animals are provided relief from their suffering and a reasonable end-point to the experiment, and the experiment's goals and potential for success are evaluated before the project is undertaken.

A research plan must be justified on the basis of outcomes that can be envisioned at the time it is initiated; it is morally questionable to engage in significant pain- and death-causing experiments with the mere hope that the experiment will produce serendipitous results that are significantly beneficial.[22]

(2) Supposing that we really do need to know the information, is experimentation on animals the only way in which the information can be obtained? In some cases, this seems to be so. Animals have played an essential role in the development of modern medicine and technology. Pig skin has served as a temporary bandage to protect patients suffering from burns. "Research on sheep with kidney failure allowed surgeons to perfect and implant the arteriovenous shunt, a device that allows patients with kidney failure to be connected to dialysis machines for long-term treatment." Because woodchucks and humans have similar hepatitis viruses, woodchucks have been used to study liver cancer. Primates have played essential roles in research for treating cancer and Parkinson's and for studying how HIV works. Dogs have provided basic information regarding cardiovascular and respiratory research.[23]

Other cases are less clear. To return to our above example, the charge against the cosmetic industry's use of the Draize test and lethal dose–50 (LD-50) — which measures the toxicity level of the chemical by determining the highest dose at which there are no adverse reactions — is that

22. "Sigma Xi," p. 76.
23. For a list of animal models of human disease, see Charles W. Leathers, "Choosing the Animal — Reasons, Excuses, and Welfare," in *The Experimental Animal in Biomedical Research,* ed. Bernard E. Rollin and M. Lynne Kesel, vol. 1 (Boca Raton: CRC Press, 1990), pp. 70-77.

they frequently were unnecessary. The tests, some argue, were conducted merely in response to the industry's desire to label their cosmetics as having been safety tested. The fact that safety testing of this sort constituted only 1 percent of experiments on animals is irrelevant; the issue concerns whether this testing was necessary because it was the only means available to obtain the information. The U.S. Food and Drug Administration did not require such tests for cosmetic ingredients (excluding color additives); only evidence of safety was required.

Public pressure, the Congressional action in 1985, and the increasing economic pressures of caring for and using live animals forced the cosmetic industry to change.[24] Though animals are still used for toxicology research, in many cases the protocols have been changed. First, initial chemical screening is done to test the destructiveness of the chemicals involved. Dermal equivalents, which are complex cultures of cells that, like human skin, have inner and outer layers of different types of cells, have been developed. Creams, oils, and ointments can be applied to this laboratory-grown piece of "skin," which releases chemical distress signals like those sent out by damaged cells in normal skin.[25] Not only are tissue cultures sometimes satisfactory for testing purposes, but they are much easier and cheaper to maintain. The dermal equivalent materials are less expensive than a monkey, for example, which costs $2,000 to buy and $100 a week to feed and house.

Admittedly dermal equivalents are not wholly satisfactory, for the ways in which complex organs and organisms will respond to the chemicals still needs to be ascertained. Thus, once substantial initial data are gathered from the use of living cells and tissues, tests on animals need to be conducted. The traditional tests have, however, been modified. For example, in the classical LD-50 test, the chemical was applied in increasing dosages until 50 percent of the subjects died. "This test, which can require as many as 200 test animals, has been replaced almost completely by a procedure called the 'range-limit' study [using only] six to ten animals. In the first part of the test, a small number of animals is used to determine a range of minimum and maximum toxicity. Following that, a reasonable limit is set on the number of doses to be tested. Doses higher than the set maximum toxicity doses are not tested." However, at times the traditional LD-50 test is still used for "certain pharmaceutical products for which the dose-toxicity relationship is

24. By 1989 the European Commission abandoned LD-50 entirely. "Alternatives to Animals," *The Economist* 313, no. 7631 (2 Dec. 1989): 97.

25. "Bunny Love," *The Economist* 318, no. 7693 (9 Feb. 1991): 74.

extremely sensitive and complex."[26] The Draize test, too, has been modified, so that the chemical is applied only to one eye, and its application is accompanied by medication to reduce the pain.

We do not mean to rule out animal research; nor are we arguing for an end to safety testing. The principle of successful animal experimentation before application to humans still holds true. Consequently, there are contexts in which experimenting on animals is justified. Toxicity levels need to be ascertained. It is necessary to know what long- and short-term effects chemicals have. Does their potency decline over time, or do they produce even more acute effects on the organism? In our litigious society, companies have a genuine stake in knowing that their products are not carcinogenic or cannot cause long-term damage, either to persons themselves or to their offspring. The specters of Agent Orange and Thalidomide still haunt us.

Animal research and use is also necessary for improving human health. For example, some have claimed that the use of chimpanzees was essential in the development and testing of the polio vaccine, in some experiments with hepatitis, and more recently with AIDS research, for chimps are genetically the most proximate to humans.[27] Or again, in research in which a person's life depends upon an organ that can be supplied by an animal (given that the problems of incompatibility and rejection have been satisfactorily resolved, and given that this is the only reasonable and significant alternative), one would be justified in killing the animal so that the vital organ can be used to save the human life. To date, a baboon heart and liver have been transplanted experimentally into humans. In 1984 an infant known as Baby Fae died twenty days after receiving a baboon heart. In 1992 a man in Pittsburgh, suffering from hepatitis, lived for ten weeks on a baboon liver. A woman whose liver was failing lived thirty-two hours after receiving a pig liver as a temporary bridge to sustain her until a human liver became available for transplant. Though these failures have made the news, in fact animal tissues have been and continue to be used to preserve human life. Animal skin is employed to cover burns, while heart valves from pigs replace defective human heart valves.

---

26. *Animal Research and Human Health: The Use of Animals in Product Safety Testing* (Washington, DC: Foundation for Biomedical Research, 1992), 4.

27. Yet even this claim can be questioned. One author writes, "Research with the animal model of polio resulted in a misunderstanding of the mechanism of infection. This delayed the development of the tissue culture, which was critical to the discovery of a vaccine." Stephen Kaufman, "Most Animal Research Is Not Beneficial to Human Health," in Rohr, ed., *Animal Rights*, p. 69.

As we argued in Chapter Four, the fact that we have stewardly obligations to nature does not commit us to a completely egalitarian model where humans and animals have equal standing. Even non-stewardship ethics note that organisms live off of other organisms; survival cannot be noninterventionist.[28] Rather, we want to stress that in ruling over creation, we must do what is best for humans *and* the creation; hence, where possible, we must take the route both that is the least destructive or harmful to all involved and that at the same time accords with our positive obligations to be benevolent and just. Thus, animals used for experimentation or research must be treated with compassion, and if it is true that in fact "as much as 40% of basic biological research can be conducted on cells rather than whole animals,"[29] we must continually reassess the role of animals in basic research to make sure that substitutes yielding accurate data are not available.

(3) Is the research program sound? Background checks of prior experiments should be a first priority, for needless duplication results in needless suffering. Sound statistical methods should be employed to guarantee that the right number of animals is being used — neither too few nor too many — so that the experiment will not have to be repeated and the experimentees wasted.[30] The experiment should also be structured to maximize the relevant information gathered and to minimize the harm and suffering to the research subjects. This suggests, as a minimum, that the information gleaned should be weighed against the harm done, so as to achieve a balance between the two. Perhaps less information would be adequate and the suffering caused to research subjects could be reduced, hence preserving our concern for the research subject. "The greater the animal harm envisaged, the stricter and more compelling the ethical justification must be, and that a principle of proportionality — inflicting the least harm for the particular good sought — ought always to hold."[31] Not everything that can be done or known is worth doing or knowing.

Minimizing the harm done to research subjects also suggests that the research program must be consciously structured to be painless where this is possible and where it does not negatively affect the research results. Singer,

28. Strachan Donnelley, "Animals in Science: The Justification Issue," *The Hastings Center Report,* Special Supplement, 20, no. 3 (May/June 1990): 10.

29. "Alternatives to Animals," p. 97.

30. See George E. Seidel, Jr., "Basic Principles of Experimental Design," in Rollin and Kesel, eds., *The Experimental Animal,* pp. 1-93.

31. Lilly-Marlene Russow, "The Troubled Middle *In Media Res,*" *The Hastings Center Report,* Special Supplement, 20, no. 3 (May/June 1990): 3.

in the same editorial mentioned above, provides the example of a research film distributed by the largest research institute funded by the Medical Research Council of England, which shows a rabbit undergoing experimental surgery with little anesthesia, struggling to get on its feet. The evidence of extreme, unnecessary suffering should cause us to roundly condemn such experimentation in the very same way we would condemn similarly conducted experimental surgery on humans. The inadequate anesthesia shows the experimenter's callousness toward the subjects studied.

The individual institutional review committees, or IACUCs, should be (and in many cases already are) given a strong hand in evaluating research protocols. This can occur in the very formation of the committees, by naming as members both those who have some independence from the institution or its research and those who will work most closely with the animals, whether laboratory technicians or veterinarians. These can and should act as advocates for the silent animals. Further, IACUCs can also assess the qualifications of the researchers, students, and technical staff to conduct a specific protocol.[32] Hospital ethics committees can provide models for their composition and conduct.

(4) What kinds of animals are being used for experimentation, and how much pain and suffering can or do they experience? Contrary to many activists, not all organisms and not even all animals are equal. Though it must be assumed that many animals can feel pain, not all animals apparently experience the same level of pain or suffer to the same extent, and many probably do not suffer at all. Hence, proper discernment of the nature of the animals used for experimentation is of vital importance. Where animals lower on the scale of sentience can be used to achieve comparable results, or especially where animals can be replaced with non-animal experimentees such as cell and tissue cultures, they should be. Where the animals can experience significant levels of suffering, our concerns for protecting their interests must be greatly increased, perhaps even to the point of prohibiting painful research on them unless it can be established that it is absolutely necessary to achieve some necessary and significant good unachievable in other ways.

(5) How do we manifest our concern for the welfare of the experimental animals? One way of approaching this is to consider the conditions under which the animals are captured, transported, and housed. Jane Goodall, the

32. See Rowan, "Ethical Review and the Animal Care and Use Committee," pp. 23-24.

noted researcher on gorillas, describes her visit to a biomedical research facility, under contract to the National Institutes of Health, that housed primates. In this facility, "young chimpanzees, 3 or 4 years old, were crammed, two together, into tiny cages measuring 22 inches by 22 inches and only 24 inches high. They could hardly turn around. Not yet part of any experiment, they had been confined in these cages for more than three months."[33] After being infected with hepatitis, they are isolated, since the disease is highly contagious, "living in severe sensory deprivation, for the next several years. During that time, they will become insane." Not only were the chimps subjected to cruel and perhaps useless experiments, since we already know a great deal about hepatitis, but the conditions in which they had to live showed little care for them. As Goodall points out, chimps are intelligent and social animals, establishing close bonds with family members. Yet the conditions under which they were housed in this facility denied them any meaningful emotional or social activity. Likewise their physical activity was limited to eating what was supplied to them. Opportunity to exercise was generally denied. Caged with bars on every side, including the bottom, they endured conditions that reflected the ease of the experimenter, not their own comfort.

Confining animals in separate cages so that they are isolated from each other often reflects our own individualism more than the natural good that best serves the interests of the animal. Frequently, without undue cost, it is possible to improve the living conditions of experimental animals so that they are provided not only with physical comfort but also with the possibility of realizing, at least to some degree, the society that is a part of that animal's normal environment. Federal regulations now specify the cage sizes, temperature, humidity, ceiling materials, and availability of proper veterinary care. The objection to these regulations by some researchers has been that they will make the use of animals in experimentation too costly. It is true that the additional cost to relieve bleak conditions might be hard to justify solely in terms of the experimenter's intended outcomes. Changing cage sizes, for example, might be costly.[34] But a stewardship ethic rejects any such simple cost-benefit analysis that takes into account only the experimenter's desires and bottom line.

We agree most strongly with what have been called the "three Rs" of animal research: *replacement*, where computer and mathematical models,

33. Jane Goodall, "A Plea for the Chimps," *The New York Times Magazine,* 17 May 1987, pp. 108-9.

34. "Winning through Intimidation," p. 49.

microbiological systems, and cell and tissue cultures are substituted for live animals wherever feasible, that is, without compromising legitimate and justified research goals; *reduction* of the number of animals used; and *refinement* of the procedures, so that the same results can be achieved with significantly less suffering and distress.[35] These "three Rs" demonstrate stewardship of the animal community and of our resources, without abandoning the need for animals to further human essential goods.

Our point is that, as stewards, we are to benefit both human beings and the rest of creation. Thus researchers, whether in the laboratory or pursuing animals in the wild, have significant and substantial obligations to the creation that they take as the subject of their knowing activities. Christians have as much obligation to call attention to abuses of other members of creation as they have to point out abuses of other human stewards. Thus we might add a fourth "R" with regard to our treatment of creation in general: *reorientation* of both our desires and our outlook on creation. With respect to our desires, we consciously decide that we can do with less, particularly where that "less" can benefit other members of creation. With respect to our outlook, we must begin to rethink our relation to the rest of creation, turning from a mere utilitarian mode to one of stewardship, in which the rest of creation, too, has goods that we can preserve and foster.

In sum, any limits to be placed on knowledge will be applied, not to the knowledge itself, but to how that knowledge is obtained and used. With regard to the former, attention continually must be directed to the means by which humans get their knowledge. In particular, attention must be paid to what is done to the subjects of the knowledge. With regard to the latter, attention must be directed to how that knowledge will be used. Will it be used to enslave other persons, or will it be used to free them mentally or physically to a life of greater fulfillment?

The church must assume the mantle of the prophet to speak to the scientific culture. It must speak the truth and speak for the truth (including a defense of what it takes to be the truth), encourage the just and proper acquisition of knowledge, and speak out for the proper and moral use of knowledge and against the abuses to which knowledge is sometimes put.

---

35. W. M. S. Russell and R. L. Burch, *The Principles of Humane Experimental Techniques* (London: Methuen, 1959).

The Ethic of Stewardship and the Justification of Knowing

Let us turn our attention to the second issue noted in the opening paragraphs of this chapter: On what basis can one justify knowing? One justification of knowing relates to its pragmatic value. It would appear that this pragmatic justification of knowing is compatible with all three modes of our steward-ship ethic.

First, knowing is justified by the command to rule or subdue the earth. Unless the research is performed, carefully and meticulously, there is no possi-bility of a controlled changing of nature. Research is necessary to tell us the structure of what we are working with (what it is like), to inform us what can be done with what we have (if we do this, that will be the result), and to contribute to making judgments about what *should* or *should not* be done. (We say "contribute to," for empirical facts alone cannot decide the matter, though they supplement the value structure by providing the information relevant for deciding which values are applicable in a particular moral situation.)

Second, knowing is justified by the command to fill the earth. We have interpreted this command both quantitatively and qualitatively, and both aspects are relevant here. If we are to help species reproduce, for example, to replenish the ecological devastation we have caused, scientific knowledge is required. For example, through hunting and killing the whooping crane for its plumage, Americans reduced the whooping crane population from mil-lions to just twenty individuals by the 1940s. To assist the remaining cranes to reproduce, scientists studied their reproductive habits, noting that although they laid two eggs, usually only one chick survived. To increase the population they removed one egg from the nest and attempted to raise it themselves. The first attempts were made in a hatchery, but although the chicks hatched, they patterned themselves after the humans who raised them and laid no eggs. The next experiment, in which scientists introduced the extra egg into the nests of sandhill cranes for incubation, also met only limited success, for again, although the eggs hatched, the whooping crane chicks' mating attempts with sandhill cranes were rejected and the lineage ended. Now, because of the knowledge they have gained about reproductive patterns and chick pattern-ing behaviors, scientists feed and raise the chicks using puppets that look like the whooping crane mother, so that when they mature they will engage in courtship with other whooping cranes.[36]

36. "Puppet 'Parents' Raise Cranes," *USA Today Magazine* 119, no. 2553 (June 1991): 9.

This use of knowledge to fill the earth also touches on the third command, the command to care for creation. Stewards should fill the earth in ways that care for the members of creation and the total environment, seeking to protect, provide for, and find ways to better the quality of existence for all. Hence, though the acquisition of knowledge might result in the knower using the knowledge for evil, knowledge still is required for any caring activity.

In short, knowledge has a pragmatic justification consistent with our stewardship ethic. It is necessary to our biological survival and cultural existence, our very filling of the earth, and the possibility of our changing the human and environmental context for the better.

A second justification for knowing seems to diverge from the stewardship motif that we have adopted. It expressly connects the human good with what it is to be human. Knowing is a valuable human activity because it fulfills human beings. So it is that Aristotle begins his *Metaphysics* with the claim that all persons by nature desire to know. His optimism might have been overblown regarding the extent of this desire, but he correctly captures the intellectual curiosity that spurs scientists and many others in their intellectual pursuits. For Aristotle, the knowledge sought is not primarily the ascertainment of specific facts but ultimately the discovery of the necessary principles, essences, and causes that govern nature, a knowledge that is both scientific and philosophical.

Though we need not follow Aristotle into the specifics of his view of science, his ethical perspective contributes in important respects to our view. According to Aristotle, the good for each thing is relative to the nature of that thing. Thus what is good for a fish is relative to what it is to be a fish; similarly what is good for humans is what fulfills human beings. Human ethics are grounded in providing for human fulfillment. Goods are valuable and are to be recommended to us because they are good *for us*. Accordingly, the goods we should seek connect with what will bring meaning and fulfillment to our lives. If we are, in part, rational beings with the unique capability of cognitively knowing our world, then human fulfillment will occur in part through the development of our knowing capabilities and their exercise in knowing both the world and what brings the highest human fulfillment. To this end both theoretical wisdom (knowledge of scientific principles) and practical wisdom (deliberation, understanding, judgment, or right discrimination) need to be cultivated through education and right habits.[37] In short,

37. Aristotle, *Nichomachean Ethics*, bks. 1 and 2.

knowledge is good because it comes from utilizing our human potential in ways that bring us happiness.

There is no reason to think that this ethic is ultimately incompatible with a stewardship ethic. The basis of the stewardship ethic is the contention that God created both us and the world. In putting us in the position of stewards over creation, God does not intend that we deny or fail to develop our humanity. We are stewards not only of the rest of the created world but also of ourselves, and in that capacity we are obligated to preserve, develop, and invest ourselves. Part of our stewardly activity is the development and investment of human potential in the realization of meaningful human existence. This realization can only occur in a context where distinctly human goods are realized on the individual human level.

However, humans do not solely determine their own good. Their good arises from God's creation of human beings to be what they are. God has given us certain drives, desires, and abilities, the fulfillment of which contributes to both personal satisfaction and self-realization. Thus, again, though good is related to human nature, it is a derived good, derived from the creative activity of the omniscient God, who knows what best fulfills his creation.

This means that the pursuit of knowledge, whether scientific or nonscientific, is a legitimate endeavor in itself, apart from any practical application. As we give free rein to our curiosity, we penetrate the creation, reveling in the ways in which we are one with and different from that creation. To understand the world can itself be a joyous, truly human event.

What can we make of these two justifications for knowing? Are they of equal weight? In our pragmatic society, where truth and value are determined by workability, it is easy to side with the first. It fits with our drive to change our world and our entrepreneurial economic structure, where the doer is the person who succeeds and there is no succeeding without doing. There are many who engage in science for pragmatic reasons, either selfish or benevolent: for their own self-aggrandizement or to make the world a better place to live. But there are other scientists more inclined to the nonpragmatic view, who see their research as personally fulfilling without necessarily feeling committed to implement their findings by themselves or see their findings implemented by others. For them, knowing actualizes the cognitive potential and satisfies the curiosity God created in us.[38]

38. It goes without saying that both justifications of knowing operate within the broader moral context. Thus scientists cannot justify causing unnecessary pain for a research

While the pursuit of knowledge for nonpragmatic reasons can certainly be justified, it may be asked of those who justify science as self-fulfilling whether their lack of concern with the practical applications of their research reveals some moral deficit in their character. If they seek knowledge for its own sake and their own self-fulfillment and not to benefit others, are they not at best mere egotists and at worst lacking moral conscience in the face of great human need? This moves us to the third of our major questions: Do scientists and other researchers have obligations to society, either in general or to specific individuals?

Obligations of the Scientist

In the opening paragraphs of this chapter we noted three questions relating to the obligations that scientists have to others: (1) What are their obligations to society in general? (2) What obligations do they have to individuals who are affected by their research? and (3) What obligations do they have to their fellow scientists?

Obligations to Society

Let us begin by considering what obligations scientific researchers may have to society in general. Consider the following adaptation of a true case. Researcher A is studying whether the possession of a certain biological property that has behavior manifestations divides along racial lines. Researcher B queries what A will do with the results and what might be the social impact of their publication. A replies that should his study reveal that there are no racial differences, he would feel free to disseminate the information. However, should his study indicate that there are racial differences, he would withhold them from publication, or only publish those results that would not have a bearing on racial differences. B queries why A would do this, and A replies that there is already enough incentive in society for racism and he does not wish to contribute further to it.

---

animal simply because what they do satisfies their curiosity. The questions raised above about need, etc., still apply.

There are many issues to consider in this adapted case. There are initial concerns having to do with the research program itself. How carefully is the research program set up? Can the protocol be set up so as to delineate sufficiently between causation and correlation or between the causes and the effects? Is the research program constructed so that it will really test for racial similarities or differences, or might the factors be so complex that racial contributions cannot be distinguished from cultural and environmental ones? That is, is the research program carefully enough constructed to insure that the results reflect true racial differences and not something else?

There are also questions concerning A's motivation. Why is A doing the research? What is it that A is trying to discover? If A is willing to publish his results only if no racial differences emerge, he is approaching his project with a significant bias. The research is being conducted to legitimate some preconceived position about both the facts and their possible relation to human behavior. But this violates the openness that characterizes good scientific research, that is, openness to whatever data emerge and to the postulation of the most reasonable and supportable hypotheses, and to the possibility that one's previous theories need to be restructured in light of the data.

But beyond these, there are questions about the obligations of the researcher to society. A sees his task as involving a possible conflict between his obligations to society at large (to inform it about the results of his research) and his obligations to a particular segment, here racially designated, of that society (to protect them from possible discrimination based upon his findings). A proposes to resolve this conflict by tailoring his report to favor one segment of society.

One might question whether A resolved this conflict of obligations properly. A should make certain that his research project is conducted properly, that the controls and variables are properly set, that the data are as clear as possible, that other reasonable hypotheses can be excluded on the grounds of incompatibility with other results, and that his results can be verified (or at least not falsified) by independent tests. If a racial difference still is indicated, he should take special pains to communicate exactly what the results indicate and what they do not indicate, and to carefully warn against potential misuse of results.[39] The fact that there are

39. The debate over the racial implications of research and the emotions it evokes have been reignited by Charles Murray and Richard Herrnstein, *The Bell Curve* (New York: Free Press, 1994).

racial differences does not entail that we should discriminate racially; to do so is to move from a thesis about human differences to a thesis about human value.

But beyond the particular way in which A resolved the dilemma, two points become clear. First, A recognizes that he does have obligations to society, both to society as a whole and to particular segments of that society. And this was our concern. How one handles these obligations is the subject of moral scrutiny.

Second, those obligations involve a host of things, from making sure that the research is competently done in the first place to careful communication of the results. Regarding the former, one never benefits society by shoddy work, no matter whether the conclusion is what is desired or anticipated or not. The fulfillment of professional research obligations forms the basis of the trust that society must be able to place in its scientists. Researchers are stewards of scientific knowledge; they hold this knowledge in trust for the community. The violation of this trust not only destroys the scientific enterprise but also places in jeopardy the scientists' role as stewards of this knowledge on behalf of the community.

Regarding the latter, the researcher has the obligation to inform society about the results. As we have said, censorship works against trust. Truth matters. But *how* the purported knowledge claims are communicated matters a great deal. Researchers have the obligation to communicate in ways that are consonant with their role as stewards of scientific knowledge. The obligation to communicate commits the researcher (1) to provide the results in ways that allow the reader/listener to have an informed opinion about the issues the research raises; (2) to qualify the results carefully, so that the reader/listener is not misled by the claims but can recognize the role that interpretation and theory are playing in stating the facts, noting their relationships, and explaining them; and (3) to both indicate the positive good that the research might make possible and warn against possible misuse of the information or hypotheses. Finally, researchers should also reveal what they personally have to gain from the research and its publication. The increasingly close linkages between the researcher, the government, and private enterprise (sometimes established by the researchers themselves) provide opportunity for significant abuse.[40]

The sensational reporting of studies or the leaking of partial and

40. Lawrence M. Fisher, "Profits and Ethics Clash in Research on Genetic Cloning," *New York Times* 143 (30 Jan. 1994): 1, 18 (N).

fragmentary results or interpretations to the media provides a ready context for abuse. Even before the results of a Canadian study on the value of mammography for women between the ages of forty and fifty were published, leaks to the press in late 1992 created a storm over whether the studies were properly conducted, whether the data were clear, and whether mammography itself was harmful to women in this age group. Instead of enlightening women on the matter, the research was reported in a way that only clouded the issues, making it difficult for women in society to make an informed judgment about the value of early mammograms.

Obligations to Individuals

Let us turn to the question regarding whether the scientist has obligations to specific individuals. Researchers are stewards of God for the benefit not only of society in general, for whom they are attempting to acquire the knowledge, but also of the individuals who are affected by the research. Thus researchers have obligations to their research subjects. The research, unless it is therapeutic, might not directly benefit the experimentees, but because they are subjects of the experiment or research, their welfare lies in the researchers' hands. The subjects have put their faith and trust in the competency of the researchers, and this trust should not be betrayed.

It is sometimes suggested that it is appropriate to distinguish between the roles of the experimenter and of the health care provider. The former has the primacy of society in mind, whereas the latter is responsible for the individual patient. Though this distinction might be helpful in providing a larger description of what is being done in experimentation as over against providing health care, the experimenter should not consider the societal good apart from or independent of the good of the research subject. The researcher is as much a steward of the research subject as the health care provider is of the patient. This is not to deny that at least two persons with differing roles should be involved in the care of human research subjects, one to provide an independent check on behalf of the subject on the activities of the other. This is probably necessary, given the human fallenness we have spoken about above. But the fact that there is a person to check for abuses does not free researchers from concern for the subject of their research.

The obligations researchers have to their subjects include (1) obtaining meaningful, free, informed consent prior to taking any actions,

(2) being competent to conduct the research, (3) making sure that the project is both well constructed and worthwhile (e.g., that it does not duplicate previous research or is unnecessary because of what else is known), and (4) not harming the subject. If harm is done, it should be done in a way that *(a)* does not go beyond the informed consent of the subject, *(b)* is not permanent, or where permanent (in therapeutic experimentation), is the lesser of the two evils (the harm done is less than the evil of allowing the disease to continue), *(c)* is necessary for the success of the project, and *(d)* is part of an experiment that the subject can end at any point. We might speak of these as being obligations of the researcher directly related to the conduct of the research.

But beyond these basic principles enumerated clearly in the Nuremberg Code, does the researcher have a steward's obligations to specific individuals who are not part of the research project? That is, can individuals demand of scientists that they provide, if not cures, at least amelioration of their illness, and if not even this, that they at least study their specific illness?

This demand resounds nowhere more clearly than in the cry raised by some individuals infected with HIV. The radical group ACT UP (AIDS Coalition to Unleash Power) has accused the scientific community of not moving quickly enough to meet the needs of HIV-infected persons. Neither a vaccine nor a cure for AIDS is in sight. The members of ACT UP *demand* that science provide and government fund research leading to vaccines and cures.

The demand for a cure for AIDS or any other disease reveals nothing short of a misconception of the powers of research and medicine. Research cannot guarantee the discovery of a cure. The most it can provide is research into the problem.

But what can be said about the less radical demand that the scientific community take up the subject of AIDS with the goal of finding a prevention or cure? Is this a legitimate demand to be made of scientists? Or is research on AIDS a supererogatory act that some scientists willingly perform? That there is a moral obligation of the scientific community in general to sufferers follows from our stewardship ethic. We are obliged to care for others. That is, we not only have obligations not to harm; we also have obligations to do others good. To care for other human beings means that we need to assess their needs and take steps to meet them when they are legitimate and reasonable. Seeking to find treatments and possible cures for AIDS meets this obligation.

However, there are two caveats. First, though research on AIDS or any other illness might be an obligation of the scientific community in general, it does not seem to be a requirement that can be assessed against any one researcher. The general principle that we are obligated to help our neighbor does not mean that just anyone can have a claim against any other person. The needy neighbor has a legitimate claim to be helped; but ascertaining who is responsible to help is another matter. Determining who is obligated has to do with the assessment of who is in a position to best help the needy person. Position can mean a number of things: being in contact with the neighbor, knowing the plight of the neighbor, possessing the resources to help the neighbor. Applied to the AIDS sufferer, many researchers are not in a position to assist the sufferer. Others, however, by virtue of their training and expertise, are in such a position. Hence, though the AIDS sufferer cannot single out any one particular researcher, the sufferer might indicate that certain individuals, by virtue of having benefited by society's training in certain areas of medicine and being in possession of certain intellectual and governmental resources, have obligations to assist them.

The second caveat relates to the matter of scarce resources. Though we are to care for others, and though the needy can legitimately make claims on us, yet where the resources are such that not all legitimate claims can be met, there must be a careful assessment of (1) the degree of suffering involved, (2) the risk that this disease or condition poses to the larger community, (3) the amount of benefit that can be gained in proportion to the cost expended, and (4) the seriousness of this condition vis-à-vis other needs (including nonmedical needs like education, defense, and public safety). Stewards must wisely invest the resources that are entrusted to them, putting them to both maximal and compassionate use. Hence, the claims of the cancer patient, the Alzheimer's patient, the AIDS patient, and so on must all be considered in deciding both the research priorities and the amount to be spent on researching causes, treatments, cures, and preventions. We will return to this subject in Chapter Ten.

Obligations to the Scientific Community

Finally, as stewards of particular sets of research data, researchers have obligations to other members of the scientific community. Their obligations include many that we have already listed: being competent to conduct the research, making sure that the project is both well constructed and worthwhile (e.g., that it does not duplicate previous research or is unnecessary because of what else is known), and communicating honestly what is learned to the scientific community.

This last obligation dispels the notion that scientific knowledge may morally be held privately, especially when it has the potential for significant social good. This, too, flows from an ethic of stewardship. What researchers have discovered or learned is not theirs to hoard, as a treasure only for themselves. As stewards of knowledge, they hold what they have discovered or learned in trust for the common good of the community. Consequently, insofar as it bears on the common good, it is not privileged but public information, to be shared both with other specialists and with the public at large.

Perhaps nowhere are the temptations of abuse by scholars and researchers more prevalent than here, for the possession of certain information or knowledge provides opportunity to aggrandize one's personal status in the profession. But what is forgotten in hoarding the information is that scientists and researchers function within a community and that they owe substantial debts to that community, beginning with their early education. Since the community has transmitted to the researcher its store of knowledge, ideas, and values, and has at various stages provided for the researcher's well-being, the researcher has a debt to repay to that very community. And this repayment is through responsible stewardship of what is learned.

Conclusion

We have covered much ground in this chapter. We have argued that a stewardship ethic provides a fundamental framework from which we can address questions concerning the dangers of knowledge, the justifications of knowing and doing, and the obligations of those who take the extension of human knowledge as their primary task. When those who seek to know acknowledge their responsibilities of stewardship and obedience to the Creator and see their role to discover and use what they know to benefit both other human beings and the rest of creation, the temptation to abuse others, to hoard information, and to fail to act justly should be diminished. At the same time, we acknowledge that all humans struggle with weakness of will and therefore need to be reminded of their place within God's creation and the obligations that result therefrom.

CHAPTER NINE

Brains, Genes, and ——— *Moral Responsibility* ———

T HE 1990s have been designated the "decade of the brain," a phrase that reflects the rapid progress that has been made in recent years in understanding the structure and function of the human brain. It is also meant to portend that new methods of investigation will produce further striking advances before the decade's end. While better means for controlling serious diseases will be discovered, interpretations of the new knowledge will challenge our views of human nature and human moral responsibility.

Who are we? Can we really make choices about what we want to be, given who we are? What will insights into the role of genes and the environment contribute to the discussion? Will the language of choice, freedom, and moral accountability disappear, to be replaced by language about the brain, central nervous system, and their functions? Or will some new language be invented that will bridge language about choice and language about brains?

Let us begin our discussion with where we stand in the 1990s. It is a place that has changed drastically from two decades ago, a place where illnesses, dispositions and tendencies, and abnormal behavior, once viewed as mysteries, are associated with genetic predispositions. The structure and function of the brain are believed to hold the key to understanding mental illness and deficiencies of mental abilities or task performance. Persons with unusual or socially unacceptable behaviors — e.g., having delusions, attention deficits, or difficulty in learning to read; or being overly aggressive, withdrawn, depressed, or drug dependent — which once were deemed, in part or in whole, under a person's control or perhaps even

250

under the control of demonic forces, are now being treated with powerful drugs. Children who are raging terrors in the evening are tamed into model behavior the following morning by their dose of Ritalin. Treatment with Clozapine slowly allows people who have withdrawn into their own private world to emerge and begin to live a life they have missed.

At the same time the voluntary dimension is not abandoned. Counselors encourage alcoholics to voluntarily adopt and follow the 12 Steps to bring about their recovery. Even children on Ritalin must learn to control their aggressive behavior when the effect of the drug wears off. So the more we learn about the brain and the stronger the drugs we develop to control abnormal and unwanted behavior, the more pressing it is that we help persons to take control of their lives and behavior. Nature and nurture, brain and mind, biological determinism and free choice — is there a bridge, a coming together, an integration of these two worlds, or will further study show that they remain as far apart and their relation remains as intractable as it has been for centuries?

We certainly make no claims to know what will happen or to resolve the dilemma. But perhaps we can contribute to the continuing discussion by putting the issues in better focus.

The Brain

A major link between genes and behavior is the brain, the most complex and highly organized organ of the body, if not the most complex structure known to us, far surpassing the most powerful Cray computer. Anatomically it consists of two large hemispheres connected by the corpus callosum and other axonal bridges. These convey stimuli, or coding information, from one side of the brain to the other. The hemispheres are covered by a thin layer, only two millimeters thick, called the cerebral cortex. Across and through its many convoluted folds move many of the nerve impulses.

Physically the brain consists of more than 100 billion neurons, which receive information along branched dendrites and send information to other neurons through unbranched axons. Short axons extend to neighboring neurons, while long axons run to other regions of the brain. Neurons do not act alone but are grouped in columnar clusters extending through the cortex. Groups ranging in number from 4,000 to as many as

251

100,000 neurons function together in these modules, providing a reasonable number of working units in the brain.

The work in the neurons is done electrochemically. Information is conducted along the axons in the form of brief electrical impulses called action potentials. When neurons are at rest, they admit potassium ions through their surface membranes but keep out sodium ions. When physical or chemical stimuli decrease the voltage gradient at the cell's surface membrane, positively charged sodium ions are allowed to penetrate. Moving from fluids outside the cell into the cell's interior, they change the electrical charge in the cell from -70 millivolts to a positive reading. This admittance of sodium ions into the cell is what creates action potentials. At a critical point the positive feedback from the cell causes the membrane that prevents entry to reverse voltage, and the penetration of sodium ions declines, so that the cell returns to its negative voltage or resting state. The entire process, though taking only milliseconds, is one million times slower than the time it takes to conduct electrical current through a copper wire. Speed of information processing thus must be achieved through many axons working in parallel through their many connectors.

Once the action potential reaches its terminus, the information must be communicated to a neighboring neuron. Chemical transmitters at the contact points (synapses) between neurons are set in motion. The action potential causes neurotransmitter molecules to be released to carry information across the gap to a corresponding point (receptor) on another neuron's (postsynaptic) dendrite.

As many as fifty neurotransmitters have been identified so far, including gamma-aminobutyric acid (GABA), serotonin, and dopamine. These transmitters bind to receptors, contact points that open or close to control the flow of specific chemicals through the cell membrane. Some stimulate the cell's response, while others inhibit it. There are two basic kinds of receptors. In one, ion channels allow ions to cross the cell membrane and proceed as described above. In the other, the transmitter interacts with a neighboring membrane protein, creating many biochemical reactions that, though slow to begin, continue for a longer period of time.

Though all the brain cells or neurons have the same genes, different cells express different genes, thus performing different functions. We have 50,000 or more genes, and half of these are active primarily in the brain, though not all at the same time. As the nerve tube develops in the fetus, certain genes help organize broad divisions of the early brain that later will have different functions. The dividing nerve cells then migrate out from the central tube, initially guided by genes and later by the environment of the target areas.

Eventually the neurons form complex patterns of connections through their slender extensions of axons and dendrites. While the information is being passed on, the neuron itself is being affected by the process. Action potentials not only convey information; they also produce metabolic aftereffects that alter the circuits through which they pass. Further fine-tuning is accomplished as neurons and connections that are no longer needed are removed by a genetic program of cell death.

A single neuron can receive signals from hundreds of other neurons that use different neurotransmitters. This communication could not take place unless there were different kinds of receptors that are sensitive to the different neurotransmitters. Thus, this single neuron must integrate a diverse set of inputs into a coordinated response.

Other types of receptors are constructed from subunits, each produced by a different gene; and each of these genes in turn may have different mutant forms. As the brain develops, neurons in different parts of the brain switch the expression of the different subunits on or off. The net result of this variability is that the brain is able to adapt to a wide range of environmental challenges.

The pathway between genes and behavior does not work in one direction only. It is now clear that behaviors may affect genes. Seizures, for example, can turn on the expression of genes known as "immediate early genes," which may be involved in the brain's response to injury. This means that the interplay among these processes is changing dynamically as the individual develops. Each step is affected by environmental factors and conditioned by the life experience up to that time. The development is not rigorously preprogramed but responsive to events as they occur.

This description suggests two things. First, because the brain structure and its processes are complex, we should be cautious in interpreting new research findings. James Watson, who shared in developing the DNA model, claimed that "the brain will be to the next century what the gene has been to the 20th century. . . . But compared to the gene, the brain . . . seems an infinitely more daunting objective."[1] Second, genes *alone* do not directly cause specific behaviors; rather, genes work through physiological systems in the brain. At the same time, these physiological systems

1. James Watson, preface to *The Brain* (Plainview, NY: Cold Spring Harbor Laboratory Press, 1991).

have been formed according to genetic instructions, so that genes are often a dominant player in the causal structure of specific behaviors. In short, very few if any behaviors are completely without genetic influence, and many are heavily influenced by genes.

A Genetic Basis for Schizophrenia

Dan began to act strangely shortly before he went off to college in the 1980s. He would "attack the refrigerator for reading his mind and threaten family members for using the word 'right'." His parents took him to psychiatrists, hospitals, and adult homes. When he was released, he returned home, only to create chaos there. His family bore the brunt of his episodes, so much so that his brothers and sisters were reluctant to bring their families to their parents' home for a visit. "As Christmas approached, two of our children called to say they would not come to visit over the Christmas holidays unless we could assure them that Dan would not be here. After all, they did have the joys of young children to consider. . . . Our two younger children wanted to be home." Yet the Christmas event was quickly brought to ruin. "Dan was stomping, slamming doors, screaming until 4:00 A.M. . . . I was up early, found a burned skillet in the sink, a broken one in the trash. Oysters were on the wall and floor. I washed the dishes and prepared the turkey. Gradually people began to get up." The younger children left or went out with friends, leaving the parents to "celebrate" Christmas alone with Dan. "My husband, John, and I have given up the hope of 'golden years' together, though he occasionally says he wants to help Dan find a place before he dies."[2]

Dan was diagnosed with schizophrenia, a disorder expressed by a variety of symptoms. Dan's case is typical; the disease has its onset often between the ages of fifteen and twenty-five. Those who suffer from schizophrenia can manifest confused or illogical thought processes. Dan, for example, wanting immediate satisfaction, would pawn his new shoes or his watch to eat in restaurants. Schizophrenics often withdraw from others and talk to themselves or to fictional beings or things. They can have hallucinations or delusions of being persecuted, as Dan had about the

---

2. Irving I. Gottesman, *Schizophrenia Genesis: The Origins of Madness* (New York: W. H. Freeman, 1991), pp. 167-70.

254

refrigerator. They seem to have no emotions or to express inappropriate ones for the circumstances. At times they are totally out of control, at other times withdrawn and unapproachable.

In some senses schizophrenia is difficult to diagnose. In the absence of any reliable pathological test, diagnosis must be made on the basis of observed behavior and reports given by sufferers compared against a priori criteria. Even these a priori criteria can vary. In the 1970s, for example, the United States reported a frequency of schizophrenia that was more than twice that of Great Britain. Subsequent tightening of the diagnostic criteria altered this figure, so that the number of cases per 100,000 in the U.S. population currently is less than in Great Britain. "Although the [diagnostic] problem has been improved by the use of standardized criteria, the lack of a defined biologic marker still makes accurate diagnosis problematic."[3]

Even with this diagnostic uncertainty, however, the occurrence of schizophrenia is widely acknowledged to be familial. About 1 percent of the general population will have developed schizophrenia by age fifty-five. But the lifetime risk is 9 percent for siblings of affected persons, 13 percent for children with one parent who suffers from schizophrenia, and as high as 40 percent if both parents suffer.[4] The lifetime risk is 48 percent for an identical co-twin of an affected person, compared with 17 percent for a non-identical (fraternal) co-twin.[5] In short, "the larger the proportion of genes shared with an affected individual, the higher the morbid risk."[6]

Yet schizophrenia does not fit a Mendelian inheritance model, where the probability of a child being affected if the parent is affected can be determined a priori. For example, only 50 percent of identical twin-pairs both develop schizophrenia, even though all co-twins presumably share a genetic tendency with the index case. This suggests that the factors involved in its transmission and in the appearance of its symptoms are complex, that possibly they are environmental, psychological, and social as well as genetic. "Genes in and of themselves do not result in schizo-

3. Roland D. Ciaranello and Andrea L. Ciaranello, "Genetics of Major Psychiatric Disorders," *Annual Review of Medicine* 42 (1991): 156.

4. Claudia Wallis and James Willwerth, "Schizophrenia: A New Drug Brings Patients Back to Life," *Time* 140, no. 1 (6 July 1992): 55.

5. Gottesman, p. 96.

6. Nicola De Marchi, "The Genetics of Schizophrenia," *Developmental Medicine and Child Neurology* 33, no. 5 (1991): 452.

phrenia, but they do establish a predisposition or substrate — an inherited vulnerability or likelihood — that, when combined with the consequences of environmental stress, tip the scale into illness."[7]

Adoption studies support this impression. Children of a schizophrenic parent who are placed in adoptive homes early in life have the same risk as those who are reared by their affected parent. In the opposite direction, adopted children who are reared in a home with a schizophrenic parent have no increased risk. The conclusion is that the environment in which a person is reared is seldom sufficient to cause the disorder.

Recent neuroscientific research has attempted to discover a biological factor that might cause schizophrenia. A likely hypothesis is that schizophrenia results from some structural or biochemical change within the brain. As we have described above, brain function involves neurons and their connectors (synapses). Communication across synapses takes place by means of neurotransmitters that one neuron releases and receptors on an adjacent neuron recognize. The most effective drugs in the treatment of schizophrenia block receptors for the neurotransmitters dopamine or (in more recent applications) serotonin, whereas agents (such as amphetamines) that exacerbate schizophrenia have been found to increase dopamine levels.

The "dopamine hypothesis" can be tested by studying dopamine levels or activity in the brain. However, the best ways to determine these levels would require invasive techniques for analyzing specific brain areas. Gene mapping strategies can circumvent this problem. For each kind of receptor there are one or more genes that could undergo mutation and thus lead to a diagnosable behavioral problem. The location of such genes on specific chromosomes can be identified if "marker" DNA segments are found that are transmitted through families as tracers of the behavioral disorder.

In 1988 Robin Sherrington and others studied five Icelandic and two English families with multiple cases of schizophrenia in two or more generations. They reported isolating a region on chromosome 5 that showed a linkage with schizophrenia in these families.[8] In the same issue of the journal, however, James Kennedy reported that there was no link between the specific areas on chromosome 5 and schizophrenia in a

---

7. Gottesman, p. 18.

8. Robin Sherrington et al., "Localization of a Susceptibility Locus for Schizophrenia on Chromosome 5," *Nature* 336, no. 6195 (10 Nov. 1988): 164-67.

northern Swedish family that had been studied for a long time.[9] In subsequent studies Sherrington was unable to duplicate his results.

Conflicting data have led researchers to conflicting conclusions. All agree that as yet there is no evidence of a specific genetic basis for schizophrenia. Yet most are convinced that genes play some sort of role, some believing that schizophrenia involves a specific gene, but most holding that the disease is polygenic (where different genetic combinations create different symptoms depending on the number of genes involved). The former conclude that schizophrenia is a single disease with a variety of symptoms, depending on other factors, including other genes and environmental factors. The latter argue that schizophrenia is not a single disease but a group of diseases manifesting a variety of symptoms, yet possessing some common core that results from the expression of genes in the brain.

A Genetic Basis for Alcoholism

As with schizophrenia, there is substantial evidence that alcoholism runs in families, that children of alcoholics have a greater predisposition to alcoholism than the general population.[10] Studies of twins raised independently confirm that the predisposition is more than environmental or social. But again, the familial pattern is not enough to resolve the question of the respective roles of genetic and environmental factors.

Animal studies have been somewhat more successful in teasing apart the biological effects of alcohol. Experimenters have used selective breeding to create a strain of rats that prefer alcohol over water. This craving seems to be based not on the taste, smell, or calories of the alcohol but on its pharmacological effect. They have developed other strains for differences in sensitivity to alcohol; for example, some sleep longer after a standard

9. James L. Kennedy et al., "Evidence against Linkage of Schizophrenia to Markers on Chromosome 5 in a Northern Swedish Pedigree," *Nature* 336, no. 6195 (10 Nov. 1988): 167-70. See also Sevilla D. Detera-Wadleigh et al., "Exclusion of Linkage to 5q11-13 in Families with Schizophrenia and Other Psychiatric Disorders," *Nature* 340, no. 6232 (3 Aug. 1989): 391-93; W. F. Byerley, "Schizophrenia: Genetic Linkage Revisited," *Nature* 340, no. 6232 (3 Aug. 1989): 340-41.

10. Gilbert S. Omenn, "Genetic Investigations of Alcohol Metabolism and of Alcoholism," *American Journal of Human Genetics* 43 (1988): 579-81.

dose of alcohol. Some strains develop alcohol tolerance faster than others, in that they can learn to handle larger doses. Finally, they have created strains that exhibit more severe signs of withdrawal, such as seizures. These results, together with observations of humans, suggest that several separate aspects of alcoholism can be distinguished: level or frequency of consumption, sensitivity, tolerance, dependence (inability to function well without some alcohol), withdrawal signs, consequences of use on behavior, and perhaps also a tendency for tissue damage (such as liver cirrhosis).

Researchers also have begun a search for biological markers for alcoholism in humans. Kenneth Blum and his colleagues examined the brain tissue from thirty-five deceased alcoholics.[11] They focused their attention on one type of dopamine receptor (D_2), which earlier studies had associated with alcoholism. Reduction in dopamine activity might help to explain the behavior of people under the influence of alcohol. The researchers discovered the A1 variant (allele) of the dopamine D_2 receptor gene (DRD2) in 77 percent of the brain tissue samples taken from the alcoholics, whereas it was absent in 72 percent of the nonalcoholic control group. On the basis of this discovery they argued that a gene related to human susceptibility to alcoholism is located somewhere on chromosome 11, near the receptor gene.

Subsequent studies yielded mixed results. Some tended to confirm Blum's studies.[12] However, Annabel Bolos and her team reported that the gap between alcoholics and nonalcoholics with regard to the A1 allele of the DRD2 gene was not as significant as first reported.[13] Whereas 30 percent of nonalcoholics had the allele, only 38 percent of alcoholics did. In a subsequently reported study of 44 alcoholics and 68 control subjects by Joel Gelernter and others, 35 percent of the control subjects and 45 percent of the alcoholic subjects were found to have the A1 allele, a difference that was not considered statistically significant. Their conclusion was further confirmed when they divided the alcoholic subjects into groups

11. Kenneth Blum et al., "Allelic Association of Human Dopamine D2 Receptor Gene in Alcoholism," *Journal of the American Medical Association* 263 (18 Apr. 1990): 2055-61.

12. David Comings et al., "The Dopamine D_2 Receptor Locus as a Modifying Gene in Neuropsychiatric Disorders," *The Journal of the American Medical Association* 266 (2 Oct. 1991): 1793-1801.

13. Annabel M. Bolos et al., "Population and Pedigree Studies Reveal a Lack of Association between the Dopamine D_2 Receptor Gene and Alcoholism," *Journal of the American Medical Association* 264 (26 Dec. 1990): 3156-61.

based on their family history of alcoholism, age at onset, and patterns of consumption and found no differences from the control group.[14]

Again, conflicting data bring researchers to conflicting conclusions. Some contend that there is as yet no evidence of a specific genetic basis for alcoholism. In any case, the A1 allele provides no argument for such, for the suspect allele is found in at least half of the nonalcoholic population and is absent in more than half of the alcoholics studied. Indeed, some who had the allele did not have the disease, and even within the same family whose members each possessed the A1 allele, not all were alcoholics.

Others appeal to these same considerations and contend that although "other genes or factors are the primary cause of alcoholism, a variant of DRD2 gene, such as the A1 allele, was acting as a modifying gene."[15] They argue that this allele, combined with a primary gene as yet undiscovered for alcoholism, exacerbates the symptoms. The modifying gene does not cause the disorder but makes the disorder more severe.

A second biochemical difference has better support. Aldehyde dehydrogenase is an enzyme that helps the body to eliminate alcohol. While a high proportion of Japanese and Chinese have a deficiency of this enzyme, among Japanese and Chinese alcoholics the proportion of those with the deficiency is much lower. This suggests that the enzyme deficiency allows toxic chemicals to build up with unpleasant effects in drinkers. That is, those with the deficiency tend to become ill when drinking too much, so that this deficiency may actually help prevent alcoholism from developing.

At present it appears that there is no specific gene for alcoholism per se. Part of the problem of attempting to trace alcoholism back to one gene lies in the significant variability of alcoholism. Some alcoholics suffer serious psychological impairment (e.g., mental illness); others do not. Some imbibe continuously, while others drink in binges. Some alcoholics use only alcohol; others move to other drugs. Robert Cloninger has suggested that there are several different types of alcoholism, distinguished by age at onset, differential expression in males and females, and presence of accompanying aggressive actions.[16] About half of alcoholics have various

14. Joel Gelernter et al., "No Association between an Allele at the D_2 Dopamine Receptor Gene (DRD2) and Alcoholism," *Journal of the American Medical Association* 266 (2 Oct. 1991): 1801-8.

15. Comings et al., p. 1794.

16. Robert Cloninger, "Recent Advances in Family Studies of Alcoholism," in

mental disorders, which might be traceable to other genetic or environmental factors. Some may have "a general susceptibility to compulsive behaviors whose specific expression is shaped by environmental and temperamental factors."[17]

What can we conclude at the present time? There appear to be several types of alcoholism, not a single condition.[18] Furthermore, genetic susceptibility is probably not specific to alcohol but may extend to other substance abuse and temperamental characteristics. That is, there "may be a gene for 'compulsive disease' and subgenes (modified genes) that dictate susceptibilities to particular substances,"[19] such that different environmental factors can lead to different kinds of compulsive disorders, such as gambling and eating disorders. Environmental influences are most likely involved, but they are not well defined and may operate only on a subset of the population.

The Question of Moral Responsibility

Traditionally alcoholism was viewed as a behavior over which a person had control. Control was a matter of will power: one had a choice to say no to drinking alcohol, or one could, out of weakness of will, yield to temptation or desire. Of course, the more one consumed, the less power one had to control one's behavior. The determining effects of the alcohol consumed supplanted volition. But the initial responsibility remained with the drinker.

Only recently has alcoholism in some of its forms been termed a disease. As we have seen, the results of recent studies indicate the possibility that a gene or combination of genes plays a significant role in the inheritance of this predisposition. Although these genetic factors have yet to be

Genetics and Alcoholism, ed. H. W. Gedde and D. P. Agarwal (New York: Alan R. Liss, 1987), pp. 47-60.

17. Constance Holden, "Probing the Complex Genetics of Alcoholism," *Science* 251 (11 Jan. 1991): 163.

18. The underlying "genetic diathesis is likely, in part, to be manifested in temperamental characteristics that in turn influence the nature of and sensitivity to critical developmental experiences." Matt McGue, "Genes, Environment and the Etiology of Alcoholism," in *The Development of Alcohol Problems: Exploring the Biopsychosocial Matrix of Risk,* ed. R. Zucker, G. Boyd, and J. Howard, *NIAAA Research Monograph Series,* No. 26, NIH Publication 94-3495, Public Health Service, Rockville, MD.

19. Holden, p. 164.

isolated, many are convinced that it is only a matter of time before the genetic puzzle will be laid bare.

Let us suppose, then, that these genetic factors will be isolated, that there is a gene or combination of genes responsible, at least in part, for alcoholism and for schizophrenia. What would such a finding imply for our understanding of human personhood and moral accountability?

Most schizophrenic persons have minimal control over their schizophrenic behaviors. Psychotherapy can help, but only to a limited extent. The genetic predisposition to schizophrenia leaves little to volition. Accordingly, we would not hold Dan morally accountable in many instances for the damage or human suffering he causes. With alcoholism the situation is more complicated, since the manifestations can be more complex. But again the issue of accountability arises: if certain people have a genetic predisposition to alcoholism, can they control their behavior? Can we any longer attribute their alcoholism to human volition? If not, what follows regarding their moral responsibility for their alcoholism-associated behavior? Can they excuse their behavior by saying, "My genes made me do it"?

Even if genes influence behavior only in part (as is most likely), so that what the genes do not cause the environment does, the same problem arises; only the language changes. Instead of attributing behaviors simply to genes, behaviors are held to flow from the combined action of genes and environment. Now someone accused of moral impropriety might respond with the excuse, "My genes and my environment made me do it."

The problem, of course, extends beyond alcoholism; it encompasses all our behavior. The part of our behavior ascribable to genes and/or environment would appear to be out of our hands or control. We inherit our genes from our parents, they inherit their genes from their parents, and so on. Accordingly, our genetic predisposition is not of our own making. Similarly, any particular environment results from prior environmental conditions, and these conditions result from still prior ones, to a point that extends prior to our own existence. But if by our conscious choices we have not contributed to either of these sets of causal factors, what room is there for moral agency and moral accountability?

Some might respond that although we have not contributed to our genetic heritage, we do create our own environment by our choices. There is mutual influence: we affect the environment, and it in turn affects us. But on what basis do we act to create or influence our environment? Are such creative acts themselves the product of our gene-environment nexus, or is there something more involved — an element of human choice and

261

agency by which we structure our environment? If the former, humans still are causally determined; if the latter, the genetic-environmental causes or influences must be supplemented by something like human free agency.

Prima facie it would appear that in speaking about genetic-environmental determination we are dealing with a continuum. At one end are actions that, given the causal conditions present, we can either perform or not perform. As agents, we can choose how to behave. Here the causal conditions are weakly conditioning, so that the individual has maximal accountability. At the other end of the continuum lie actions that, given the causal conditions present — genetic and environmental — we have no choice but to perform. Here individual accountability is absent. The agent has not acted but has been acted upon to behave in a certain way. Since persons cannot do otherwise in these circumstances, where these actions are moral ones persons are not morally accountable for their behavior. Our actions fall somewhere on this continuum, probably at some distance from either extreme. Our genetic heritage and our environment in various ways and to various degrees condition our actions, so that we are partially free and partially determined.

Given our present knowledge, schizophrenia seems to lie near one end of the continuum. Alcoholism, too, lies somewhere on that continuum. Without a clear indication that alcoholism is unavoidable in the same way that cystic fibrosis or Huntington's disease is unavoidable, it would seem that at least partial control can be exercised over alcoholic behaviors. These behaviors, influenced by brain predispositions, environmental conditions, and human choice, can be altered by persons choosing to make changes in their environment.

But though we have this prima facie account of human action, the methodological and ontological presuppositions of the various scientific disciplines having to do with the brain appear to leave little room for the personal contribution of human choice. The neuroscientist R. W. Sperry, in laying out the position he wishes to attack, puts it this way:

> Science insists that there is no real freedom of will or choice, nor any actual moral right and wrong — that ours is a deterministic universe in which the flow of events is causal and inexorable. Science tells us, further, that the entire conscious content of the life experience is merely an accessory artifact, a superfluous by-product of brain activity, with no effect whatever on the sequence of events, either in the brain or in the real world.[20]

20. R. W. Sperry, "Search for Beliefs to Live by Consistent with Science," *Zygon* 26, no. 2 (June 1991): 241.

Clearly, Sperry has a particular view of science in mind, one that he thinks needs replacing. Neuroscientists frequently show a tendency to want to see consciousness as a mere function of the electrochemical processes within and between the brain cells. "Just as sensing, moving, adapting, learning, and feeling can ultimately be explained in terms of the structure and workings of the brain, so will consciousness be ultimately explained." The mind is what the brain does; it is "the working of the brain as a whole. Those brain events that we experience as *conscious* are only those events that are processed through the brain's language systems."[21]

Certain biological methods and theories can also be viewed as portending causally deterministic scenarios. For example, as we shall see shortly, behavioral genetics ascribes the causes of our behavior to a combination of genetic and environmental factors. "The actions of genes underlie all types of behavior, including mechanisms of learning, since many of the molecules that control the function of neurons are direct products of genes."[22] What is not a product of the genes is ascribed to the environment, which is characterized as "outside factors." But what role does this leave for the contribution of human choice? Can a scenario that ascribes our actions to genes and environment allow for persons to be agents, controllers of their behavior?

These approaches, if followed out, have significant consequences for understanding what it is to function as a human being. We maintain that being human involves more than simply having human parents or a body of a certain size and shape, genes and a brain with certain capacities. Being human, in its full-blown richness, invokes dimensions of embodied personhood: the ability to be self-aware; to know something about oneself and the world; to have beliefs and desires; to make meaningful choices or decisions about actions; to employ percepts, concepts, and reasoning in support of those choices; and to be able to communicate these to others. At the heart of this description lies the matter of agency. Persons choose to be themselves or to hide behind facades, to hold views or to reject them, to act and behave according to certain values, etc. If what we think, do, say, feel, and communicate is the product merely of genetic and environmental causes, and these are causes over which we have no control,

21. Floyd Bloom, Arlyne Lazerson, and Laura Hofstadter, *Brain, Mind, and Behavior* (New York: W. H. Freeman, 1985), pp. 206, 207.

22. Jeffrey Hall, Ralph Greenspan, and William Harris, *Genetic Neurobiology* (Cambridge, MA: MIT, 1982), p. 3.

then who we are is determined by conditions that lie outside ourselves. We cannot take charge of our individual acts, let alone of our lives. Moral ideals and norms no longer have any bite, for not only are we incapable of choosing and acting to realize them, but we also cannot be held morally accountable for failing to meet them. Moral accountability does not apply to that over which we have no control.

So where do we go in this "decade of the brain" to resolve this dilemma? To begin to address this problem, we will explore two questions. First, we have hinted that the problem arises from certain scientific perspectives on being human. What is there about the methods and conclusions of certain of the biological sciences that creates the perception that personhood and moral responsibility are in jeopardy? And second, what are the options for reconciling the two sides of the dilemma — the neurological and biological side and the humanistic side with its emphasis on knowing, choosing, and moral accountability? We will address these two questions in turn.

From Genes and Environment to Behavior

Neuroscience

Neuroscience attempts to trace cognitive processes to brain structure and function, and these mechanisms to the cell and molecule. As one neurologist puts it, "The ultimate goal . . . is extraordinarily ambitious. Eventually researchers such as myself hope to be able to analyze higher mental functions in terms of the coordinated activation of neurons in various structures in the brain. It should also be possible to identify the cells that mediate the activity of those structures. Such research will help explain the origin of mind."[23] The basis for memory is sought in the long-term potentiation at certain synapses in the brain; the basis for drug addiction is looked for in a neural pathway, the nucleus accumbens. This reductionism, as a methodological approach to the question of the workings of the brain, is appropriate (though perhaps not fully adequate).[24]

23. Patricia S. Goldman-Rakic, "Working Memory and the Mind," *Scientific American* 267, no. 3 (Sept. 1992): 111.

24. Some have suggested that there is another side to scientific methodol-

However, some want to go beyond making reductionistic claims about the brain and its processes to making such claims about the mind and mental processes and the causal relations that hold between neural activity and minds, beliefs, and volitions.[25] They hold that mind and its processes are higher-level structures composed of lower-level patterns, and these in turn are composed of still lower patterns. At the base lie the individual brain cells and their firings. Individually, neural firings are unpredictable; however, statistical regularity reveals that these firings occur in patterns that encode the information necessary for the organism to respond interactively to other information in the environment. When these firing patterns are interpreted at the highest levels, the level of mind, they are given meaning. These higher-level patterns do not possess their own, independent reality. Rather, they emerge from lower-level patterns and these in turn from the fundamental, nonintelligent firings of the neurons. Viewed from the level of the individual cell activity (that is, viewed from the bottom up), what gets done is accomplished without questions of meaning and interpretation being raised or considered. The "emergence of various mechanisms [are] natural responses to external pressures."[26] Mind need not be invoked to account for individual neural activity, though without higher levels the coherence necessary for the existence of an organism is absent. However, invoking these higher-level patterns, as humans are wont to do, provides an easier way to explain what goes on in and around us. These patterns help us understand the transmission of information within the brain and hence to make sense of the organism's

---

ogy. Michael A. Savageau argues (in "Reconstructionist Molecular Biology," *The New Biologist* 3, no. 2 [Feb. 1991]: 191) that

> any respectable reductionist is also a reconstructionist. If you ask a reductionist what his or her objective is, you will find that it is to reduce complex systems to their elemental units in order to characterize them, and once this is accomplished, to use this knowledge for reconstructing an understanding of the intact entity with which the investigation started. The problem is that the reconstructionist phase of this program is seldom carried out.

25. See, e.g., Daniel Dennett, *Brainstorms* (Cambridge, MA: MIT Press, 1978); Douglas R. Hofstadter and Daniel C. Dennett, *The Mind's I* (New York: Basic Books, 1981); Paul Churchland, *Neurophilosophy* (Cambridge, MA: MIT Press, 1986); Patricia Churchland, *Matter and Consciousness* (Cambridge, MA: MIT Press, 1988); and Daniel C. Dennett, *Consciousness Explained* (Boston: Little, Brown and Co., 1991).

26. Hofstadter and Dennett, p. 173.

operations. This type of explanation coheres with our penchant to use language that ascribes purpose to events. Thus, language about persons simply becomes one way of speaking about the higher-level patterns that carry coded information, but it should not be thought that such "top-down" explanation corresponds with any causal activity or "that higher-level laws actually are responsible, and govern the system above and beyond lower-level laws."[27]

For example, an explanation of why a person took a particular action — say, of going to the grocery store and purchasing a loaf of bread — is rooted in, indeed is reducible to, lower-level events. Making a decision to go to the store and acting on that decision involves brain processes that can be analyzed further in terms of the electrochemical discharges and transfers in various parts of the brain. Ultimately to understand causally why persons have the ideas of food, a grocery store, and hunger or why they act as they do, one must focus on the neural processes and the states in the central nervous system associated with those processes. However, for purposes both of simplicity and to make sense of the behavior, it is appropriate to invoke purposive explanatory language, so long as we are not misled into thinking that in using this purposive language we are describing something over and above the basic levels of brain activity — that is, misled into ascribing causal powers to purposive systems.

On this view, language about meaning and purpose, although it fulfills an important role by allowing us to speak about events in ways that satisfy our desires for meaning and make sense of our situation, ultimately does not point to anything other than the biological and environmental. Choice itself refers to nothing but a complex event within the brain, an event that is determined by prior neural structures and present input from the environment. Freedom or volition is nothing more than a complicated, internal, causal configuration that enables the object to react to its environment as a representational system.[28]

In short, neuroscientists adopt a *methodological* reductionism in accounting for brain processes: to be understood, the brain must be analyzed into its component parts and the relations that hold between them. Those

27. Hofstadter and Dennett, p. 195.

28. A robot, designed to operate on the rocky terrain of Mars, is constructed without any central control system, but rather with modules that can interact in layers. Here one gets "apparently purposeful" behavior without any higher level, organized control system. Alun Anderson and Joseph Palca, "Who Knows How the Brain Works?" *Nature* 335, no. 6 (Oct. 1988): 490.

who argue for an *ontological* reductionism believe that what is real, what exercises causal influence, occurs at the micro-level of neurons and their firings. Human choice and agency ultimately can be understood in terms of the causal processes and information transfers within the brain. Does this then mean that the concept of person is dispensable? Some claim that, in certain contexts which concern origins and which "trace events to their ultimate causes," it is dispensable, for when I forget "about evolution and seeing things in the here and now, the vocabulary of teleology comes back: the *meaning* of the . . . . distribution and the *purposefulness* of signals. . . . However, with some effort I can always remember the other point of view if necessary, and drain all these systems of meaning, too."[29]

We would be remiss if we left the impression that all neuroscientists are ontological reductionists. Some argue that it may be more appropriate to consider bottom-up and top-down explanations between the molecular, biological, psychological, and sociological levels as complementing each other. Some years ago Gunther Stent noted that in the years prior to 1953 physical and chemical methods had been successful in determining the *structure* of the DNA molecule, but these methods could not explain the *functions* of DNA. Then Watson and Crick introduced two fundamental genetic questions: (1) How can DNA be copied accurately and faithfully? and (2) How can DNA carry information for synthesizing proteins? Thereupon they realized that DNA would function in these two ways if the "bases" formed complementary pairs at the center of the double helix. In other words, the phenomena could be explained from the bottom up (in terms of physics and chemistry) if the questions were posed from the top down (in terms of biological function).[30] This "breakthrough" from one level to the next did not hurt biology at all. A parallel would be that problems in cognitive psychology have the potential of being explained by biology to the benefit of psychology and cognitive philosophy, but only if behavioral scientists, philosophers, and neuroscientists develop and maintain strong collaboration and recognize the legitimacy of approaches from both top and bottom. At the same time, however, *how* they correlate remains problematic.

29. Hofstadter and Dennett, pp. 195, 174.
30. Gunther S. Stent, "That Was the Molecular Biology That Was," *Science* 160 (26 Apr. 1968): 390-95.

Behavioral Genetics

Behavioral geneticists study variations in behavior and estimate *heritability*, the proportion of the variability that results from genetic differences. Whatever does not result from genes is ascribed to the environment. In studies of a number of types of behavior, behavioral geneticists calculate the heritability to be in the range of 40 to 60 percent.[31] Seldom is the heritability close to 100 percent or to 0 percent. This means that genetic factors account for about half of the individual differences in these behaviors.

One must be careful how one uses this percentage, however. The heritability for a specific kind of behavior is not constant; it only describes the situation in a specific sample of the population at a given time. Under different circumstances or at a different time the percentage could be considerably different. Thus this information cannot be used to predict how much a certain individual would be affected by a specific environment. That is, it is not very helpful in trying to decide what type of intervention might be beneficial in any particular case. For example, it is true that good health and nutrition are likely to improve the measured IQ of children, but the amount of improvement that might be seen in a given child by altering the environment is uncertain.

It is readily apparent that behavioral genetics and neuroscience agree significantly. Indeed, many scientists working in these areas promote a closer integration of the research conducted in the two fields. Molecular research is not finished when a gene is mapped, identified, and sequenced. The work must continue in order to determine the gene product, to find out how the gene is regulated (when and where it is expressed), and to explore the interaction with other genes and the environment in the pathogenesis of a disorder. Behavioral research (approaching from the opposite end of the gene-to-behavior pathway) must include reliable definition of the behavioral trait, an effort to understand the brain systems involved, and exploration of the interactions between genes and environment throughout development. This marriage of neuroscience and behavioral genetics provides more comprehensive insight into the biologically determining conditions of human behavior.

If there is a basic agreement between neuroscience and behavioral

31. Robert Plomin et al., "The Role of Inheritance in Behavior," *Science* 248, no. 4952 (13 Apr. 1990): 183-88.

genetics, will one find a similar reductionism present in the latter? Prima facie, it appears so. The behavioral geneticist ascribes the cause of human behavior to a combination of genetic and environmental factors. What is not genetic or heritable derives from the environment. But what room then remains for human agency? We have seen that without choice there cannot be moral responsibility. But choice means that persons contribute something to the action over and above what derives from their genetic heritage and environmental input. Thus, while behavioral genetics contributes to our understanding of the causal factors at play in our behavior in that it helps us comprehend the significant contribution that genes make, it raises the critical question of human responsibility.

Behavioral geneticists might reply that, although it appears that dividing the sources of variability into only two categories (genes and environment) exhausts the possibilities, this is not in fact true. A description of the way in which estimates of heritability are calculated may clarify this point.

IQ test scores are adjusted for age, so that about 95 percent of the general population (at each age) will score in the range of 70 to 130, with the average at 100. If the heritability for IQ score is 50 percent (which is about the average from many studies), individuals selected to have an average *genetic* potential would (95 percent of the time) have test scores in the range of 80 to 120.[32] In this theoretical example, fixing the genetic effects will narrow the range but will not *determine* the performance of any specific individual. The remaining variability results from the chance events in the individual's development (prenatal and postnatal), nutrition, health, family circumstances, and everything else. For convenience, the sum of everything that is not genetic is labeled "environment." This example illustrates the point that genetic relationship can be defined more precisely, while the environmental factors relevant for a given trait cannot be exhaustively identified and studied.

The critic, in reply, will appreciate the range left open to the non-genetic in any individual case. Behavioral geneticists are not necessarily committed to a freedom-denying reductionism. But what is to be understood by "environment"? That it refers to "everything else" besides heredity

---

32. Heritability studies include twins with different degrees of relationship, some sets reared together, some apart. The similarity in test scores is then analyzed with the knowledge that genetic relationship (genes shared) is 100 percent for identical twins, 50 percent for nontwin siblings, and 25 percent for half siblings. The 80-120 range includes the effects of individual choice to the extent that choices are not indirectly constrained by genetics.

would allow for the contribution of individual choice. At the same time, the term "environment" generally refers to what surrounds and affects the individual, to what is outside it.[33] But agency has to do with what is "in" the organism but not ascribable to heredity. Thus, to say that behavioral geneticists leave the nature of the other, nongenetic causes open still raises the question of the nature of those other causes. What is worrisome is their labeling of the other dimension as the "environment," for in its most obvious sense this label excludes what might be termed human agency — purpose, considered choice, and deliberation. If this dispute is thought to be merely verbal, perhaps a more felicitous term to describe the nongenetic might be employed.

Sociobiology

A third perspective joins hands with the two just mentioned. Sociobiology is "the systematic study of the biological basis of all social behavior."[34] Though originally applied only to nonhuman animal behavior, Edward Wilson, its chief proponent, extended it to humans. For him evolutionary biology holds the key to understanding human nature and behavior; to fully comprehend human societies we must understand their biological and cultural evolution.

Wilson reviewed data about the social behavior of a variety of animal species and concluded that genetic factors contributed significantly to the neural basis for such behavior. He extrapolated his findings to humans, concluding that all identifiable human traits and all social behavior have a biological basis.[35] They are all gene constrained, environmentally adaptive, and naturally selected.

For Wilson, natural selection is the major mechanism of development. It is the driving force behind the development of all traits, "the agent that molds virtually all of the characteristics of species."[36] Natural

33. Hall, Greenspan, and Harris, p. 3.

34. Edward O. Wilson, *Sociobiology: The New Synthesis* (Cambridge: Cambridge University Press, 1975), p. 4.

35. This assumption is widely shared by sociobiologists. For documentation see Howard L. Kaye, *The Social Meaning of Modern Biology* (New Haven: Yale University Press, 1986), p. 138.

36. Wilson, p. 67. "The pervasive role of natural selection in shaping all classes of

selection provides the means by which "certain genes gain representation in the following generation superior to other genes located at the same chromosome positions."[37] From this he concludes that genes, not organisms, are basic.[38] The goal of organisms is to carry and reproduce those genes that are in effect their masters. Already, then, we have hints of a reductionism. Wilson repeats an aphorism: as "the chicken is only an egg's way of making another egg, . . . [so] the organism is only DNA's way of making more DNA."[39] But why, one might ask, make the DNA more basic than the organism? Cannot we say that as the egg is the only way a chicken can make more chickens, so DNA is the only way an organism can make more organisms?

Ethics assumes a prominent place in Wilson's discussion. For example, he holds that altruism constitutes the central theoretical problem of sociobiology, since altruism cannot be readily explained by natural selection. Altruism, which he understands biologically in terms of actions that promote the genetic fitness of kin, is necessary for the possibility of the existence of society, for its harmony, and for the development of culture. But since our genes are prone not to altruism but to selfishness, in that they work for their own survival, and since vertebrate societies "favor individual and in-group survival at the expense of societal integrity,"[40] there is a built-in tendency, not toward evolutionary social progress, but toward social devolution.

Wilson's concern is how to provide for a postreligious, scientifically informed society on a new basis. Religion at one time provided for "genetic advantage and evolutionary change. . . . When the gods are served, the Darwinian fitness of the members of the tribe is the ultimate if unrecog-

---

traits in organisms can be fairly called the central dogma of evolutionary biology. . . . [B]ehavior and social structure, like all other biological phenomena, can be studied as 'organs,' extensions of the genes that exist because of their superior adaptive value" (pp. 21-22).

37. Wilson, p. 3. Natural selection is interpreted in terms of population genetics; it is defined as "the change in relative frequency in genotypes due to differences in the ability of phenotypes to obtain representation in the next generation" (p. 67).

38. "In a Darwinist sense the organism does not live for itself. Its primary function is not even to reproduce other organisms; it reproduces genes, and serves as their temporary carrier. . . . [T]he individual organism is only their vehicle, part of an elaborate device to preserve and spread them." Wilson, p. 3.

39. Wilson, p. 3.

40. Wilson, p. 382.

nized beneficiary."[41] However, science has destroyed the "myths of traditional religion and its secular equivalents." "Religion . . . can be systematically analyzed and explained as a product of the brain's evolution." Science is superior to religion by virtue of "its repeated triumphs in explaining and controlling the physical world; its self-correcting nature open to all [who are] competent to devise and conduct the tests; its readiness to examine all subjects sacred and profane; and now the possibility of explaining traditional religion by the mechanistic models of evolutionary biology."[42]

This victory has not been without cost, however. "The price of these failures [of the myths of traditional religion and its secular equivalents] has been a loss of moral consensus, a greater sense of helplessness about the human condition and a shrinking of concern back toward the self and the immediate future."[43] The altruism that is essential to society and culture, which religion fostered, lies in jeopardy. To begin the rescue of altruism, ethics should be "removed temporarily from the hands of the philosophers and biologicized." What needs to be discerned is the biological basis for ethics. Our emotive centers guide our moral actions. Hence "only by interpreting the activity of the emotive centers as a biological adaptation can the meanings of the [moral] canons be deciphered."[44] Our ethics, our general sense of right and wrong, is thus guided by genes that have an evolutionary history. Moral philosophers have failed to appreciate the genetic and neural basis for their ethics, so that their ethics lack the grounding that only evolutionary biology can provide.

Not all moral rules have significance; some are mere relics left over from what the genes prescribed as the appropriate behavioral patterns for humans' hunter-gatherer days. The task of sociobiology is to uncover the "rules by which human beings increase their Darwinian fitness through the manipulation of society."[45] This set of rules, as a "genetically accurate and hence completely fair code of ethics,"[46] can then be used to construct a more harmonious social order. The ethic will be both altruistic and

41. Edward O. Wilson, *On Human Nature* (Cambridge, MA: Harvard University Press, 1978), pp. 172, 184.

42. Wilson, *On Human Nature*, p. 201.

43. Wilson, *On Human Nature*, p. 195.

44. Wilson, *Sociobiology*, p. 563.

45. Wilson, *Sociobiology*, p. 548.

46. Wilson, *Sociobiology*, p. 575.

pluralistic, with different moral codes applied to different segments of the population in terms of their particular age or sex role within the society.

But how is such an ethic — the goal of whose biology is to design future societies by genetically engineering characteristics, and which includes obligations that might run counter to present genetic prescriptions — possible, when the evolutionary-derived "anatomical, physiological, and behavioral machinery . . . carries out the commands of the genes,"[47] where the very explanation of ethics is found in the basic, unconscious neural elements that are the products of selfish genes and natural selection? Wilson's sociobiology faces a dilemma. On the one hand, he holds that by an "exercise of will" we can alter the human predicament. Human biology can "fashion a biology of ethics, which will make possible the selection of a more deeply understood and enduring code of moral values,"[48] taking into account the genetic pool that has enabled humans and their society to survive. On the other hand, the mind and the human will are "an epiphenomenon of the neuronal machinery of the brain,"[49] and since the brain is genetically programmed for spreading genes, it is not their master.

Thus sociobiology, with its scientific materialism, continues the reductionism of molecular biologists such as Francis Crick and Jacques Monod. Crick argues that "apart from the principle of natural selection itself, all [life] can be explained in terms of the ordinary concepts of physics and chemistry, or rather simple extensions of them. The ultimate aim of the modern movement in biology is in fact to explain *all* biology in terms of physics and chemistry." "Eventually one may hope to have the whole of biology 'explained' in terms of the level below it, and so on right down to the atomic level."[50] This applies equally to consciousness and mind.

47. Wilson, *Sociobiology*, p. 23.

The biologist . . . realizes that self-knowledge is constrained and shaped by the emotional control centers in the hypothalamus and limbic system of the brain. These centers flood our consciousness with all the emotions . . . that are consulted by ethical philosophers who wish to intuit the standards of good and evil. What, we are then compelled to ask, made the hypothalamus and limbic system? They evolved by natural selection. That simple biological statement must be pursued to explain ethics and ethical philosophers . . . at all depths. (p. 3)

48. Wilson, *On Human Nature*, p. 196.
49. Wilson, *On Human Nature*, p. 195.
50. Francis Crick, *Of Molecules and Men* (Seattle: University of Washington Press, 1966), pp. 10, 14.

Consciousness "deals with the exact behavior of the brain," while "mind is simply a way of talking about the functions of our brain."[51]

For Wilson, too, the ultimate goal is reductionistic in nature. It is to explain ourselves in "mechanistic terms: of neurons and genes." He writes:

> Only when the machinery can be torn down on paper at the level of the cell and put together again will the properties of emotion and ethical judgment come clear. . . . Cognition will be translated into circuitry. Learning and creativeness will be defined as the alteration of specific portions of the cognitive machinery regulated by input from the emotive centers. Having cannibalized psychology, the new neurobiology will yield an enduring set of first principles for sociology.[52]

Wilson at times tries to avoid a human-demeaning reductionism. He notes that the possibility is open "that a highly intelligent species can abandon genetic programs and come to depend more or less exclusively on cultural evolution." He is even willing to talk about an "individual free will [that] probably will remain forever invulnerable." Yet only a few paragraphs earlier he suggests that the "chain of causation runs from genes to physiological process to species-specific behavioral responses to a limited array of possible social organizations,"[53] and that the biological research program will eventually be able to map the mind on the DNA sequence, though not necessarily point-by-point.

Wilson sees human behavior as emerging "as the outermost phenotypes following behavioral development, the range and scope of which is constrained by the interaction of polygenes with the environment."[54] This would, on the surface, allow for the input of human agency. Yet, when he goes on to say that "with reference to the interaction [between polygenes

---

51. Crick, pp. 17, 87.

52. Wilson, *Sociobiology*, p. 575. This reductionism is also present elsewhere. Richard Dawkins (*The Selfish Gene* [New York: Oxford University Press, 1976], p. ix) writes that human beings are "survival machines — robot vehicles blindly programmed to preserve the selfish molecules known as genes." Kaye (chapter 5) traces this in others, such as Richard Alexander *(Darwinism and Human Affairs* [Seattle: University of Washington Press, 1979]) and David Barash (*Sociology and Behavior* [New York: Elseview, 1977]).

53. Edward O. Wilson, "The Ethical Implications of Human Sociobiology," *The Hastings Center Report* 10 (Dec. 1980): 27-28.

54. Edward O. Wilson, "Biology and the Social Sciences," *Zygon* 25, no. 3 (Sept. 1990): 253.

and the environment], there is no reason to regard most forms of human social behavior as qualitatively different from physiological and nonsocial psychological traits," the contribution of human agency seems precluded, for human choice plays little or no role in producing physiological traits.

Moral Responsibility

The upshot of the methods and presuppositions of neuroscience, behavioral genetics, and the paradigm of sociobiology is that, although ontological reductionism is not entailed, nor is a role for human agency precluded, certain features of these approaches lend support to this preclusion. The methodological reductionism of neuroscience easily slides into a metaphysical reductionism. The ascription by behavioral genetics of whatever is not due to heredity to the environment, combined with our common understanding of environment as what surrounds us, raises questions about what this discipline is willing to ascribe to human agency and how it would understand this feature. Sociobiology's appeal to natural selection as the primary, if not sole, guiding mechanism in human evolutionary development raises questions concerning the role of a purposive, deliberative human moral agent. Even the potency of culture is carefully circumscribed: "culture relentlessly tests the controlling genes, but the most it can do is to replace one set of genes with another."[55]

Typically, defenders of moral responsibility have maintained that for persons to be free and hence capable of assuming moral responsibility for their actions, they must be able to transcend to some extent their causal determinants, whether internal or external. Persons' ability to choose, though conditioned by their biological structure and the environmental circumstances in which they find themselves, enables them to act in ways that do not strictly follow from these causal factors.

We are not invoking absolute freedom; the total absence of causal conditions is a chimera. What those who hold to human freedom deny is that the internal and external causal conditions are completely determinative of human moral action. Our choices and actions lie at various points along a continuum between absolute freedom and total determinism and never at either end. Absolute freedom is impossible, for then we have

55. Wilson, *On Human Nature,* p. 178.

neither a context in which to act nor a finite agent with a particular biological heritage to perform the act. Absolute determinism is also impossible, for then there is neither choice nor agency; the person might behave or operate (as the planets behave or as an engine operates) but cannot act.

How can we reconcile these two perspectives — that of free agency, necessary for personhood and moral responsibility, and that of the respective scientific disciplines we have discussed? Several possibilities come to mind. The first is that we have wrongly posed the issues in the first place, that we have wrongly pitted freedom against causation. The second is to suggest that our problem is part of the larger, traditional mind-body problem, and that what is required are new ways of conceiving the problem. The third is to suggest that a new era of understanding is upon us, that neuroscience, genetics, and artificial intelligence will be able to contribute positively either to resolving the problem or to making it disappear as a pseudoproblem, similar to the way in which the discovery of DNA and its mechanisms put an end to the treatment of life as a property supervenient on the biological. We will consider the first two of these possibilities here; the third would call for a substantial treatment far beyond the limits of this volume.

Compatibilism

One possibility, attractive to many — especially in biology but not limited to it — is that our dilemma arises because we have wrongly posed the issues in the first place. We have pitted freedom against causation and so have failed to see that, properly understood, freedom and causal determinism are not incompatible after all. They really address different issues, so that to compare them or hold that they are contradictory is to relate apples to ping-pong balls. Determinism addresses questions about the causes, such as genes and environment, of biological or physical states. Freedom addresses questions about human choices stemming from our motivation and desires. Compatibilists hold that all events follow from their respective causal conditions (though not necessarily in a mechanistically determined way; descriptions of these events must be compatible with quantumly inspired probabilistic accounts of causation). We are free, however, when we are not coerced by external forces to choose or act contrary to the way in which we want to act. We are free when, in a given instance, we could have done other than we did if we had wanted to or tried.[56]

In moral contexts, we ascribe responsibility to those individuals who can be brought to act differently by, for example, reward or punishment. Where persons act in the same way no matter how the conditions are altered — for example, no matter what punishment is threatened or rewards offered — they are not morally responsible, for presumably such persons could not have done anything else even if they had wanted to or tried. What matters with respect to ascribing moral responsibility is not whether the decision made or action taken is causally determined but whether it accords with what the person wanted and whether they could have acted differently on the basis of those wants.

For example, suppose that in our above account Dan engaged in destructive behavior in the kitchen that Christmas morning despite various threats from his parents and despite his desire to be an obedient and properly behaved son. In such a case, we would not hold Dan morally accountable for his behavior, not because it was caused by his genetic predisposition to schizophrenia or because his brain was malfunctioning, but because he could not act differently even if he tried. On the other hand, we would hold an alcoholic morally responsible for his behavior if in a particular situation he could have refused the first drink, knowing that once he took it he could not stop. We say that he is free in this situation, regardless of his genetic predisposition to alcoholism. At the same time, we might not hold the alcoholic morally accountable for his first drink in a situation where the environmental circumstances (for example, social pressures) are such that he could not have avoided taking the drink even if he had wanted to.

There is much to be said for this view. In particular, it captures why we are concerned for the freedom issue. To be free is to be able to act in accord with our wishes and desires. Hence, it gets to the heart of our motivation and to the acts we perform out of that motivation. Choices that stem from our beliefs and desires and that accord with moral norms and principles — these are what morality is about, not genetic predispositions or environmental influences.

56. "We can accept the view that the full meaning of the human self as a moral agent is wholly structured by the interplay of genes and environment, and still maintain that the actions of the agent are not fully predictable from its genes and environment. What is structured is free to act in morally responsible ways." R. Cole-Turner, "The Genetics of Moral Agency," in *The Genetic Frontier,* ed. M. S. Frankel and A. Teich (Washington, DC: American Association for the Advancement of Science, 1994), pp. 161-74.

The problem with compatibilism, however, is that it fails to provide a sound basis for our freedom of choice. Compatibilists argue that we are free when we could have chosen to act otherwise if we had wanted to. That is, we are free when we can do what we want. But the problem has to do with our *wants*. If we are causally determined by our genetic makeup and our environment, we cannot *want* to act differently from the way we do. For someone to want to do otherwise, different causal conditions in the brain and/or environment must be present. But since, according to causal determinism, every behavior and internal state results from genetically and environmentally caused conditions, nothing can be different from the way it is (given the current causal conditions). Thus, being able to act or to want to act otherwise than we do is not in our power. Ultimately the compatibilist's causal determinism means that there is no genuine human freedom, for the very wants that it holds to be variable cannot vary within a causally deterministic system apart from variation of the causal conditions themselves.

Here again, however, it might be contended that the critic has confused language about causes with language about freedom. It is inappropriate to speak about biological causes of our wants. Genes and environment affect our biological structure; what we want is affected by our past actions and experiences and present beliefs. We should speak about our wants in the context of the latter concepts, not in terms of genes or brain cells.

Whether or not this reply succeeds depends upon how the biological relates to the psychological. Are they causally related or independent? Clearly, the compatibilist approach to resolving the problem of moral accountability points to an incommensurability between causal accounts about brains and agent accounts of psychological or mental states such as wanting, choosing, believing, and desiring. This in turn indicates that the freedom-necessity issue is symptomatic of a deeper problem, namely, the question of the relation of mind to brain, of mental states to brain states.

Mind and Brain

Moral language fits with language about minds or mental activity: choices, motivation, beliefs, desires, etc. Language about causes fits with language about the biological causes of our behavior: our body and its brain. Thus the question of moral accountability ties up with one of the most intractable philosophical problems: the relation between mind and body.

It would employ too much hubris to think that we can resolve this larger problem, let alone doing so in this limited space. In fact, it would be too much to think that anyone at present possesses a clear, indisputable resolution. What we can do, however, is to sketch some possible options, along with their strengths and weaknesses.

(1) Reductionism

One approach is to affirm some form of reductionism. We have already observed that reductionism understood as a methodological approach comports well with neuroscience. As a solution to the mind-body problem, however, neuroscience's methodological reductionism is transformed into an ontological reductionism, into the contention that mental language at best translates what occurs on the neurological or molecular level. Because it has provided an unprecedented understanding of the workings of the brain, some predict that neuroscience ultimately will resolve the problem of human consciousness, in the same way that molecular biology has resolved the problem of the nature of life by revealing its molecular basis.

Reductionism, however, threatens the very concept of the person. Where persons' actions and beliefs are ultimately explainable in terms of unpredictable neural firings and chemical transfers, those acts and beliefs are no longer the purposeful product of human choice.[57] On this view, one need not appeal to minds, concepts, and reasons in order to explain human behavior. We *can* do so because this language fits more easily into our teleological approach to things; we see things happening for a purpose. But ultimately, we can eliminate this type of explanation as we become more and more able to trace these higher pattern levels to the basic levels of neural firings.

This means that reductionism is particularly disastrous for morality, not to mention our concept of personhood itself. In a reductionistic system, moral norms can be nothing but conventions or, if the sociobiologist is correct, derivations of genetically conditioned emotions that have been naturally selected. They provide the rules by which society operates,

57. Gareth Jones, "The Human Brain and the Meaning of Humanness," in *The Reality of Christian Learning*, ed. Harold Heie and David L. Wolfe (Grand Rapids: William B. Eerdmans, 1987), p. 174.

placing restrictions on behaviors and designing societal responses to them. But there is no ground for preferring one set of norms to another. Preferences have causes, but the causes need not be reasons, let alone good reasons, for adopting that particular set.

Where norms are mere conventions, ethical relativism results. What each society or person thinks is right *is* right for them. Each society's norms may or may not apply to others, depending on the others' relationship to that society. Should anyone in a society not opt for those norms, the society can only compel them either to conform or to leave that society. It cannot attempt to persuade them by moral arguments, for there are none. Conformity must be compelled by force.

(2) Interactive Dualism

A second approach moves to the opposite end of the spectrum. Suggesting that it is illicit to reduce minds to brains and moral choices to neurological events, it proposes an interactive dualism between mind and brain. As brain language is referential, so is language about the mind. These constitute two ontological realities that somehow interface. Hence, any explanation of human moral choice must include not only the biological predispositions and the environmental factors but also the actions of a nonbiological agent or self that, with varying degrees of freedom, chooses between alternatives.

The neurologist John Eccles and the psychologist Daniel Robinson contend that if the materialist theory of the mind were correct, if consciousness had no causal powers but was merely a physiological by-product, we would be mere "passive spectators of the performances carried out by the neuronal mechanisms of the brain. Our beliefs that we can really make decisions and that we have some control over our actions would be nothing but illusions."[58] But to the contrary, we are true agents, selves with self-consciousness, beings with wills that respond to reasons as well as to causes.

Further, on the purely materialist schema there would be no evolutionary reason for us to have consciousness at all. "According to biological evolution, mental states and consciousness could have evolved and developed *only if they were causally effective* in bringing about changes in neural happenings in the brain with the consequent changes in behavior. That can occur only if the neural machinery of the brain is open to influences

58. Sir John Eccles and Daniel N. Robinson, *The Wonder of Being Human* (New York: Free Press, 1984), p. 33.

from the mental events of the world of conscious experiences, which is the basic postulate of dualist-interactionist theory."[59]

Persons are not the ensemble of their body parts operating under the hegemony of the brain, for even if the bodily parts are damaged, destroyed, or exchanged, the person's essential identity remains. Even most of the brain is not intimately related to our self-identity. "For example, removal of the cerebellum gravely incapacitates movement, but the person is not otherwise affected."[60] We are, Eccles concludes, "a combination of two things or entities: our brains on the one hand; and our conscious selves on the other."[61] Our uniqueness does not come from our genetic constitution; otherwise identical twins would be the same self. It is not dependent upon our accumulated experience and memories; no matter what their content, they belong to our unique soul, which existed before we had any memories. Eccles therefore concludes, "We are constrained to attribute the uniqueness of the psyche or soul to a supernatural spiritual creation . . . which is 'attached' to the growing fetus at some time between conception and birth."[62]

Following Karl Popper, he contends that we belong to several different worlds. World 1 is the matter-energy world that includes the brain. World 2 consists of our mental experience: our outer sense (of light, color, sounds, etc.), our inner sense (thoughts, feelings, memories, imaginings, etc.), and what is referred to as the psyche, self, soul, or will. Eccles also identifies World 3, which is the world of our cultural heritage; it includes our theoretical structures, language, and values. He argues that the development of human personhood comes about through interaction between Worlds 2 and 3, whereas the brain develops on a track indicated by its genetic structures in interaction with environmental input, all of which takes place in World 1.

Events in World 1 affect World 2, and vice versa, not by any physical interaction, since the mental is nonphysical, but through transfer of information. That is, there is a constant information flow between the brain and the mind. The brain is the liaison between the world of our mental

---

59. Eccles and Robinson, p. 37.

60. Eccles and Robinson, p. 29. They distinguish between the self (which arises from our awareness that we exist and have experiences), self-identity (the knowledge, arising primarily from memory, of who we are), and personal identity (which refers to how others perceive us) (p. 41).

61. Eccles and Robinson, p. 33.

62. Eccles and Robinson, p. 43.

experience and the world of our motor and bodily events. Thus, in voluntary movement, mental intentions initiate a sequence by transferring information to the brain. Eccles locates the point of interaction between mental intention and the body in the nerve cells of the supplementary motor area (SMA), located at the top of the brain. He notes experiments in which nerve cells in the SMA discharged before cells responsible for motor activity. From this he concludes that mental intentions act on cells in the SMA, which contain an "inventory of all learned motor programs."[63] This information, in turn, is transmitted to the cells responsible for initiating movement in the body.

Although it is an ancient tradition in Western and Indian thought, dualism faces severe difficulties. First, how does the mind and its intentions cause the neurons in the SMA to fire? The dilemma that interactive dualism faces is this: If the point of interaction is physiological (as the SMA is), then how does that physiological point relate to the mental intention? Or if the point of interaction is mental, then how does that mental point relate to the physiological? In either case we have an infinite regress. Eccles's solution is that what passes between them is not matter-energy but information. But must there not be something that conveys the information?

Second, though interactive dualism admits that the biophysical affects the mind to some extent, it fails to account for the significant degree to which it does so. World 2 is conditioned by World 1. Illustrations from recent genetic research reveal just the tip of the iceberg. For example, Robert Plomin and his group collected data from cognitive tests given to six hundred children ranging in age from six to twelve years.[64] Now they are typing the children for more than one hundred genetic markers that are thought to be related to brain function. Their goal is to identify genes that are common in children at the top of the IQ scale. A clear presupposition of this endeavor is that genes bear on individuals' mental capacities.

Similar intriguing results, which may lead to new insights into learning and memory, come from studies of mice.[65] Researchers grew mice from embryos in which a single gene was "knocked out" or inactivated. The missing gene normally codes for a kinase, an enzyme involved in

63. Eccles and Robinson, p. 161.
64. Peter Aldhous, "The Promise and Pitfalls of Molecular Genetics," *Science* 257 (10 July 1992): 164-65.
65. Marcia Barinaga, "Knockouts Shed Light on Learning," *Science* 257 (10 July 1992): 162-63.

intracellular signaling. Defects in this and several related kinases block long-term potentiation, a process thought to be important in spatial learning. In a further test, researchers found that "knockout" mice perform worse than their littermates in a water maze. The fact that several different kinases are involved in long-term potentiation (needed for memory and learning) suggests that there is a network of interacting regulatory mechanisms. These results will be tested further in mice and then presumably can be applied to human studies.

One can multiply examples that show that the brain and its genes condition the mind. Studies of twins reveal that the closer their genetic identity (identical versus fraternal), the closer are their IQs. The highest correlation (85 percent) exists between identical twins, 60 percent between fraternal twins, 40 percent between first-degree relatives, and 20 percent between adoptive parents and their children or between children adopted apart.[66] Other twin studies show correlations between features measured on personality tests. Between identical twins correlation is 50 percent, while between fraternal twins it is 30 percent, suggesting that heritability of personality is somewhere between 20 percent and 50 percent.[67]

Studies of persons who have experienced brain damage show how much this damage affects awareness, consciousness, memory, and conceptual abilities, all of which are features of Eccles's World 2.[68] For example, persons whose corpus callosum (connection between the brain hemispheres) has been severed, when presented on the left with an object that is sensed only by the right hemisphere of the brain, cannot tell what they saw, though they can point to it with their left hand. The right brain, which controls the left hand, knows the object, but the left brain, which is the main area controlling speech and communication, does not know of the object because no information passes through the severed corpus callosum.

Most intriguing is the research of Wilder Penfield conducted on patients suffering from epilepsy.[69] Penfield wanted to know the function of particular brain areas before he destroyed any part of the brain in his attempt to control patients' seizures. Accordingly, while patients were

66. Plomin et al., p. 185.

67. Plomin et al., p. 186. See also Thomas J. Bouchard, Jr., et al., "Sources of Human Psychological Differences: The Minnesota Study of Twins Reared Apart," *Science* 250 (1990): 223-28.

68. For two interesting cases, see Jones, pp. 164-67.

69. Wilder Penfield and Theodore Rasmussen, *The Cerebral Cortex of Man: A Clinical Study of Localization of Brain Function* (New York: Macmillan, 1950).

under a local anesthesic, he stimulated parts of their brain and inquired what they experienced. He discovered that stimulation of particular neural locations can cause particular perceptions or recollections. When he stimulated the rear of the brain, patients reported visual sensations, and when he stimulated the nonspeech parts of the temporal lobes, sometimes (7.7 percent of the time) the patients, in flashbacks, "relived" earlier experiences. One woman said "she was suddenly aware . . . of being in her kitchen listening to the voice of her little boy who was playing outside in the yard. She was aware of the neighborhood noises, such as passing motor cars, that might mean danger to him."[70]

The use of positron emission tomography (PET) scanning allows similar but noninvasive exploration of specific brain areas. Donald MacKay describes the PET process. "Oxygen containing a short-lived radioactive isotope is injected into the bloodstream of the subject. Within the following 2 minutes most of the radioactive atoms disintegrate, each emitting two positrons in opposite directions. A ring of coincidence counters around the subject's head allows a back-computation to be made of the position of each atom when it disintegrated."[71] By measuring the radioactivity in various areas, one can measure where oxygen brought by the blood to the brain increases. In this manner one can determine in general which areas of the brain are more involved in particular mental processes, such as perception, thinking, and memory. For example, visual activity correlates with activity in the occipital areas at the rear of the brain, auditory activity with areas in the side of the brain (Herschel's gyrae), cognitive activity with the frontal lobes, and memory with the hippocampus.

These data do not refute dualism, for a savvy dualist would not deny that there is a causal relation between the two. Eccles records this information, and Penfield came to hold a dualist position. The mind, Penfield suggested, may be "a distinct and *different essence*"; the brain may supply the mind both with energy and with its "basic neuronal mechanisms."[72] But in the absence of any reasonable account of how the mind and brain interrelate, these data certainly make it seem less likely that the dualist account is true.

---

70. Wilder Penfield, *The Mystery of the Mind* (Princeton, NJ: Princeton University Press, 1975), pp. 21-22.

71. Donald M. MacKay, *Behind the Eye* (Oxford: Basil Blackwell, 1991), pp. 119-20.

72. Penfield, *The Mystery of the Mind*, pp. 62, 72.

(3) Emergentism

A third approach tries to mediate between reductionism and dualism. Emergentism, like the theories we have considered above, comes in diverse forms. One of the more well-known advocates of emergentism is the philosopher John Searle. Against the reductionists Searle argues that first-person mental experiences ("I am in pain") cannot be reduced to claims about neural firings, for in doing so important first-person features like subjectivity are lost. Knowing that I am in pain is different from knowing that someone else is in pain. Consciousness, as manifested in particular subjective mental states, is as basic a feature as any other irreducible physical feature.[73] Against the dualists he argues that the strict dichotomy between mental and physical properties should be discarded. Mental properties are one kind of property that physical things can have. *"Pains and other mental phenomena just are features of the brain (and perhaps the rest of the central nervous system)."*[74] Consciousness is simply a higher order feature of the brain. Just as solidity and liquidity are not separable properties but are both causally explained by atomic elements and realized in the same physical system, so consciousness arises from the causal relation between fundamental physical things (here, neurons) and is realized in the same micro-system.

How then are we to understand agency? Searle argues that there are clear differences between human actions and other events in the world. First, actions involve more than human bodily movements. Though one can give many descriptions of the action, only one describes it as my action. That is, actions have a preferred description that invokes, in large measure, what I think I am doing. When I am walking down my street, one could say that I am getting closer to Chicago or that I am wearing out my thongs. But that would not describe *my action*, what I am doing. What I am doing is taking Rachel to the beach. Second, I know this not because I observe myself doing it but because I am in a special or privileged position to know why I am doing what I am doing, to know the principle that explains the action.[75]

Consequently, my actions consist of both mental and physical com-

---

73. John Searle, *Minds, Brain and Science* (Cambridge, MA: Harvard University Press, 1984), p. 25.
74. Searle, p. 19.
75. Searle, pp. 57-59.

ponents. The mental states from which I act have both content (what it is I believe, desire, or intend) and a psychological mode (belief, desire, intention). Such states are intentions, that is, are about something. As such, they both provide a mental representation of the event and bring it about. That is, because I want to take Rachel to the beach I am both aware of what I am doing and performing that action. Intentions make things happen.

But how do intentions do this? They must do this through that in which the mental states are realized — namely, the brain.[76] Searle denies that consciousness transcends the physical, that it has causal powers that cannot be explained by the causal interactions of the neurons. Consciousness has no life of its own apart from that in which it is realized. But because of this, Searle's emergentist view leaves no room for free moral agency, and the dilemma we have posed above remains. Searle himself describes the difficulty he cannot reconcile.

> The top-down causation [by mental events] works only because the mental events are grounded in the neuro-physiology to start with. So, corresponding to the description of the causal relations that go from the top to the bottom, there is another description of the same series of events where the causal relations bounce entirely along the bottom, that is, they are entirely a matter of neurons and neuron firings at synapses, etc. As long as we accept this conception of how nature works, then it doesn't seem that there is any scope for the freedom of the will because on this conception the mind can only affect nature in so far as it is a part of nature. But if so, then like the rest of nature, its features are determined at the basic micro-level of physics.[77]

Yet as Searle himself argues, we are incurably wedded to the fact that we are free human agents who can engage in voluntary, intentional acts.

Several different types of experiments suggest a necessary role for conscious agency. James Jones notes three of them. First, Jones points to biofeedback, which enables persons to "develop conscious control over . . . the function of their central nervous system by regulating their own brain waves; their peripheral nervous system by changing their heartbeat and respiration rates; the mediation of conscious experience by overruling their

---

76. Searle, p. 64.
77. Searle, p. 93.

pain centers; physiological function by . . . raising and lowering their levels of white blood cells."[78]

Second, Jones appeals to the experiments of Penfield noted above. When Penfield stimulated a particular region of the brain of a fully conscious person on an operating table, the person's arm moved and the person was aware of it moving, but the person reported that Penfield, not he, moved it. That is, the patients distinguished between having the arm moved when the proper neuronal region was electrically stimulated and moving it themselves. Yet Penfield was unable to locate any region in the brain where this experience of self-awareness and volition was located. Jones concludes that "the mind-brain system appears organized in a hierarchical way in which there is direct action (moving the arm) and awareness of the action (noticing my arm is moving) and also awareness of the awareness (this is not my act but yours). The first two are directly linked to neuronal activity; Penfield's findings suggest that the third may not be."[79]

Third, Jones points to experiments by H. H. Kornhuber and L. Deecke, who found that as much as a second prior to persons' randomly moving their finger, there was a gradual buildup of electrical activity in the brain. This "readiness potential" was associated with volitional activity but had no apparent neural cause.[80] The upshot of these experiments, as well as of our common or ordinary experience, is that an emergentism that has no role for the agency of consciousness or the mind cannot adequately account for significant experimental data or for our sense not only of performing a particular action but also of choosing to perform one action rather than another.

Other emergentist views attempt to accommodate moral agency. William Hasker advances the view that "the human mind or soul is produced by the human brain and is not a separate element added to the brain from the outside. . . . [At the same time], the mind is distinct from the brain and its activities are not completely explainable in terms of brain

---

78. James W. Jones, "Can Neuroscience Provide a Complete Account of Human Nature? A Reply to Roger Sperry," *Zygon* 27, no. 2 (June 1992): 190.

79. James W. Jones, p. 191. From his experiments with one patient, whose speech area he blocked, Penfield concludes that what the mind does in searching out other options cannot be accounted for by neuronal mechanisms (*The Mystery of the Mind*, pp. 51-54).

80. James W. Jones, p. 192. MacKay, however, argues that this experiment does not establish that the mind or agent has control over his brain (pp. 122-33).

function."[81] That is, on the one hand, the mind is brain-based; thus neurology properly informs us about the necessary conditions for mental activities. On the other hand, at the same time the mind "emerges out of the physical," in that it is not fully explained by the physical, allowing for human agency and freedom. The brain and the emergent mind interact and mutually influence each other.

How does the mind emerge, and what is the status of this emergent mind? Hasker provides no straightforward account but instead appeals to analogies to assist our understanding. He suggests that the relationship between the mind and the brain is like the relation between a field — electrical, magnetic, gravitational — and the object that generates that field. The magnetic field would not exist without the magnet that generates it; at the same time the two interact, the field affecting the magnet and vice versa. Consciousness is the field of the brain, emerging out of it and influencing it.

Hasker's view of consciousness differs not only from Searle's but also from the dualist's. He agrees with the dualist that consciousness has mental properties, like "feeling, choosing, and imagining." However, consciousness also has properties more akin to material objects, for example, "spatial location and extension."[82]

Hasker, unlike Searle, provides room for moral agency. The problem with his view is that his explanatory analogy is flawed. Whereas Hasker takes the magnet as fundamental and the field as secondary or caused, the reverse is true. What is primary in magnetism is the magnetic field, which is a manifestation of moving charges. In fact, one can have a magnetic field without there being a magnet. Magnets are the things that need to be explained; they are magnetic because they have atoms in which electric charges are moving around a nucleus. In short, magnets do not help us understand magnetic fields; rather, they must themselves be explained in terms of those fields. Since Hasker uses his field analogy to explain his emergentism, and since his analogy is flawed, we are no farther advanced in our understanding of the relation of mind to brain.

81. William Hasker, *Metaphysics* (Downers Grove, IL: InterVarsity Press, 1983), p. 73.

82. Hasker, p. 74.

(4) Complementarity and Bridge Principles

A fourth approach recommends a complementarist thesis. There are, it is asserted, two accounts or stories: the story of the person who speaks about his or her own experiences (I think, I hear, I judge, I believe) and the story about the physical activity in the central nervous system (about what goes on, for example, in certain groups of neurons). On the one hand, these two accounts should not be confused. Brains don't think; people do. Language appropriate to the one account should not be transferred to the other account. On the other hand, surely the two stories relate; they are not independent of each other. The complementarist holds that they are two equally acceptable accounts of reality, somehow interdependent.

For example, the neurologist Donald MacKay contends that we have two different explanations of a unitary phenomenon, one given in terms of brain events, the other in terms of the person's own experience. Neither can be dispensed with or reduced to the other. The accounts are neither identical (the one is not a translation or a renaming of the other) nor parallel, for "our conscious experience is *embodied in* our brain activity."[83] Nor does one explanation take primacy over the other.[84] "They are harmoniously complementary. The one is spelling out the *personal* significance of a unitary situation, another aspect of which is dealt with in the brain-story."[85] We have, in effect, a dualism of complementary descriptions of human agency.

MacKay's dualism differs from the interactive kind described above. Mental acts are "efficacious in determining, sometimes, the form of our action," though the field of their action is not our brains or neurophysiology. The mind does not interfere with or control the brain;[86] rather, it affects the brain by affecting the field both have in common, namely, human behavior and the external world. Similarly, the brain does not exercise causal activity on the mind, but on that same common field. In short, mental activity and brain activity do not interfere with each other;

83. MacKay, p. 9.
84. Though MacKay does not want to give primacy to either description, that of the mind or that of the brain, he does accord a kind of primacy to the mental, since we understand both accounts through the mind. At the same time, he also ascribes a kind of primacy to the physical, for it is "in and through" the brain that we conduct our mental activity (pp. 59-60, 133).
85. MacKay, p. 60.
86. MacKay, p. 61.

they are not functioning on the same level.[87] At the same time, however, though mental events do not causally interact with the brain, he sees the two as "interdependent."

MacKay's description of interdependency is inadequate, however. Simply because A and B are both efficaciously related to field C does not establish that A and B are interdependent. Think, for example, of a jogger in a park and a dog who makes use of the park's trees. They both produce effects on the grass, yet, as MacKay would suggest, there is no reason to think that the dog and the jogger are causally related to each other. Neither might directly, causally affect the other. But contrary to MacKay, the fact that they inhabit and use the same park does not show that they are interdependent. The dog does not need the jogger, nor does the jogger need the dog. Interdependency rests on something more than simply affecting the same field — namely, on some intrinsic or causal connection between what they both efficaciously produce.

To help us understand the relation between the two accounts, the brain-account and the "I-account," MacKay provides the analogy of complementary descriptions of a house plan. "In the case of the plan and elevation drawings of a house, for example, it would be silly to say that the plan drawing is 'nothing but' and identical with the elevation drawing, and yet they are both projections of one and the same house which has more dimensions . . . than plan or elevation alone."[88] Here the contention seems to be that the two accounts are related in that they are both descriptions of something that can be captured only in part by each alone. But not only does this fail to show that they are interdependent; it also raises the further question whether this "something" might be describable in another way or by another language.

It is evident that a major question facing MacKay concerns whether there are any bridge principles between these two accounts. While recognizing the difficulty of discerning such principles, MacKay believes that approaching both the mental and the neurological as information systems provides a good start in linking the respective categories, for both persons and brains, as goal-directed systems, run on information. "The language and ideas of information theory have the right kind of hybrid status to

87. MacKay's descriptions are not always clear. He speaks of searching for the "seat" (his quotation marks) or localization of brain activity in mental activity, of mental activity determining the form of brain activity but not controlling it (pp. 118, 132-33, 200).

88. MacKay, p. 63.

provide a vocabulary of (rigorous) concepts which latch on to the physiological concepts of the brain-story on the one hand and also to the characteristic concerns of the I-story on the other."[89] MacKay uses the example of a thermostat that has an effector system, a receptor system, an organizing system, and a comparator as an evaluative system. To this he adds a supervisory system that is necessary for more complex, goal-setting processes. Self-adapting and self-regulating information systems such as the thermostat are possible without being conscious; consciousness becomes present when the evaluation system becomes its own evaluator.

MacKay's account of human freedom is complex and controversial. He contends that human freedom, which belongs to the I-account, is compatible with the strictest determinism of brain activity (though he denies that brain activity is best presented in these terms). Briefly, his argument is that, although an outside observer might be able to provide a causal account of what action a person will take from an analysis of all present physiological and environmental states, it would be inappropriate for the person (the possessor of the I-account) to believe that account. The reason is that believing or not believing something has a physical correlate, so that whether the person believes the observer's report or not will correlate with an alteration of the extant physiological conditions. The correctness of the observer's report for that person is determined by whether the person believes it or not, not vice versa. It is not that the observer is wrong; the brain state of the person at that time is as the observer describes. But the observer cannot provide a *complete* description, for any description of the brain state only says what is the case at the present moment; it does not include whether the person believes the observer's report or not, for that is yet to occur. In short, since the observer cannot give a complete description, because the observer lacks the description of the brain correlate of the person's belief, there is no reason for the person to believe the observer's account, accurate though it may be. Hence, the person should not believe that he or she is determined in the choices to be made.[90]

But MacKay's analysis is, in a sense, trivial, for if the brain states (what he calls "cognitive mechanisms") operate deterministically (as sup-

89. MacKay, p. 41.
90. MacKay, pp. 134-36 and chap. 9. "Even if your brain were as deterministic as a clockwork model of the classical universe, there does not exist and cannot exist a complete account of its immediate future such that if only you knew it you would be correct to believe it and mistaken to disbelieve it" (p. 135).

posed for the argument), then the observer, in knowing the present brain states and other relevant data, can know whether the agent will believe the report, and that specification will be part of the future description of the brain state. Should the person not believe it, the observer could have predicted that from his knowledge of the cognitive mechanism and other relevant data, since either believing or not believing correlates with causally conditioned changes in the brain. Hence, MacKay's account amounts to the claim that future changes in the cognitive mechanism allow for changes in the person's mental state; but this is not an account of freedom, let alone one sufficient to provide for moral accountability.

MacKay seems to be providing a version of the compatibilist account of human freedom noted above. And the problem he faces is similar to that faced by the compatibilist. If the observer can know from the physiological causes what brain states will result, then in reality the person or agent does not have a free choice in the matter (though of course they might *think* that they do, from their perspective). If the person had chosen differently, this would have been reflected in the appropriate causal structure in the brain, since the choice correlates with a certain brain activity. The choice to believe or not believe, act or not act, is causally determined.[91]

MacKay's response to this objection is similar to the response we gave above to compatibilism, namely, that to apply causal categories to language about human choices is a category mistake. Persons could have done otherwise, not brains.[92] Hence, there are two accounts that are not to be confused: the causally deterministic account that applies to brain states, theoretically accessible to the observer; and the first-person account accessible to the person entertaining the I-story, the account that allows for no complete specification, that allows persons the freedom to determine future states.

Whether this suffices remains to be seen. That is, it remains to be shown that logical indeterminacy is any more relevant to human freedom than the physical indeterminacy described by quantum theory (MacKay denies that the latter is relevant), or whether it is an illusory freedom resulting merely from the ignorance that the person entertaining the I-story has about his or her brain states.

---

91. See C. Stephen Evans, *Preserving the Person* (Downers Grove, IL: InterVarsity Press, 1977), pp. 111-14.

92. MacKay, pp. 203-4.

What Direction Might One Go?

Our brief account cannot do complete justice to the rich complexity of the above approaches, but it does lay out the issues at stake and the questions that need to be addressed. Which position is most adequate is left to the reader to explore in more detail. But perhaps it is appropriate at least to suggest a direction to move in resolving the question. The cue, we suggest, comes from how we now view a concept that is intriguingly parallel to consciousness — namely, life.

At one time people believed that there must be in living organisms a separate principle responsible for their being alive. This principle, interestingly enough, was the soul, the same thing that many postulate to account for human agency and self-consciousness. With the discovery of DNA and understanding of its functions, vitalism has to all intents and purposes disappeared. To be alive is not to possess a life-making immaterial entity; rather, being alive results from our complex strands of DNA. The DNA encodes information biochemically; when communicated to proteins, this information sets in motion processes that are essential to something being alive. As we noted earlier, to explain life one has to appeal to both bottom-up explanations in terms of physics and chemistry and top-down explanations in terms of biological functions.

One might contend that something similar occurs with human consciousness. Consciousness results from our complex neural activity, likewise conditioned by our DNA. To explain consciousness we have to appeal to both bottom-up accounts in terms of electrochemical actions in the neurons and top-down explanations in terms of mental functions. And just as biological functions are not wholly different from the physical-chemical structures that underlie them, so the mental functions are not wholly different from the biophysical structures in which they are rooted.[93]

Conclusion

It may be thought that we have wandered a long distance to come to only a suggestion, not any definitive conclusion. In one sense this is true. We

93. John Searle, *The Rediscovery of the Mind* (Cambridge, MA: MIT Press, 1992), pp. 89-90.

have proposed no ground-breaking hypothesis that will make the philosophical and biological worlds stand breathless. On the other hand, an ancient proverb says that a journey of a thousand miles begins with one step. We have directed our one or two steps in this chapter toward showing why the problem of the relation of mind to brain and genetic structure is important, how it is a particularly trenchant problem for contemporary science, what the options are, what difficulties must be surmounted by those who adopt each position, and where one might find a parallel that suggests a direction in which to proceed.

We want to affirm several things. First, a full concept of human personhood must embody moral responsibility. Ethics is concerned with actions, that is, with behavior to which the person contributes. In this book we have explored the meaning of three motifs of a stewardship paradigm for ethics: ruling, filling, and caring for. Since these would have no meaning in the absence of meaningful choice, moral responsibility presupposes a denial of complete determinism. As we noted above, human actions generally are distanced somewhat from the extremes of absolute freedom and complete determinism, though to different degrees (as evidenced by our earlier examples of schizophrenia and alcoholism). Thus, in most cases it is reasonable to expect us to *rule over* our actions and be held morally accountable for failure to do so. Moral accountability rides along with agent involvement.

Second, human personhood cannot be reduced to a purely immaterial or spiritual notion, as sometimes occurs when people identify themselves with souls. We are essentially and fundamentally embodied beings, with a genetic heritage from our parents, their parents, and so on. This genetic heritage affects all parts of us, including our brain, and thereby plays a significant role in determining who we are, the choices and decisions we make, and the actions we take. Our consciousness and mental functions are firmly grounded in our neurology, just as our being alive is grounded in our biology.

Third, we must be aware of the strong influences that biology and the environment, relating dynamically, have over our actions. Both the brain, structured by the genes, and the environment play significant roles in our decision making, in dimensions and ways stronger than perhaps we admitted in the past. Humans must be understood, in large measure but not completely, as selves molded through their biology and environment. These provide the base upon which we build, through our choices, both moral and amoral. Hence, biological (including genetic) and environmental factors must be taken into greater account when in particular cases we consider degrees of moral and legal responsibility for actions performed. In *caring for* others, we must go to pains to address not only the behavioral symptoms but also the root causal conditions of their behavior. Insofar as these conditions manifest biological dysfunction, mere moral blame often will be a less adequate approach to getting persons to alter their behavior than will therapies that couple treatment of the underlying biological condition with moral and motivational components. A new realization of the biological rootedness of behavior will make us more understanding of and sympathetic to the person's condition and more willing to find multidimensional ways of addressing the underlying problems, without at the same time completely bypassing human agency and moral responsibility.

Fourth, though personhood involves the biology and the environment, persons should not be reduced to these in ways that dehumanize and remove the real contribution of human agents. We must guard against any schema, including therapy, that treats persons or clients as less than human. A schema can do this by claiming an unwarranted role for genes, the brain, or the environment. That is, though we recognize their essential contributions and hence can conceive of them as mitigating factors in evaluating human actions, in many cases they cannot be taken as an excuse.

In all our scientific endeavors, we must not ignore the fact that those whom we study or treat are human beings who must be viewed holistically, as embodied persons. Neither can we overlook the fact that those who do the study, interpret the results, implement the conclusions in patient treatment, and report them to others are, first and foremost, human beings who themselves contribute to their findings through their personal knowledge and choices.

CHAPTER TEN

Stewardship and ———— *Human Sexuality* ————

C LAIMS that the size of the human skull and its component brain correlates with individual human behavioral abilities have a long and checkered history. In the nineteenth century two German physicians, Franz Joseph Gall and Johann Spurzheim, advanced phrenology as a strict science. They believed that by measuring the circumference of the skull and noting the dimensions and location of its bumps, a person's intellectual potential and diverse abilities and tendencies could be determined. In general, the larger the skull, the greater the intellectual potential.[1] Phrenological charts or maps were composed, showing the precise brain locations of affective propensities or sentiments and intellectual abilities. The ability to perceive individuality, for example, was thought to lie between the eyes; behind the ear were the propensities to be combative and destructive. Though the chart included areas for acquisitiveness, self-esteem, hope, wit, perception of size, and causality, missing was any location accorded to human sexuality.

The brain is exceedingly more complex and integrative than the phrenologists thought. Human personality traits are not located in one part of the brain but result in some measure from the interaction of various

---

1. Identification of brain size with ability continued into the last part of this century. For example, as late as 1962 the *Encyclopaedia Britannica* maintained that "averages calculated from groups . . . show statistically that there is a small, though measurable correlation between large size of head and a high intelligence. The relation becomes still more apparent when the heads of the intellectual classes are compared with those of the lower classes." *Encyclopaedia Britannica*, 1962 ed., s.v. "brain."

parts of the brain. Yet the phrenologists left a legacy to contemporary neurology. Neurologists continue to attempt to locate primary brain areas that govern certain mental and physical processes and to measure the brain and its segments (as opposed to the entire skull).

Contemporary neurologists have expanded interest in brain size to wonder whether there are significant structural differences between male and female brains. The average male brain at maturity is larger than the average female brain by five ounces, though this absolute difference in size is tempered by the fact that the proportion of brain to body weight and size is approximately the same in both sexes. The size differences are not present in humans initially; up to age three, brains of males do not differ from those of females. The major change occurs between the ages of three and six, when, some speculate, male sex hormones become active and stimulate brain growth.[2]

But although there is an overall difference in brain size, do the differences between the sexes extend to local areas of the brain? Some researchers point to findings that there are structural differences between male and female brains in animals and in humans. "In the human brain, such differences have now been observed in three major structures — the hypothalamus, the anterior commissure, and the corpus callosum — as well as in lesser areas."[3] Others see the evidence as more ambiguous.[4] Nevertheless, even if one acknowledges the presence of local differences, the significance, if any, of these differences in size remains in dispute.

Whatever the outcome, researchers have concluded that the brain is much more than an intellectual organ. Indeed, some wish to call it a sex organ, if not "our most significant sexual organ."[5] It plays significant roles in determining gender differences and sexual identity, in stimulating hormone production, in sexual arousal, and perhaps in sexual orientation.

Simon LeVay, a brain scientist at the Salk Institute in San Diego, was intrigued with studies done at UCLA by a team headed by Roger Gorski, which indicated that regions (called nuclei) of the anterior hypothalamus of men were twice as large as women's. He wondered whether

2. Ann Gibbons, "The Brain as 'Sexual Organ'," *Science* 253 (30 Aug. 1991): 958.
3. Gibbons, p. 958.
4. Doreen Kimura, "Sex Differences in the Brain," *Scientific American* 267, no. 3 (Sept. 1992): 123-24.
5. David Myers, *Exploring Psychology* (New York: Worth Publishers, 1990), p. 262.

a similar difference might hold between the hypothalami of heterosexuals and of homosexuals.[6] The area of the brain he chose to examine was the medial zone of the anterior hypothalamus, which is known to play a role in sexual activity. In previous studies, monkeys that suffered lesions in this area had impaired sexual drives.[7] Other studies with rats showed that the size of the nuclei in the anterior hypothalamus depended on testosterone levels. Rat pups were castrated at birth, lowering their levels of testosterone production. This in turn reduced the size of the sexually dimorphic nucleus (SDN) in the anterior hypothalamus. The researchers correlated this reduction with a decrease in male-type sexual behavior, such as mounting females. On the other hand, when testosterone was injected into female pups, the SDN was enlarged and the females demonstrated sexual behavior more characteristic of males.[8]

LeVay studied brain tissues from forty-one cadavers, nineteen of whom were homosexual men who died from complications of AIDS. Sixteen men were assumed to be heterosexual, six of whom died from AIDS. Six others were assumed to be heterosexual women. The absence of lesbians in the study meant that no conclusions could be drawn about the respective hypothalami of heterosexual and homosexual women. In order to rule out changes in the size of the nuclei due to age, the brain tissues studied came from populations with approximately the same age. The mean age of the homosexual men was thirty-eight; that of the heterosexual men was forty-three.

LeVay found that, although three of the nuclei studied showed no differences in size between the two groups of men, one nucleus (called INAH 3) showed marked differences. The volume of this nucleus in the heterosexual men was twice that of the homosexual men.

But could the difference in volume be the result of AIDS? LeVay compared the volume of INAH 3 in the heterosexual males who had died of AIDS with those who had died from other causes and found no difference. On the other hand, the volume of INAH 3 in homosexual males was more like that of the women studied. He concluded that the difference in volume had nothing to do with AIDS or its complications.

6. Simon LeVay, "A Difference in Hypothalamic Structure Between Heterosexual and Homosexual Men," *Science* 253 (30 Aug. 1991): 1034-37.

7. J. C. Slimp et al., "Heterosexual, Autosexual and Social Behavior of Adult Male Rhesus Monkeys with Medical Preoptic-anterior Hypothalamic Lesions," *Brain Research* 142, no. 1 (Feb. 1978): 105-22.

8. Marcia Barinaga, "Is Homosexuality Biological?" *Science* 253 (30 Aug. 1991): 957.

LeVay's conclusion, put most tentatively, was that sexual orientation is possibly influenced by certain biological structures in the brain.[9] Although he could not rule out that the structural differences he discerned were the result and not the cause of sexual practices of certain kinds, the results of the studies on rats indicated that the production of androgens influenced the size of the nuclei in the hypothalamus, which in turn resulted in behavioral differences.

A number of questions about the study arise. (1) How good was the sample? Were all the brains labeled heterosexual really from heterosexuals? Or could the subjects have been closet homosexuals or bisexuals? No such determination could be made from the death records. (2) Further, in what sense were they homosexual? Were they homosexual in orientation only, did they have only periodic encounters, or did they frequently practice homosexuality?[10] (3) Perhaps what matters is not the volume of the INAH 3 nucleus but its density or number of cells. Because the area measured, which is smaller than a snowflake, is difficult to delineate precisely, LeVay rejected the approach of determining cell density. Yet cell density might provide a more accurate way of determining the significance of this brain area. (4) Could the features noted be correlative with different kinds of sexual activity but not be a causal factor, just as the trees growing leaves in the spring is correlative with birds laying eggs but not a cause of it? If the relation were merely correlative, it would not lead to the conclusion that homosexual behavior is rooted somehow in biological structures but only that certain features correlate with homosexual behavior.[11]

The issue raised by LeVay is critical to an understanding of human sexuality: What is the role of the biological in influencing our sexuality? The question must be posed on two levels. First, to what degree, if any, does genetic heritage influence sexual *orientation*? (By "orientation" we mean the basic

9. Reporting on a correlative study by Dean Hamer, LeVay and Hamer conclude, "The most straightforward interpretation of the finding is that chromosomal region Xq28 contains a gene that influences male sexual orientation. The study provides the strongest evidence to date that human sexuality is influenced by heredity." Simon LeVay and Dean H. Hamer, "Evidence for a Biological Influence in Male Homosexuality," *Scientific American* 270, no. 5 (May 1994): 49.

10. Anne Fausto-Sterling, "Why Do We Know So Little About Human Sex?" *Discover* 13, no. 6 (June 1992): 28. Robert T. Michael et al., *Sex in America* (Boston: Little, Brown, 1994), p. 171.

11. William Byne, "The Biological Evidence Challenged," *Scientific American* 270, no. 5 (May 1994): 55.

drives, instincts, desires, and dispositions that a person has.) Second, to what degree, if any, does one's genetic heritage influence particular sexual *behaviors?* Is the genetic factor in any way predominant? Or does the environment so condition sexual behavior that, though the biological provides the necessary conditions for sexuality, the primary input informing how we act comes from the nurture we receive and the environment in which we live?

This issue might be of merely biological interest were it not for the fact that many of our moral judgments focus on human sexuality. In condoning certain sexual behaviors and condemning others, we maintain that sexual practices or orientations are in part, if not wholly, controllable by the individual's will. There are moral norms to be followed; if we deviate from them, we are subject to moral judgment. And should that deviancy harm others, some form of active intervention by others to change our behavior is appropriate. Thus, in dealing with both sexual orientation and behavior, we are concerned not merely with human biology and the environment but also with what individuals by their choices contribute.

To be morally accountable requires that we are able to act otherwise than we do, that given the biological and environmental conditions present, we still can control our own behavior. If we cannot control our behavior by our choices, if our sexual orientation and/or behavior results *merely* from our biology and/or environment, then it becomes inappropriate to raise moral concerns about persons' character and behavior, just as we cannot raise moral questions concerning whether a person is male or female or tall or short. Condoning or condemning would be as irrelevant to human sexual orientation and behavior as they would be to a person's height. Such a conclusion would be unwelcome for any attempt to develop an ethic of sexuality, let alone a Christian one. If we are to consider the ethics of sexuality, we must consider the critical question of the extent to which our sexual orientation and behavior are affected by our choices. To prepare for this discussion, we must first attend to the biology of sexuality.

The Biological Heritage of Sex

Gender Development

Biology plays an extremely important role in human sexuality. Sexual development is a gradual process that begins in the fetus and continues well into postnatal life. Prior to the fifth week the sex of the human fetus is phenotypically indeterminate. Between the fifth and seventh weeks, the fetus develops two gonads that eventually become either ovaries or testes. If the fetus's twenty-third pair of chromosomes contains a Y chromosome, by the eighth week the gonads begin to develop into testes. If the twenty-third pair of chromosomes are both Xs, at the thirteenth week the gonads begin to develop into ovaries.

In the male the testes produce testosterone, which controls the development of other features of male sexuality (the penis and scrotum) and stimulates the Wolffian ducts in the fetus to develop into the epididymis, the vas deferens, and the vesicles that will carry the sperm from the testes. The testes also produce a testicular protein that inhibits the other pair of ducts in the embryo, the Müllerian ducts, from developing. In the female the absence of testes-produced testosterone allows the Müllerian ducts to develop into the fallopian tubes, uterus, and inner part of the vagina, while the Wolffian ducts atrophy.

The genetic and chromosomal story is not simple, however. Although the X and Y chromosomes contain the primary sex genes, they also control other factors of human development (including the ability to recognize red and green). Further, while the Y chromosome controls the development of the testes, genes on other chromosomes control the development of other sex-related characteristics and the ability of the body to use testosterone. To put it simply, our sexual anatomy is significantly influenced by a variety of genes and the chemical-biological systems through which they work.

Biological Errors

Since human sexual heritage and development have a substantial biological component, biological errors can cause abnormalities in sexual development. Normally persons have forty-six chromosomes, divided into twenty-three pairs. Some persons, however, have more than the normal complement of forty-six chromosomes. Persons affected by Klinefelter syndrome have an extra X chromosome, making them 47,XXY. Either an XX egg was fertilized by a Y sperm, or an X egg was fertilized by an XY sperm. Although such persons have male genitalia, they can manifest male secondary sexual traits (facial hair) that are poorly developed and female secondary characteristics (enlarged breasts). Generally they are sterile and have a somewhat higher frequency of mental retardation. Other persons are 47,XYY. They are generally taller, with a range of IQs from 80 to 118. As we noted in Chapter Seven, earlier reports that such persons were "supermales," prone to aggression and violence, have now been discounted. Other, more rare combinations of chromosomes — 47,XXX; 48,XXXY; 49,XYYYY; etc. — are possible; persons with such chromosomal structures may suffer from mental retardation and sexual abnormalities.

Some other persons do not receive the complete complement of sexual chromosomes. Persons with Turner syndrome, which affects about one in every 2,500 liveborn females, have only one X chromosome (45,X). They have female genitalia, but since the ovaries are absent or poorly developed they are sterile, and secondary sexual characteristics such as breasts and pubic hair are underdeveloped. Most fetuses with this disorder are not viable and die before birth.

But as we noted, the X and Y chromosomes are not the only chromosomes responsible for sexual development. Even when the normal complement of chromosomes exists, defective genes on other chromosomes may produce abnormalities.

In one type of anomaly, the defect lies in the androgen (sex hormone) receptor. The Y chromosome is present, so that the gonads develop into testes. The testes produce both testosterone and the protein that blocks the development of the internal female ducts and uterus. At the same time, however, a defective gene blocks the body cells' ability to respond to the testosterone produced by the testes, so that the external male sexual organs fail to develop. The affected persons thus are genetically males in that they have the Y chromosome and the resulting (undescended) testes, but at the same time they manifest the external bodily characteristics of a female. Since they look like females, they are raised as females; in fact, it is generally only when they do not menstruate or are unable to conceive that the genetic condition is discovered. They can function well in life as females, except that they have a predisposition to testicular cancer and cannot reproduce.

Another case presents more of a dilemma. Persons with a 5-alpha-reductase (5AR) deficiency have a normal Y chromosome that produces testes, which in turn produce testosterone and the protein that suppresses the development of the Müllerian ducts into the uterus and fallopian tubes. In normal males the 5AR enzyme converts some testosterone into DHT (dihydrotestosterone). In persons with a 5AR deficiency, the gene governing this enzyme is defective, so that testosterone is not converted into

DHT. Since DHT controls the early development of the external male genitals and the prostate gland, someone born with 5AR deficiency has the external genitals of a female in early life. Nevertheless, since testosterone is produced and in part processed, at puberty male traits, such as the external male sex organs and body muscle build, begin to develop. Thus, those afflicted by the 5AR deficiency have a mixture of female characteristics — a vagina and diminished bodily hair growth — and male characteristics — a penis and no breasts.

This condition can be psychologically devastating to persons who begin to develop male traits after they have been raised as females. "Most known examples are clusters of related cases in remote, isolated Third World villages, where marriages between close relatives are common and children can inherit many of the same genes from both parents. . . . In a rural, inbred village . . . physicians identified a total of 38 pseudohermaphrodites, all descended on at least one side from the same now-deceased woman and many of them traceable to that woman through both their mother and father."[12]

Finally, in a few very rare cases some persons can have the normal sex chromosomes and yet turn out to be true hermaphrodites in that they possess both ovaries and testes, though usually in a rudimentary state. Though they are generally sterile, there are at least four recorded cases of pregnancy and two births from such persons.[13]

Diverse Ways of Understanding Gender Identity

It is obvious that biology plays a formative role in determining an individual's sexuality. Our sexual organs and the production of the sex hormones result from our genetic structure. Fortunately, in the majority of live births the number of chromosomes is normal and the genes function properly to produce a person of a distinct gender.

However, the presence of sexual abnormalities suggests that there are diverse ways of classifying persons sexually. We might do this *genetically:* persons with a Y chromosome are males; those without it are females. In most cases this is adequate; yet as we have seen, in rare instances persons

12. Jared Diamond, "Turning a Man," *Discover* 13, no. 6 (June 1992): 74.
13. Arthur P. Mange and Elaine J. Mange, *Genetics: Human Aspects,* 2nd ed. (Sunderland, MA: Sinauer Associates, 1990), p. 94.

can be genetically males (in that they have a Y chromosome) and yet genitally females. This suggests a second method of classification, namely, according to a person's *genitalia*. Or again, some persons with a Y chromosome, because of a genetic abnormality, possess sex-related female body traits (absence of facial hair, size of pelvis, muscle and breast development). Here a sexual classification could be *somatic*, given in terms of sets of bodily traits. This diversity of possible classifications raises the question of which criterion or combination of criteria best serves to categorize humans sexually.

One might think that this categorization ambiguity is insignificant; yet it achieves particular importance when, for example, being a female is a necessary condition for participation in a particular activity. Participants in athletic events frequently are divided according to gender. Normally the division is not problematic. Yet should a person suffer from blocked androgen receptors, as was found to be the case in 1985 with Spain's best hurdler, Maria Jose Martinez Patino,[14] how should that person's sex be determined? According to the person's genitalia or body type, the person is a female, since the person cannot respond to male testosterone; tested chromosomally, however, the person is male. In 1985 it was decided that Patino could not participate in the female track events at the World University Games in Japan. After appealing her case to the International Olympic Committee (IOC), three years later she was allowed to compete. But as late as the 1992 Olympic Games in Barcelona, the debate continued as to which criterion — genetic or genital — the IOC should use to determine whether a person is eligible to compete in women's events. Rejecting genital examination because it invades privacy and is traumatic to women, yet keeping the above type of exceptional case in mind, the IOC decided to use a more reliable test "based on amplification of the *Sry* gene, . . . the absence of [which] seems to be a key element in determining the female sex."[15]

Thus, the data suggest that human sexual physiology cannot always be divided cleanly into the distinct categories of male and female. Sexual identity lies on a continuum shaped much like a barbell, with the majority of persons located at one end or the other of the gender spectrum and a few scattered in between. A very small number of persons are her-

14. Denise Grady, "Sex Test of Champions," *Discover* 13, no. 6 (June 1992): 81.
15. B. Dingeon et al., "Sex Testing at the Olympics," *Nature* 358 (6 Aug. 1992): 447.

maphrodites (having both male and female organs), and some are pseudo-hermaphrodites (having female and sometimes male external genitalia and internal male sexual organs but being incapable of reproducing as a female). In such cases the biological determination of sexuality allows for ambiguity.

Since there can be biological ambiguity regarding how to classify persons who are genetically males and yet somatically females, in some cases a more appropriate categorization of sexuality might be *psychosocial.* A psychosocial understanding of sexuality has to do with how one understands one's own body, identity, and role in society. For example, in the case noted above of persons with a 5AR deficiency, who are genetically males but look like females early in life and at puberty begin to develop external male sex organs, the question thus arises how these persons will understand themselves and how others in their society will understand them sexually as they mature. Individuals who are transsexuals likewise might fall under this category.

In short, being a male or a female is a social and psychological phenomenon as well as a genetic or somatic one. Fortunately the psychosocial and the genetic normally coincide. People who are genetically and biologically females normally function psychosocially as females; similarly with males. But the fact that there are genetic and biological abnormalities shows that the biological categorization cannot be taken as completely determinative.

The role of the psychosocial becomes significant in understanding persons who have difficulty with their sexual identity. Where the genetic conflicts with the somatic (including the genital), the production of certain hormones will be a factor in how persons feel about themselves. In some cases surgery on the genitals and on secondary sexual characteristics might be necessary to bring the somatic into line with the psychosocial. Invoking our stewardship ethic, caring for such persons means helping them to bring into congruence the totality of sex-determining factors, so that they have the best unity of sexual traits possible. Doing this might require a combination of surgery, therapy, hormone treatments, and counseling.

The upshot of this is that, although our biology is generally the most important factor in determining our sexual identity, sexuality has a psychosocial dimension as well. This means that who we are and how we act sexually depends on a variety of factors, including our genetic heritage, our hormonal components and their action on the brain, how others react socially to us, how physicians surgically decide ambiguous cases, and how we understand ourselves and act from that understanding.

The Ethic of Stewardship:
Filling and the Purposes of Sexuality

To this point we have considered sexuality largely in terms of gender. Before we turn to the questions of orientation and behavior, we need to address the question of the purposes of sexuality. The consideration of purpose will bring us back to our stewardship paradigm and its locus in Scripture.

It hardly needs to be pointed out that sexuality has *reproduction* as its primary function. The male and the female are physically structured to reproduce. Both contribute in their unique way to the continuation of the species. Sperm joins with egg; the half-complement of chromosomes carried by each merge. Biologically, humans are sexual *complements*.

The relations among biological complementarity, gender pairing, and reproduction are reflected in the Genesis account from which we have derived our ethical paradigm: "male and female [God] created them. God blessed them and said to them, 'Be fruitful and increase in number; fill the earth and subdue it'" (Gen. 1:27-28). The divine command to fill sanctions sexuality in its primary, reproductive role.

But it would be a mistake to restrict sexuality to the reproductive role, for sexuality serves functions other than biological ones. This, too, is consonant with our stewardship paradigm. We have contended that the command to fill is to be treated both quantitatively and qualitatively. The reproductive role fulfills the quantitative dimension of the command to fill. What can be said about its qualitative aspect?

At the end of the second creation story, the writer of Genesis not only recounts the creation of the female from the male but also specifies the reason for the female's creation and notes the resulting relationship between them. The man is alone, and this is found unsatisfactory. So out of the man God creates the woman, for whom the man leaves all. "For this reason a man will leave his father and mother and be united to his wife, and they will become one flesh" (Gen. 2:24; repeated by Jesus in Matt. 19:5). The ideal of personal fulfillment here is found in the unity of man and woman. Thus the qualitative dimension is asserted: sexual union brings a *bonding* of persons, a union of being. What once were two become one.

The biblical understanding of this unity emphasizes several critical features. First, the union is *exclusive*. In the Genesis paradigm the man joins or cleaves to the woman, excluding all others, including his birth family, in order to enter into union with his wife. In marriage the couple engages in exclusive sexual bonding. Second, the union involves *commitment*. Jesus focuses on this feature when he quotes this passage from Genesis in his strong stand against divorce. He allows divorce only in those cases when the cleaving has been severed by one marriage partner entering into sexual relations with someone outside that marriage bond. Where there is adultery, the commitment is broken and the union can be formally dissolved, for without commitment it has in actuality already been dissolved.

The procreative and bonding functions of sexuality have traditionally played a central role in Christian sexual ethics. What is often overlooked, however, is another qualitative dimension. Ideally, sexual relations also bring *psychological and physical fulfillment* to the partners. Sexual intimacy can bring mental pleasure, as the participants enjoy the close presence of the other. And surely it can bring varying degrees of physical pleasure. Indeed, both of these aspects help account for the frequency with which humans engage in sex.

Some have attributed the failure to include the pleasure aspect in a healthy attitude toward sex to a long tradition of Christian asceticism. There are hints of this in Paul's first letter to the Corinthians (7:1-9); it is developed by Justin Martyr in the second century, and it is well entrenched by the time of Augustine. Saint Jerome, for example, writes that "all sex is impure." Elsewhere he writes, "I praise the marriage bond but I do so because it produces virgins for me. I gather roses from the thorns."[16] Augustine also sees procreation as redeeming sexual intercourse within marriage; though marriage is proper, refraining by mutual consent from intercourse makes marriage better.[17]

The inclusion of the Song of Songs in the Christian canon gives adequate evidence against a Christian asceticism that diminishes the goodness and pleasures of sex. In the Song the delights of sex are abundantly and unashamedly evident.

> Beloved: Let him kiss me with the kisses of his mouth —
> for your love is more delightful than wine. . . .
> Take me away with you — let us hurry!
> The king has brought me into his chambers. . . .

16. St. Jerome, *Against Jovinian* 1.20; *Epistle* 22, 19.4. Quoted in Raymond J. Lawrence, *The Poisoning of Eros* (New York: Augustine Moore Press, 1989), p. 123.

17. Augustine, *Treatises on Marriage and Other Subjects* (New York: Fathers of the Church, 1955), quoted in Lawrence, p. 127.

> Lover: How beautiful you are, my darling!
> Oh, how beautiful!
> Your eyes are doves.
>
> Beloved: How handsome you are, my lover!
> Oh, how charming!
> And our bed is verdant.
>
> (1:2, 4, 15-16)

In giving pleasure to each other, the couple fulfills biologically endowed sexual yearnings and desires. Mutually and legitimately they satisfy each other's needs. However, sexual relations that are directed only to filling one's own personal sexual needs or desires, that treat the partner as a thing to be used for one's own benefit, lack the requisite mutuality and commitment. The pleasure is stolen at someone else's expense.

In short, the stewardship command to fill gives us the obligation to fill the earth (to reproduce) and to fulfill others as well as ourselves. Sexual relations within a context of biological procreation, psychological bonding where exclusivity and commitment are present, and the mutual, free sharing of psychophysical pleasure that seeks the other's good provide *a* legitimate way to fulfill that obligation.[18]

Population Control

This discussion provokes two questions regarding the command to fill. The first of these is this: If we are commanded to fill the earth, can humans limit that command by controlling population?

The recent dramatic increase in human population is indeed a serious concern. As late as the early seventeenth century, the world population was less than 500 million (less than twice the current population of the United States). In the next three centuries it had doubled twice to more than two billion, and by 1950 it had increased to two and one-half billion. After World War II the population rate skyrocketed, so that the world

---

18. It goes without saying that sexual relations are neither necessary nor sufficient for mutual fulfillment. We can meet others' needs quite apart from sexuality, as in friendship and parenting relations. Furthermore, Christians maintain that humans cannot be truly fulfilled apart from the redemptive, reconciling act of God's grace in Christ and the relational possibilities with God that this creates.

population took not a century but only forty years to double to 5 billion. It is projected to increase by more than another billion by the year 2000, eventually to reach 8.6 billion by 2025. Even should the rate of increase decline, as it has in the last decade, the world population will still increase, for population increase is exponential, building upon an ever-increasing base.

Not only are the raw population figures disturbing, but so is the very context of the increase. "Of the projected increase of some 3.2 billion [from now to 2025], . . . 95% will be in the less developed countries."[19] Whereas the annual natural growth rate in developed countries is 1.9 percent, in developing countries (excluding China) it is 4.4 percent. In developed countries the fertility rate is 2.7; in developing countries it is 6. This means that, unless there is a dramatic change in the world economy, the majority of those who are born will increasingly find survival a struggle. For example, they will confront an increased demand for diminishing resources. While the population increase builds on an already increased base, the farmer starts each year with an empty field. Thus the quantity of food produced provides a check to population growth. They also will face a labor market with an imbalance between increased production due to labor-saving technologies and the diminished need for labor.

It would be a mistake to appeal to the biblical command to fill to justify continuance of uncontrolled population growth or failure to take steps to control it. This follows from our function as stewards. As stewards we are to care for humans. To continue to bring children into the world for whose health and general well-being we cannot adequately provide is to abandon our role as conscientious stewards.

We also have stewardly obligations to the environment. Each person exacts a toll on the environment, whose resources are needed for food, clothing, housing, and employment. We have already seen how population pressures have contributed to deforestation because of the basic need for firewood and to erosion because the need for additional farmland caused people to move into ecologically fragile areas. Population pressures have had similar consequences for air and water, as people in less developed countries are forced to move into crowded urban areas that lack basic sanitation but increasingly pollute through the combustion of fossil fuels. Urban populations will have increased sixfold between 1950 and 2020,

19. Nathan Keyfitz, "The Growing Human Population," *Scientific American* (Sept. 1989): 119.

from "only 17 percent of the population of the less developed countries, . . . to well over 50 percent."[20]

If we deny an egalitarianism between humans and the environment, then we have a prima facie responsibility to provide for the basic needs of the humans we have procreated. Clearly there comes a point at which the actions taken to meet those needs impinge so severely on the environment that viable ecosystems and the diversity of organisms are jeopardized. To locate that point — the number of people the earth can support while preserving both human and ecological quality — is difficult, for it depends upon not only population figures but also how the natural resources are parceled out and the location of the populations. In some areas the population pressures on the environment have increased to such an extent that not only is population control needed but also carefully thought out plans for population reduction are called for; such programs, however, must be carried out with moral justice, sensitivity to economic and social needs, and concerns for the integrity of human beings.

These demands are not easy to fulfill. For one thing, population control programs must be sensitive to the need for social security in old age. Family planning must be coupled with programs for economic development, so that people can be assured that they do not need to have additional children in order to make a decent living and that they will be provided for later in life and not have to rely on the earnings of their (many) children. Population programs must also pay attention to gender roles in the society, though it is difficult to change gender roles in many cultures. Improvement in the lives of women has led to declines in fertility, as women begin to have a say in their education, their occupational opportunities and careers, and their procreative activities. Finally, programs addressing procreative issues (e.g., dissemination of birth control information and contraceptives) must be sensitive to the wishes of the individuals;[21] compulsory tactics have led to serious abuses, including female infanticide, in countries such as India and China.

There are reasonable limits to human population growth and density, limits having to do with the ability to produce enough food, the environment, and the psychological, physical, and spiritual health of the inhabi-

20. Keyfitz, p. 121.
21. Tim Stafford, "Are People the Problem?" *Christianity Today* (3 Oct. 1994): 54-55.

tants.[22] The command to fill is not the command to overfill or to destroy the quality of life of the earth's inhabitants.

Singleness

The second question raised by our discussion of the purposes of sexuality is this: If marriage is the ideal, is there room in a Christian ethic for the single person, whether he or she is unmarried, divorced, or widowed? Does the Genesis ideal of marriage make single persons into second-class persons?

The biblical emphasis upon the norm of marriage clearly resounds in Scripture, starting from the opening pages of Genesis. God states (is it an observation or a declaration?) that it is not good for the man he created to be alone. So God resolves to make for him an appropriate companion, for whom the man will leave all. Oneness between complementary opposites is thus established as the ideal. The woman, once symbolically one with the man as his rib, is created as distinct from him; yet this separation, where each has his or her own identity, needs to be overcome. What God separated in creation he intends to be joined together. Marriage is to consummate that original ideal.

What, then, of singleness? Part of the answer, at least, is supplied in the apostle Paul's letter mentioned above (1 Cor. 7:7). Paul sees human sexuality as touched by God's gifts. Each of us, he notes, has special gifts from God. But what is often unnoticed here is that human sexuality and its relation to singleness and marriage form the context of Paul's remark. For Paul, whether we marry or remain single should incorporate our recognition of the relevant gifts God has given us. Marriage is itself a good thing; it answers to the God-given, stewardly obligation to fulfill the needs of another. But Paul also sees that God has given single persons gifts to be used in their stewardly roles. Continence and self-control are divine gifts that allow individuals to devote themselves as stewards fully to the work of the Lord. Those with the gift of continence have the God-given ability to control their sexual urges and to concentrate that energy creatively on other endeavors in which they find personal fulfillment. What

22. The current state of the debate on food production limits and environmental impact is sketched in John Bongaarts, "Can the Growing Human Population Feed Itself?" *Scientific American* 270, no. 3 (March 1994): 36-42.

is advocated is not sexual repression but the recognition that self-control and singleness can be a gift for the individual tasks of stewardship. Thus, although the biblical norm for human relations between men and women is marriage, it is not a norm imposed upon all. As the stewards differ, so do the divine gifts for carrying out the assignment. Singleness and celibacy, too, can be ways of living out the divine gift.

The Basis for Sexual Orientation

It is time to return to the issues surrounding the question of sexual orientation and behavior. Given the above understanding of sexuality, what can a stewardship ethic say in response to the difficult questions of sexual orientation and behavior? The usual human sexual orientation is heterosexual. Are homosexual and bisexual orientations and behavior as morally acceptable as heterosexual ones? Let us first address the question of orientation.

One of the oldest theories attributes homosexual orientation to environmental factors. The psychoanalytic theory holds that "homosexuality is a symptom of neurosis and of a grievous personality disorder. It is an outgrowth of deeply rooted emotional deprivations and disturbances that had their origins in infancy. . . . [It] is generally caused by a frustrating and non-nurturing mother-infant relationship. . . . It is an oral, pre-genital disturbance that only secondarily utilizes sexual organs."[23] Put simply, it results from arrested development at the oral stage.

However, this psychoanalytic theory faces serious opposition. For one thing, the homosexuals studied by clinical psychologists were undergoing therapy, so there is legitimate doubt that they constitute a representative sample. For another, research yields contrary results. For example, in 1969-70, 979 homosexual and 477 heterosexual men and women in the San Francisco Bay area were interviewed as part of a study to determine whether there were determinable environmental causes of homosexuality. Alan Bell concluded from his survey that there was nothing in the environment or family background that correlated causally with a ho-

---

23. Robert Kronemeyer, "Homosexuality Is Not Biologically Determined," in *Human Sexuality,* vol. 1, ed. Linda P. Cushman (St. Paul: Greenhaven Press, 1985), pp. 187-88.

mosexual orientation. There was no evidence that parents played a role in causing a homosexual orientation. "Our findings indicate that boys who grow up with dominant mothers and weak fathers have nearly the same chances of becoming homosexual as they would if they grew up in 'ideal' family settings."[24] Although there was some correlation between homosexual orientation and poor relations with the father, the correlations were weak, and it was unclear whether this feature was a cause or an effect of the person's sexual orientation before adolescence. The researchers likewise found no evidence of peer influence or of atypical experiences with persons of the opposite sex in the development of a homosexual orientation. They concluded that sexual preferences were established prior to adolescence. These findings pushed the authors in the direction of looking for a biological basis for homosexuality.

We have already referred to LeVay's attempt to ground a homosexual orientation biologically by correlating this orientation with a particular area in the hypothalamus. A recent study of twins lends support to LeVay's thesis. Michael Bailey and Richard Pillard studied fifty-six pairs of monozygotic (identical) twins, fifty-four pairs of dizygotic (fraternal) twins, and fifty-seven pairs of adoptive brothers. They found that whereas 52 percent of the monozygotic twin pairs were both homosexual when one was, only 22 percent of the dizygotic pairs and 11 percent of the adoptive brothers were both homosexual. They conclude that in the absence of evidence to show that social factors (including parental behavior) affect a child's sexual orientation, genes are the most likely candidates to predispose someone to homosexuality. They argue that genes may account for 31 to 74 percent of male sexual orientation.[25]

These findings, however, are contradicted by another study by Michael King and Elizabeth McDonald of forty-eight pairs of twins. In King and McDonald's study, as in the previous one, one homosexual twin was interviewed, while the sexual orientation of the co-twin was determined from the knowledge of the interviewee. Only five of the twenty monozygotic twins (25 percent) and three of the twenty-five dizygotic twins (12 percent) were both homosexual. The authors concluded that "discordance

24. Alan P. Bell, Martin S. Weinberg, and Sue Kiefer Hammersmith, *Sexual Preference* (Bloomington: Indiana University Press, 1981), p. 184.

25. Bruce Bower, "Gene Influence Tied to Sexual Orientation," *Science News* 141, no. 1 (4 Jan. 1992): 6. See also Michael Bailey and Richard Pillard, "Are Some People Born Gay?" *New York Times,* 17 Dec. 1992, p. A-21.

for sexual orientation in the monozygotic pairs confirmed that genetic factors are insufficient explanation of the development of sexual orientation."[26]

In addition to the fact that King and McDonald's sample is very small, their study suffers from a difficulty similar to that facing the study done by Bailey and Pillard. In both cases the sexual orientation of the co-twin is determined through interviews with the homosexual twin. King and McDonald, like Bailey and Pillard, feel confident about the knowledge that one twin has of his co-twin, though in King and McDonald's study only slightly more than half of the interviewees claimed that their co-twin had actually discussed their sexual orientation with them. However, given the privacy of sexuality and the small size of the sample, lack of direct determination creates the possibility of a significant error.

Other recent research has concentrated on various biologically influenced features, such as hand preference and cognitive patterns. Cheryl McCormick and her co-researchers found an increased preference for left- or mixed-hand preference (non-right) in homosexual men and women, as compared with the general population.[27] In a more recent study of homosexual men and heterosexual men and women, McCormick and her co-researchers measured spatial ability and fluency. They concluded that "for three spatial tests and one of the fluency tests, the mean performance of homosexual men fell between those of the heterosexual men and women. The pattern of cognitive skills of homosexual men was different from that of heterosexual men: homosexual men had lower spatial ability relative to fluency. The cognitive pattern of homosexual men was not significantly different from that of heterosexual women." They concluded that their findings were compatible with a "neurobiological component in the etiology of sexual orientation."[28] Though they could not rule out early environmental factors, they appealed to the results of other studies to discount such factors.

26. Michael King and Elizabeth McDonald, "Homosexuals Who Are Twins," *British Journal of Psychiatry* 160 (Mar. 1992): 407.

27. Cheryl M. McCormick, Sandra F. Whitelson, and Edward Kingstone, "Sinistrality in Male and Female Homosexuals: Neurobiological Implications," *Society for Neuroscience Abstracts* 13 (1987): 851; "Left-handedness in Homosexual Men and Women: Neuroendocrine Implications," *Psychoneuroendocrinology* 15, no. 1 (1990).

28. Cheryl M. McCormick and Sandra F. Whitelson, "A Cognitive Profile of Homosexual Men Compared to Heterosexual Men and Women," *Psychoneuroendocrinology* 16, no. 6 (1991): 459, 470.

The best one can say from the research to date is that the evidence is incomplete. No specific genetic or neurophysical factors have been identified as playing a causal role in sexual orientation; what has been suggested to date are statistical correlations of sexual orientation with certain physical and behavioral features.[29] More systematic investigation is required to determine whether sexual orientations are genetically determined or not, and if so to what degree.

It is worth noting that in many of the recent studies the researchers may have an unstated agenda to discover a biological cause of homosexuality.[30] What role this agenda plays in the conduct of the research and the interpretation of the results is difficult for an outside observer to determine, though in light of what we argued in Chapter Two, no scientific study is value free. The studies themselves, including the questions asked and the categories used to classify the results, reflect in various ways the researchers' own worldviews. This dependency on worldviews is increased rather than decreased when there is a significant subjective element in reading the results (as in LeVay's study).[31] This should not be taken to debunk the recent reports but rather to heighten awareness of the need for more investigation and an awareness of the role that worldviews play in setting up study protocols and interpreting results.

Leaving the question of agendas aside, let us suppose that in the future researchers discover substantial evidence correlating genetically based biological features with sexual orientation.[32] This is not unthinkable;

29. William Byne and Bruce Parsons, "Human Sexual Orientation," *Archives of General Psychiatry* 50 (March 1993): 228-39.

30. LeVay and Bailey are gay activists; Pillard is a homosexual. LeVay subsequently took a leave from the Salk Institute in San Diego to direct the West Hollywood Institute for Gay and Lesbian Education, an advocacy foundation for homosexuals in Los Angeles.

31. Eliot Marshall, "When Does Intellectual Passion Become Conflict of Interest?" *Science* 257 (31 July 1992): 620-21.

32. After this was written, the results of a study to determine whether homosexuality might have a genetic basis were reported. In tracing the pedigrees of an initial sample of 76 homosexual men, the researchers found homosexuality more common among males on the maternal side (maternal uncles and sons of maternal aunts) than the paternal side. They took this to suggest that if a genetic basis were to be found it would be on the X chromosome. Proceeding to study 40 homosexual sibling-pairs whose fathers were not gay, they found that in 33 pairs, each sibling in the pair shared a set of five markers on a portion of the long arm of the X chromosome. Since not all of the homosexual pairs had the same set of markers, they concluded that although there is a link between homosexual orientation and a certain gene or genes in the region studied (an area of four million base pairs and

in light of what we know about hormones, the brain, and their role in sexual activity, it seems very likely that genetic factors are involved to a significant degree.[33] What would this evidence establish?

First, it should not be taken to establish by itself that a biological predisposition provides a norm for moral behavior for particular individuals so predisposed. The fact that something is biological does not entail that that is the way it should be, that it is good, or that it is beneficial.[34] That our genes are responsible for certain diseases provides ample evidence that the biological is not simply to be equated with the good or beneficial. As for sexuality, we have already seen that biology produces abnormal as well as normal sexual features. If persons have indeterminate sexual features, this is neither normal nor desirable, and often some corrective action must be taken to assist such persons. Study of the biological contributes to determining a norm that lies within certain ranges, but the mere fact that something is biological does not mean that it is normal and not to be resisted, controlled, altered, or removed.[35]

Second, it cannot be concluded that the presence of a strong genetic disposition to behave in a certain way diminishes or removes moral blame or praise for the behavior. It is important to recall that genetic determination lies along a broad continuum. At one end of the continuum, genes determine a person's destiny. Persons who have the gene for Huntington's disease, for example, assuredly will, if they live long enough, be afflicted with the disease. Similarly, persons who suffer from Angelman syndrome will manifest certain kinds of behavior: jerky limb movements, unprovoked

---

several hundred genes), other genes and/or environmental factors are involved in developing such an orientation. Dean H. Hamer, Stella Hu, Victoria L. Magnuson, Nan Hu, and Angela M. L. Pallatucci, "Linking Some Instances of Male Homosexuality to a Small Stretch of DNA on the X Chromosome," *Science* 261 (16 July 1993): 321-27.

33. For an excellent summary of the current state of the research, see Chandler Burr, "Homosexuality and Biology," *The Atlantic Monthly* 271, no. 3 (March 1993): 47-65.

34. Neither does the fact that a certain sort of behavior is natural to other animal species provide any basis for holding that it ought to be a human norm. For one such claim derived from observing a sexually promiscuous type of ape called bonobos, see Meredith F. Small, "What's Love Got to Do with It," *Discover* 13, no. 6 (June 1992): 46-51. One can find similar arguments based on studies of the homosexual behaviors of certain animals.

35. To give one example of the working assumption against which we are arguing, at the end of their report Hamer et al. note, without any argument or justification, that it would be unethical to use information about the genetic basis of a homosexual orientation to assess or change a person's sexual orientation (p. 326).

bursts of laughter, and behavior associated with mental retardation. At the other extreme, genes seem to have relatively less to do with certain aspects of human behavior: how I eat, what specific foods I enjoy, the way I brush my teeth, talk, and so on are probably determined more by my environment and choice than by my genes.

It is the vast area in between the two extremes that is of the greatest interest and poses the more important puzzles. Here one might say that our genes in part and to varying degrees create dispositions or tendencies to act or behave in a certain way. The dispositions are of varying strengths, but research indicates that in most cases heritability accounts for less than 50 percent of the variation in our behavior.[36] Consequently, the fact that someone is to some degree (but not totally) predisposed does not prescribe how that person should act, nor does it make it impossible for that person to control his or her behavior. One cannot condone or excuse behavior simply on the basis of predisposition.

At the same time, the degree to which or the ease with which we can control or direct our behavior depends, in part, on how rooted in our genes the behavior is. (It goes without saying that environmentally developed tendencies can also prove exceedingly difficult to control or alter.) The fact that someone has a strong disposition might, in certain circumstances, constitute a mitigating factor. This will be the case whether the disposition is of biological or environmental origin. As a mitigating factor, it will play a role, not only in the degree to which we blame, praise, or excuse the person for the behavior (that is, hold that it reflects on their moral character), but also in our understanding those persons and selecting the type of education, help, or treatment we provide for people who engage in undesirable behaviors, whether these behaviors are self-destructive or destructive for the community.

Thus, assuming that sexual orientation is to some extent genetically determined, the question arises whether sexual orientation is determined by genetics wholly or minimally. If it is the former, the degree to which one can alter, control, or direct behavior related to it is lessened. If it is the latter, in which case persons can control their behavior significantly, persons *might* have significant moral obligations with regard to behavior that should be controlled; in contrast to the former type of case, their sexual behavior cannot be explained or excused away. Based on present information one can safely conclude that the answer lies somewhere in the middle; our genes play a

36. Robert Plomin et al., "The Role of Inheritance in Behavior," *Science* 248, no. 4952 (13 Apr. 1990): 184.

significant role in determining sexual orientation, though precisely where their influence sits on the continuum awaits further research.

If we are correct about this, it follows that persons are not morally responsible for their genetic predisposition to a particular sexual orientation. Appropriate study of genetic determinants provides helpful means for understanding sexual orientation, both our own and that of others. Furthermore, this means that we need to direct our ethical questions, not to sexual orientation per se, but to matters regarding human sexual behavior based upon genetic constitution. To what extent, if any, is control over the behavior that stems from or is affected by our sexual orientation appropriate, warranted, or required? To this question we now turn.

The Ethic of Stewardship:
Ruling Sexual Behavior

To address the question of the ethics of sexual behavior, let us return to the injunction to rule over, given in our Christian ethical paradigm. To this point in the book, we have identified what we rule over as being something other than ourselves. As stewards we are obligated to rule over the rest of God's creation. But the obligation to rule over extends also to ruling over ourselves. That we have a certain nature — or better, set of dispositions — does not imply that we are free morally to act on those dispositions. To the contrary, we have a moral obligation to control our dispositions and tendencies where it is morally indicated that we do so.[37]

The question, then, is whether homosexuality is the kind of behavior for which control of predispositions or orientation (if such predispositions exist) is morally indicated. The answer, we believe, depends on invoking the factors identified earlier as guiding motifs for implementing our sexuality: *complementarity, exclusivity,* and *commitment.*

The biblical model and biology concur that human sexuality invokes complementarity. It is manifest in our sexual or procreative relationship with another person, though it is present in nonsexual relationships as

---

37. Self-mastery, often put forth as a Christian virtue, is emphasized in the New Testament letters (Gal. 5:23; 1 Thess. 5:6, 8; 1 Tim. 3:2; Tit. 1:8; 2:2, 5, 6). God's grace enables us to put away the characteristics and desires of our sinful nature and to be renewed by Christ in righteousness and holiness (Eph. 2:1-10; 4:22-24; Col. 3:10).

well. In sexual intercourse the male takes the female and the female the male. Each gives of himself or herself to the other and receives in return. The greater the fullness with which each gives and receives, the richer will be the union. Thus, the result of their union is a paradox: although the two become one, there still are two. The two bring the richness of their differences to the union; united, they draw on those differences to raise their union to new heights.

If we are correct in maintaining that sexuality properly involves complementarity, the next question is whether complementarity must involve gender. One way to answer this is to look at the procreative systems themselves. One does not have to be a biologist to see that humans are delineated along gender lines and that the male and female reproductive systems complement each other. The biological structure encourages — indeed, demands — gender complementarity for the continuation of the species.

The Natural Law Ethic

This observation lies at the heart of the natural law sexual ethic, which holds that nature provides norms for our actions. The natural law ethic, advocated by the Roman Catholic Church, is frequently misunderstood. The natural is *not* to be understood simply as what happens naturally, apart from the interference of other creatures or of humans. The move simply from what is natural to what is normative is fallacious. As we noted above, not everything that happens in nature is good, beneficial, or what ought to be.

Rather, natural law ethics argues that the good for something can be determined by discerning the ends for which that thing was made, and these ends can be determined in two ways. First, the end can be determined by studying what happens when that thing attains its fulfillment. Observation of the life cycle of deciduous trees reveals that one of the ends of a deciduous tree (if not the end) is to reproduce, and that its seeds provide the means by which this naturally occurs. Of course, not every instance is successful; most seeds will never achieve their purpose to germinate and develop into a tree. Yet when the seeds do germinate and develop, the tree achieves its end or purpose. The fact that many events do not lead to a successful outcome suggests that a second way is needed to supplement observation to ascertain something's end. To discern something's proper end one can study the essential structures and patterns that God has instilled or that have evolved in natural things. Here one can look at something's physical or biological structures and determine to what end they are the proper means. Once something's ends are discovered and the means for achieving these ends are understood, this particular means-ends ordering is taken to be normative for that thing.

Natural law ethics applies this analysis to humans as well as to the rest of creation. Humans, however, differ from the rest of creation in their relationship to their ends. Whereas natural objects cannot choose whether or not to achieve their end but do so necessarily, with many human actions the achievement of the end depends on the choices of the individual person. Humans are uniquely able to decide whether or not they will act to achieve their proper ends by the required means. The fact that humans have this choice becomes the basis for moral obligation. Humans are obliged to seek those ends for which God made them and to follow the proper means in doing so.

The natural law theorist goes on to note that procreation is one of the ends for which we were made, and that procreation requires gender complementarity through sexual intercourse as its means. Consequently, since procreation is our end, and since we have a moral obligation to attain our ends (for this is where our good lies), and since sexual intercourse employing gender complementarity is the necessary means, sex apart from gender complementarity is immoral.

We have not adopted the natural law paradigm as our paradigm. In particular, its appeal to formal and final causation presupposes a metaphysics rooted in Aristotelian biology. The decline of essentialism and rise of evolutionary theory have, for many, terminated appeals to formal and final causation. Nature's apparent teleology, it is held, results merely from natural selection.

At the same time, our stewardship ethic is not inimical to natural law ethics. Whether this theory and its emphasis on teleology can be separated from Aristotelian metaphysics and viably reconstituted in a post-Darwinian environment is worth exploring, though we will not do so here. Indeed, even among Darwinists the presence, nature, and possible significance of teleology in nature remain a matter of considerable discussion.[38] Even where it is obviously not intended to be taken literally, teleologically descriptive and explanatory language is often used.

38. Francisco J. Ayala, "Teleological Explanation," in *Philosophy of Biology*, ed. Michael Ruse (New York: Macmillan, 1989), chap. 18.

This ethical paradigm can contribute to our moral understanding by helping us to see that the human good must be understood in terms of what we are as human beings. In this regard biology is relevant to determining norms, for what we are is, to a great measure, biologically conditioned. Biology is relevant, not in the sense we rejected above — namely, that what is biological is good — but rather in assisting us to discern human structure and functions and how they are best promoted or fulfilled. Both structurally (in terms of our biological constitution) and functionally (in terms of preserving the species), biology indicates that gender complementarity does and should play a central role in understanding human sexuality and regulating sexual behavior.

Some have responded that gender complementarity must be abandoned as a criterion of proper sexuality, because there are no traits of personhood (as over against biological traits) that fall into the male-female delineation. Traits that are often suggested, such as male activity versus female passivity, are merely culturally determined. A concept of complementarity along gender lines

> depends on a sadomasochistic concept of male and female relations. It covertly demands the continued dependency and underdevelopment of women in order to validate the thesis that two kinds of personalities exist by nature in males and females and which are each partial expressions of some larger whole. Such a view can allow neither men nor women to be whole persons who can develop both their active and their affective sides. Once women reject this psychology of dependency and that repression of their active and intellectual traits that is implied by the ideology of femininity, the myth of complementarity is overthrown. This concept of complementarity must be recognized as a false biologism that attempts to totalize on the level of the whole human existence a limited functional complementarity that exists on the level of procreative systems.[39]

The view here is radical yet significant. It raises the critical question whether there are significant differences between the two genders. Some hold that, though there are differences between the genders, the differences are neither significant in themselves nor morally significant. Others believe that there are no natural differences, only culturally imposed ones. Culture, not

---

39. Rosemary R. Ruether, "From Machismo to Mutuality," in *Homosexuality and Ethics,* ed. by Edward Batchelor, Jr. (New York: Pilgrim Press, 1980), pp. 29-30.

nature, has forced us to treat each other along distinct gender lines. For example, culture has made some behaviors taboo for one gender (men do not cry) and forced us to treat people unequally (women can be nurses but not physicians; they do not belong on road construction crews driving bulldozers or in pulpits). But obligations that are merely cultural mandates are relative and cannot provide any universal moral norms. Both views claim that nothing of moral significance rides on any differences, actual or alleged, between genders. To claim otherwise is to promote "false biologism."

If gender complementarity is to be rejected as a criterion for proper sexuality, then it is hard to require complementarity *at all* for engaging in sex. There would seem to be no rational requirement that those who engage in sex have differing or complementary psychological or behavioral traits. That one partner is sanguine does not mean that the other should not be.

If we abandon complementarity as a necessary criterion or guide for sexual relations, what is to replace it? The suggested alternative is love or affection.[40] So long as two persons love each other and seek each other's good, the argument goes, sexual relations between them are proper. The persons need not be complementary in any way; they might have the same qualities of character or physical and mental traits. What is important is that they seek to provide a healthy expression of their caring for each other. In their love they should build a relationship that is not oppressive, destructive, or exploitive.

Sexuality and Love

On the surface, Christians cannot quarrel with this ethic. Love is the central (though not the sole) ethical virtue in Jesus' teaching. He makes love of God and neighbor the summation of the law and the prophets (Matt. 22:37-40). Love is uppermost in the teaching of Paul, who understands it to be the highest gift and foremost of the three virtues, faith, hope, and love (1 Cor. 13), and exhorts his readers to adopt it as their primary motive (Eph. 3:17). It predominates in the first letter of John, where the love the Father lavished on us, and the love the Son showed in his death on our behalf, prefaces his message that we should love one another (1 John 3:1, 16; 4:7-12). However, while as Christians we rightly

40. Ruether, p. 30; "Report of the Commission on Homosexuality of the Episcopal Diocese of Michigan," in Batchelor, ed., *Homosexuality and Ethics,* pp. 128, 134.

322

concur with the emphasis on love, we must be careful to spell out what precisely is meant by love as a controlling motif in Christian sexual ethics.

The variety of meanings is reflected in the three prebiblical Greek words for love. *Eros* refers to sensual love; it is the passionate, sexual, impulsive love that seeks to appropriate the other person for oneself. It invokes a sensual intoxication in which self-control is lost; for those under its spell, sexual freedom leads to ecstasy.[41] *Philia* is the kind of love that holds between friends, neighbors, parent and child, spouses, political cohorts, etc. Here the lover is not overcome by ecstasy but experiences the warmth of friendship that underlies close human relationships. The word *eros* does not occur in the Bible. *Philia* is used on a number of occasions: for example, in John 5:20 it describes the Father's love for the Son; in John 11:36 it is used of Jesus' love for Lazarus, and in John 20:2 it describes Jesus' love for one of his disciples; and in 1 Corinthians 16:22 it is used of our love for Christ. However, the unique emphasis in the New Testament is on *agape* love. "[Jesus] demands love with an exclusiveness which means that all other commands lead up to it and all righteousness finds in it its norm."[42] In contrast to *eros*, *agape* is often selective. It carefully chooses the individuals who are its objects and commits itself to them in sacrificial ways.[43] It seeks not its own good but the good of those who are loved. *Agape* extends not only to those who like us[44] but even to our enemies.[45]

41. "What the Greek seeks in *eros* is intoxication, and this is to him religion. To be sure, reflection is the finest of the gifts. . . . More glorious, however, is the *eros* which puts an end to all reflection, which sets all the senses in a frenzy, which bursts the measure and form of all humanistic humanity and lifts man above himself." "ἀγαπάω," *Theological Dictionary of the New Testament*, vol. 1 (Grand Rapids: William B. Eerdmans, 1964), p. 35. Plato, for example, treats *eros* suprasensually; persons in love ascend from the sensual love of the body to the love of the soul and its beauty, until finally they achieve ecstasy through contemplation of wisdom, the good, and beauty itself (*Symposium* 210a).

42. "ἀγαπάω," p. 44.

43. "Ἀγαπᾶν is a love which makes distinctions, choosing and keeping to its object, . . . a free and decisive act determined by its subject. . . . Ἀγαπᾶν relates for the most part to the love of God, to the love of the higher lifting up the lower, elevating the lower above others." "ἀγαπάω," p. 37. It should be noted, however, that the meaning of *agape* is at times imprecise, that it is used sometimes in extrabiblical literature as a synonym even for *eros*. Even the Septuagint uses it in Ezekiel 16 for the love of sexual desire.

44. This is a problem that Plato struggles with when he treats love in *Lysis*.

45. We are *not* claiming that these words are always used with these distinctions in mind. The distinctions are conceptual and not always manifested in a given text; at times

If we are to replace complementarity as a criterion for proper sexuality with the criterion of love within a Christian ethic, it must be a rich love that embodies all three elements. As we have seen above, one of the functions of sex is to produce psychophysical pleasure. *Eros* is the drive that brings persons together in a union of ecstasy that transcends the mere production of physical pleasure. Insofar as the erotic plays a proper role, the Christian will not lose sight of it nor denigrate it. But the erotic must be treated like other basal desires: desires for sleep, food, pleasure, recognition by others, etc. Stewardship morality demands that basal desires be controlled, that we rule over them. They are not to be repressed, lest we starve from lack of food, become joyless from asceticism, or wither from a poor self-image. Rather, ruling over means that our desires are to be properly channeled, used for the purposes for which we have them, and monitored. In sexuality, *eros,* then, should not substitute for but be supplemented by *agape.* Sexual love should be exercised in ways that create a bond of unity between the lovers, a unity that leads to a life of sacrificial, exclusivistic commitment to each other.

Philia should also play an important role in sexuality, for the erotic is only one of the complex human relationships (or ways of living) that obtain between lovers. Our lover should at the same time be our best friend.

The distinctive aspect of love in the Christian ethic is *agape;* it brings persons to the height of their mutual relationship. *Agape* does not come alone but is surrounded and supported by the companion virtues of patience, lack of envy, humility, kindness, understanding, delight in the truth, hope, and perseverance in seeking the other's good (1 Cor. 13:4-7). In the Genesis account that provides the basis for the biblical paradigm of sexual relations, the man leaves all else, including his birth family, to be joined to his wife (Gen. 2:24). In *agape* love the lovers are exclusively committed to each other.

Those who make love the sole moral ground for having sexual relations often omit the *agape* aspect. Love is interpreted as a free expression of one's personality,[46] which seeks a mutual good, the fulfillment of both parties, but often lacks the commitment, exclusivity, and companion vir-

---

the words for love can have broader or overlapping meanings. For example, Luke 11:43 speaks of the Pharisees who love *(agape)* the most important seats in the synagogues.

46. Robert Wood, "Christ and the Homosexual," in Batchelor, ed., *Homosexuality and Ethics,* p. 166.

tues that are central to *agape*. There is a sense in which, in the name of love without exclusivity and commitment, everything (and hence nothing) can be justified. Love can be used to justify incest or refraining from incest, adultery or refraining from adultery,[47] disbanding the marriage or maintaining it. And without the companion virtues listed above, commitment can become a loveless ordeal in which everything festers.

Charles Curran points out that with regard to commitment, homosexual behavior is frequently deficient. He writes, "Interestingly, those who argue that sexuality is neutral and all sexuality should be judged in terms of the quality of the relationship fail to come to grips with the accepted fact that most homosexual liaisons are of a 'one night stand' variety. Thus there is not a sexual union as expressive of a loving commitment of one to another."[48] Though Curran's thesis constitutes a sweeping generalization, a study by Bell and Weinberg indirectly confirms it in part. In their study of homosexuals in the San Francisco Bay area, they found that 41 percent of the males surveyed reported having more than five hundred partners lifetime, and 82 percent reported having more than fifty.[49] Though females are more likely to show greater sexual fidelity in their affairs (56 percent of the lesbians surveyed reported fewer than ten partners lifetime), males "are more apt to engage in sexual activity with persons who are virtual strangers to them."[50] Though to some degree the AIDS epidemic, in creating a climate of fear in the homosexual community, apparently altered patterns of sexual behavior, some see the pendulum beginning to swing again in the opposite direction. While the threat of AIDS prompted many communities to close down gay bathhouses and bars, in places "the era of reckless abandon never ended, and at others it is coming back."[51]

47. Joseph Fletcher, *Situation Ethics* (Philadelphia: Westminster Press, 1966), pp. 73-74, 164-65.

48. Charles E. Curran, "Homosexuality and Moral Theology: Methodological and Substantive Considerations," in Batchelor, ed., *Homosexuality and Ethics,* p. 177.

49. Alan P. Bell and Marin S. Weinberg, *Homosexualities: A Study of Diversity among Men and Women* (New York: Simon and Schuster, 1978), p. 308. The recent study of sexual practices in America had insufficient data on homosexuals to draw any conclusions. Michael et al., p. 170.

50. Bell and Weinberg, p. 101. A more recent study confirms the conclusion that homosexual men have more partners than heterosexual men, though the published data do not allow one to draw any more detailed conclusion. John O. G. Billy et al., "The Sexual Behavior of Men in the United States," *Family Planning Perspectives* 25, no. 2 (March/April 1993): 55.

51. William A. Henry III, "An Identity Forged in Flames," *Time* 140 (3 Aug. 1992):

This lack of commitment, which tragically also colors many heterosexual relations, might result in part from our social structure, which reserves marriage for heterosexual relations only.[52] Homosexual marriage is generally not condoned or legalized, so that there is no social recognition given to permanent homosexual relationships. Only the very recent move by some cities (Seattle, Berkeley) and some companies (Levi Strauss, Lotus Development Corporation, Microsoft) to grant health benefits to live-in or domestic partners gives a nod in this direction.[53] State recognition of homosexual marriages might be on the horizon, if the precedent set by the Hawaii Supreme Court in May 1993 is followed. The court ruled that state regulation of marriage on the basis of sex violates equal protection; the state may prohibit homosexual marriages only if it can show compelling reasons for doing so. Beyond this, if one is going to require commitment and exclusivity as conditions for having moral sexual relations, then society will have to institute structures that parallel heterosexual marriage whereby long-term commitment between homosexual persons can be established and fostered.[54]

It is sometimes responded that one cannot require exclusivity and commitment in homosexual relations, for people who engage in heterosexual relations likewise fail to manifest exclusive commitment. But this reply falls prey to the *tu quoque* ("you also") fallacy. The fact that someone else is doing something does not justify my doing it. That one group of persons manifest lack of commitment and exclusivity in their sexual relations does not justify accepting the same lack in others. Christians deplore such a lack no matter what sexual form it takes. This follows the example of Jesus who, though he refused to join the crowd in stoning the woman caught in adultery, commanded her to go and abandon her life of sin (John 8:1-11; see also 1 Cor. 6:9-10).

We are intentionally avoiding the slippery slope fallacy. On the one hand, we are not arguing that homosexuality is immoral because it allegedly leads to promiscuity. This would be a weak argument, for not only

37. "New York City's Hetrich Martin Institute, a counselling agency for young gays, reports that more than half of 136 respondents to a survey admitted having sex without condoms."

52. "Sex without commitment or much involvement may reflect an even greater commitment to the reality of their circumstances, given the 'homoerotophobic' society in which they live." Bell and Weinberg, p. 101.

53. Katrine Ames, "Domesticated Bliss," *Newsweek* 119 (23 March 1992): 62-63.

54. Dennis O'Brien, "A Case for Gay Marriage," *Commonweal* 118, no. 20 (22 Nov. 1991): 681-85.

would it not cover cases of homosexual fidelity, but by parallel argument it would raise questions about heterosexuality. Our argument in this section has been that an ethic that is constructed solely on love often adopts a shallow view of love, one which allows for promiscuity. On the other hand, we also reject the argument that if homosexual behavior is permitted in monogamous relations on the basis of love, then there must be similar permission for any sexual relations outside marriage when there is love and fulfillment between the non-married persons. It is not uncommon to find such a move made in the literature that defends homosexuality, and it is instructive to note it. Tom Driver, for example, moves from rethinking marriage to the view that "life outside marriage is not necessarily or even normally virginal, . . . [and that] life inside marriage is not to be construed as forbidding sexual relations with other persons." "Is not this a denial of the marriage commitments?" one might respond. He continues, "If this seems to strike at the very foundation of marriage, that is because we have insisted on viewing marriage as a sexual contract, with the result that we do not care what sacrifices of personhood it requires."[55] One might reasonably suggest that Driver tragically has lost sight of *agape* almost completely.

Paradigms for Sexual Behavior

The issue, then, is not what approval of homosexuality might lead to, though at the same time one must guard strenuously against denigrating the family as the structure for giving meaning to interpersonal relations within the context of procreation and child rearing. Neither is it the accusation that homosexual encounters, at least among males, often violate the exclusive commitment required of *agape* love, for the same could be said about some heterosexual relations. Rather, the issue concerns the *fundamental basis* we will use for making moral judgments about sexual behavior.

One frequently suggested option is to adopt a worldview that is centered around human personhood, one in which being fulfilled as persons is the highest ideal, taking priority over all else. In such a worldview, one's sexuality is an instrument "in the fulfillment of [one's] subjective

55. Tom F. Driver, "The Contemporary and Christian Contexts," in Batchelor, ed., *Homosexuality and Ethics*, p. 19. Other examples can be found in Ruether, p. 31, and in more detail in Raymond, pp. 262-72.

aim or integrating purpose."[56] This might be found in either heterosexual or homosexual relationships, within or without marriage. There is no necessity that it manifest the exclusivistic commitment characteristic of *agape* love. It depends on each person to find what is personally rewarding and fulfilling. Sexuality simply provides one way to express oneself; no intrinsic moral differences exist among the various modes of expression, so long as they are not exploitative, cruel, impersonal, obsessive, or irresponsible.[57] "A new world view . . . makes it clear that each individual is uniquely capable of turning every interpersonal encounter into something new, something creative. We are not forced to limit our encounters to recapitulation either of our own past encounters or of the patterns of the majority. We can instead vivify each encounter with creativity. . . . Tolerance and vision are perhaps at once the most uniquely Christian, the most self-fulfilling, and the most important aids to authentic spiritual discovery."[58]

When sex is merely an instrument of personal expression, free creativity, and fulfillment of one's self and perhaps of others as well, subject only to the moral mandates of nonexploitation and tolerance, the question arises whether we are properly ruling over our desires, whatever our sexual orientation, in living according to such a worldview. We have argued that the mere fact that we have wants, desires, or dispositions does not mean that it is moral to fulfill them. We are to exercise control to fulfill only those desires that are moral, and to fulfill them in a moral way. Unlimited sexual gratification, the "robust" sexual life, is not part of that pattern.

Contrary to the "new" worldview that incorporates a personal fulfillment ethic, we have argued for a worldview that embodies a stewardship ethic in which ruling over ourselves and others plays a central role.[59] A stewardship ethic does not deny personal fulfillment, but it sees personal fulfillment within the larger context of stewards' responsibilities both to

56. W. Norman Pittinger, "The Morality of Homosexual Acts," in Batchelor, ed., *Homosexuality and Ethics,* p. 144.

57. James B. Nelson, "Gayness and Homosexuality: Issues for the Church," in Batchelor, ed., *Homosexuality and Ethics,* p. 201.

58. Michael F. Valente, "A New Direction," in Batchelor, ed., *Homosexuality and Ethics,* pp. 152-53.

59. It is important to note that we have not placed sexuality here in the context of ruling over others. Sexuality becomes rape when it is treated as a tool of dominance and subjugation. The ethnic rapes in Serbian-dominated Bosnia-Hercegovina in 1992-93 are no worse a failure of stewardship than the daily rapes in our own society.

God and to those for whom they act as stewards. The paradigm we have chosen places sexuality in the context of the command to fill, which has both quantitative and qualitative aspects.

In the narrower sense of specific acts, sexuality exercised in lovemaking is directed toward procreation; without this the world could not be filled, nor could the species survive. Sexuality so understood does not require that every act of lovemaking be directed toward procreation. Procreation is neither a necessary nor a sufficient condition for legitimate sexual acts. As we argued earlier in this chapter, all filling has limits imposed by various restraints, including economic, environmental, medical, and psychological. Thus, engaging in sexual intercourse requires that the procreative purpose be balanced by considerations such as the ability of the couple to assume financial, psychological, and emotional responsibility for rearing the child; concern for population growth in light of the world's limited resources; the potential negative genetic heritage parents who are carriers of a serious disease or genetic defect might bestow on the child; and the possible medical risks that the woman would assume if carrying a child to term might endanger her health. Hence, in emphasizing the procreative context, we do not deny the need for and appropriateness of birth control. By drawing attention to the filling aspect, we are, however, specifying the primary *biological context* for sex. Normally, the conception of children grows out of the fundamental union that occurs in the sexual act. Should there be no procreation, either by chance or by choice, the requirement of a biological context is not thereby annulled. It remains an essential, normative part of the biblical paradigm for sexual relations.

In its broader sense of sexual identity, sexuality specifies a primary relation we have to other persons and to the role we play in the larger social institutions concerned with engendering and raising human persons. In the primary social institution of the family, we function as spouses bonded in relation to each other, as parents and children when sexual acts are procreative, as parts of extended or complex families. Ideally, these enriching, biologically ordered family relations provide psychological and physical wholeness (though again we are not suggesting that wholeness requires that children result from the marital bond, though surely they can enrich the relationship).

We have also argued that sexual fulfillment has a qualitative aspect, revealed in the intimate bonding that sexual union brings about. Here complementarity plays an important role in fulfilling our sexuality. In the intimacy of sexual union, two persons join to become one. The biblical paradigm makes gender complementarity the norm, not only in the narrowly procreative act, but also in the broader sexual context of family life.

In ruling over ourselves, we control our actions. As we have seen, mere orientation, mere genetic predisposition, does not justify our behavior. If less than 50 percent of most of our behavior is genetically conditioned, then much remains subject to our choice. The same holds true for environmentally related dispositions. Whatever our sexual orientation, we ought to transform our improper dispositions by our actions so that we will be stewards worthy of being endowed with a sexual nature.

Finally, we must return to a point made earlier in this chapter, namely, the point that sexuality is more than genetic, genital, or somatic; it is psychosocial as well. It involves more than the physical acts of sex; it involves our life-style, how we look at ourselves, and how others look at us. This takes us beyond the morality of sexual acts, which (perhaps unfortunately) is often the focus of our sexual ethics, to the matter of sexual identity. Just as the brain is a sexual organ, so we are sexual creatures. Just as the particular sex act is only a part of human sexuality, so the morality governing those acts will be only a part of the broader moral concerns that govern our total sexuality. From the Christian perspective, the love that pervades sexual relations must not be a truncated version of love, but rather a love that incorporates all of love's dimensions: erotic love *(eros)*, love that enriches a multitude of human relationships *(philia)*, and love that grows out of and develops exclusivistic commitment to the other and that seeks the welfare of the other, even in preference to one's own welfare *(agape)*. This rich concept of love must be cultivated in ways that incorporate and transcend, that sometimes do but often do not evoke, the genital expressions of love. Love will define an entire life-style of enjoyment of the presence of other persons, fully employing the constellation of companion virtues in seeking their good and being committed to them.

The Ethic of Stewardship: Caring for Others

No contemporary discussion of sexuality can be complete without some attention to sexually transmitted diseases and our attitude and actions toward persons afflicted with them. In another era we would have spoken about syphilis, gonorrhea, herpes, and the like — and indeed, perhaps not to speak about these diseases today is a disservice, since they are becoming more prevalent. The 1980s saw the rise of syphilis to pre-1940 levels, despite the availability of penicillin.[60] Today, however, AIDS (Acquired Immune Deficiency Syndrome) demands our greatest immediate attention. Although AIDS is unique in the way it operates, it is merely the latest in a long line of debilitating, sexually transmitted diseases.

In fact, some have drawn attention to the striking parallels between AIDS and syphilis. Both diseases are transmitted by unprotected sex, usually heterosexually. Worldwide, AIDS is believed to be primarily a heterosexual disease; 70 percent of those afflicted worldwide are heterosexual, while only 10 percent are homosexual. In the United States, however, AIDS is still primarily transmitted by homosexual activity;[61] 59 percent of males with AIDS are homosexual or bisexual.[62] The Center for Disease Control estimates that 4 percent of all AIDS cases are contracted through heterosexual contact, the majority of these cases being women.[63] Before World War I, surveys conservatively estimated that 12 to 15 percent of the males in Berlin, London, and Paris were infected with syphilis.[64] In New York and San Francisco it is estimated that 70 percent of the

60. Mick Hamer and Phyllida Brown, "Those Old Jelly Roll Blues," *New Scientist* 132, nos. 1800-1801 (21 Dec. 1991): 54.

61. Kathleen McAuliffe et al., "AIDS: At the Dawn of Fear," *U.S. News and World Report,* 12 Jan. 1987, p. 62. Out of approximately 184,000 cases of AIDS reported to the Center for Disease Control through July 1991, only 10,279 were heterosexual cases. James Allen and Valerie Setlow, "Heterosexual Transfer of HIV: A View of the Future," *Journal of the American Medical Association* 266 (25 Sept. 1991): 1695.

62. "Of the first 100,000 persons with AIDS, 5% were attributed to heterosexual transmission, compared with 7% among the second 100,000." "The Second 100,000 Cases of Acquired Immunodeficiency Syndrome — United States," *Journal of the American Medical Association* 267, no. 6 (12 Feb. 1992): 788. At the same time, however, the number of heterosexual cases in the U.S. is growing so fast that it is expected to double by 1995. In young heterosexual adults between the ages of 13 and 24, cases of AIDS increased by 60 percent between August 1989 and July 1991. The problem of heterosexual transfer of HIV is particularly troubling for women, for not only is "heterosexually transmitted" the only category in the U.S. in which cases of AIDS in women exceed the number in men, but the percentage of cases of AIDS among heterosexual women increased from 30 percent in 1988 to 37 percent in 1991. Women accounted for 61 percent of all heterosexually transmitted cases. Mary E. Guinan, "HIV, Heterosexual Transfer, and Women," *Journal of the American Medical Association* 268 (22 July 1992): 520.

63. Billy et al., p. 52. "Recent evidence suggests that male-to-female transmission of HIV may be at least 20 times as efficient as female-to-male transmission."

64. Hamer and Brown, p. 54.

homosexual males carry HIV (human immunodeficiency virus).[65] It is estimated that in 1991, ten million persons worldwide had HIV.[66] Up to March 31, 1993, 289,320 cases of AIDS in the U.S. had been reported to the Center for Disease Control.[67] AIDS is the fifteenth leading cause of death in the U.S.; 182,275 deaths had been reported to the Center for Disease Control by the same date. As many as 4,880 children are reported to have acquired the disease before the age of thirteen; it is the ninth most frequent cause of death in infants between the ages of one and four.

Both AIDS and syphilis bring a period of pain and disability before death. In syphilis, a type of bacterium, *treponema pallidum,* travels through the bloodstream to the heart, brain, spinal cord, and other internal organs, where it causes lesions. Eventually lesions on the brain and spinal cord lead to insanity and paralysis, while those on the internal organs, including the heart, cause them to weaken, bringing about the patient's death. HIV destroys the body's immune system, making it vulnerable to various diseases that the normal person is able to resist. Like those with syphilis, people can have HIV for ten years or more without symptoms. Prior to the employment of Salversan, a silver and arsenic pill developed in 1907 by Paul Ehrlich, and then penicillin in 1943, syphilis was incurable; it brought almost certain death. Scientists have yet to discover the "magic bullet" to cure AIDS.

There are also similarities relating to the social dimension of the diseases. In both cases the cause of death from these diseases is sometimes not reported, either because it is not properly diagnosed or to avoid embarrassment to the family.[68] The diseases claim more of the poor than of the upper social strata, though the latter are not immune. Women generally get the diseases from men and can pass them on to the children they bear. In both cases there were initial beliefs that the disease could be caught in a variety of ways, including from toilet seats, eating utensils, drinking cups, and ordinary physical contact. Foreigners were blamed for introducing both syphilis (the Italians blamed the French) and AIDS (the U.S. ostracized Haitians) into countries.

65. Bonnie Steinbock, "Harming, Wronging, and AIDS," in *Intervention and Reflection,* ed. Ronald Munson, 4th ed. (Belmont, CA: Wadsworth, 1992), p. 237.

66. *World Almanac and Book of Facts '92,* ed. Mark S. Hoffman (New York: Pharos Books, 1991).

67. Quarterly report by the Center for Disease Control.

68. Allan M. Brandt, *No Magic Bullet* (New York: Oxford University Press, 1985), p. 10.

However, AIDS holds a unique place for us, for with the discovery of penicillin and its application to syphilis in 1943, syphilis is no longer the incurable, deadly scourge it used to be. Between the 1940s and our own day there has occurred what is often referred to as "sexual liberation" or the "sexual revolution." Whatever the precise understanding of these terms, they refer at least to the increasing openness with which people approached their sexuality and sexual relations with others. Many heralded a guilt-free, birth-control protected, uninhibited sex life without fear of death[69] as a panacea for many of our psychological ills, which were traced to sexual repression.

But AIDS casts a pall over the desirability of that revolution, for now again we come face-to-face with a disease that, given the present state of our knowledge, eventually is always fatal. Sexual promiscuity is dangerous, not only for the sexually active person, but also for each of his or her partners. It can exact a steep price with any of the sexually transmitted diseases; those who contract AIDS pay the ultimate price.

Attitudes toward Persons with AIDS

One question that a stewardship ethic raises is how to care for individuals stricken by sexually transmitted diseases, especially AIDS. For many in the United States, the initial focus was on the fact that AIDS was a homosexual or a drug culture disease. Some took it to be a punishment for immoral life-styles. When it was passed on to others heterosexually, through blood transfusions, or through birth, distinctions were made between those for whom it might be a punishment and the innocents affected directly by others or through their tainted blood.[70]

Our life-style commitments have consequences for us — for our physical health, mental health, relations with others, etc. Whether these consequences are to be understood as punishments depends upon a whole network of concepts and issues. Is God someone who punishes people, or is he too transcendent to take this kind of interest in his creation? Is God's bringing of evil into our lives inconsistent with a proper understanding of

69. Richard D. Mohr, "AIDS, Gays and State Coercion," *Bioethics* 1, no. 1 (Jan. 1987): 48.

70. Similar distinctions were made with syphilis before the turn of the century. See Brandt, p. 11.

a good and loving God? If he does punish, does God use the outcomes of our actions to punish or reward us, to give us what we deserve? Are outcomes of actions properly understood as punishments, or do punishments require the conscious application of judicial principles by one person to another? Is punishment to be understood in terms of deserts or rehabilitation? This is not the place to take up these important questions.[71] Our interest is with distributive, not retributive, justice.

Ethical reasoning must continually struggle with the matters of equality and inequality that stand at the heart of distributive justice. On the one hand, justice demands that we treat persons equally. No matter how people get into a traffic accident, whether they are drunk or sober, their injuries must be treated like anyone else's. A person who got lung cancer from excessive smoking must be treated with the same care and compassion as someone who got it from years of working in the coal mines to support his family. Persons suffering from AIDS should receive care and treatment regardless of whether they got the disease through using dirty needles in the back alley drug culture or from a blood transfusion. Each and every one is a human being for whom we are stewards and for whom God has told us to care. From the perspective of worth, a stewardship ethic will hold to egalitarianism: all human beings are equally valued by God and deserve equal consideration.

On the other hand, justice applied to the treatment of persons demands considerations of inequalities. Were our resources unlimited, the question of inequalities would be irrelevant. But where the resources are limited, caring cannot be divorced from justice, and justice must consider both the equality of persons as persons and the relevant inequalities. In medical care in general, those inequalities concern such things as the likelihood that the treatment will or will not succeed for that person, the probability of recidivism, and the responsibility people assume when they knowledgeably assume risky behaviors. For example, health care providers must allocate scarce resources based upon their best estimate of how successful the application of those resources will be. If there is one heart to transplant, one should give it to the person who has the greatest chance to survive. Or again, to take resources from one person, who we have reason to believe can benefit permanently from them, and use them instead for another person, who we have equal reason to believe will return to the

---

71. These questions are dealt with in part in Bruce R. Reichenbach, *Evil and a Good God* (Bronx, NY: Fordham University Press, 1982), chap. 5.

life-style that brought about the condition in the first place, is unjust. Similarly, where there are many requests for resources, we must take into account the life-styles that created the needs. We have argued that we have stewardship responsibilities for ourselves; we are to rule over or master ourselves. Persons' failures to control their behavior — whether it relates to smoking or drinking, injecting drugs, risky health and sexual practices, or risky recreational practices — have implications for how resources should be allocated.

Health care involves not merely someone treating the sick; it also involves the sick taking responsibility for their own treatment and the well taking responsibility for their own preventive health care. Society's obligations to continue paying for treatment are reduced for persons who knowingly and willfully refuse to take prescribed medications or to allow actions that can cure their illness. Similarly, society has limited responsibilities for persons who knowingly and willfully endanger their health through risky practices. Such persons must assume the primary responsibility for their predicament. In short, considerations about how one got into or why one continues in a health care predicament are relevant to the questions concerning the allocation of scarce resources.

Contrary to how it might appear, this does not deny that persons are of equal value. What it means is that distributive justice is a complex matter. It requires that we take into account both equality and inequality. Considerations of relevant inequalities provide the reason why more of society's resources go to the ill than to the well. At the same time, it also provides grounds for determining allocations of society's scarce resources in terms of life-styles.

We must accept responsibility for our own actions and their consequences. Willful disregard of warnings and engagement in risky health and sexual practices diminishes public obligations. Thus there is a proper tension between two demands of stewardly caring for others. Love demands that we care for others no matter how they got ill, simply because they are our neighbors. At the same time, love seeks the good of others and thus justly demands that those who engage in risky health practices take responsibility for their life-style. Concerns of love and justice interact when we speak about attitudes toward AIDS patients.

Hence, there remains a true sense in which we can speak about there being guilty and innocent persons with regard to sexually transmitted diseases. That persons are guilty (they knew the dangers and proceeded anyway) or innocent (they did not know the possible consequences of

335

their own actions or those of others) does not affect their humanity; each is equal to the other, and both deserve our love and care. What it does affect is resource allocation within the context of scarce resources.

AIDS and the Use of Scarce Resources

Caring for HIV infected patients is a costly matter. Fred Hellinger, of the Agency for Health Care Policy and Research, a part of the Department of Health and Human Services, estimates that "nearly 1% of the annual health care budget of the U.S. — $5.8 billion — [was spent in] treating all Americans with HIV infection in 1991,"[72] and that this will increase to $10.4 billion by 1994. In 1991 the treatment of HIV-infected individuals who had not yet developed AIDS cost approximately $5,100 annually; treatment of those with AIDS cost about $32,000 annually, adding up to a total lifetime cost of $85,000.

The question arises concerning who will pay for this treatment. Currently about 40 percent is paid for by the government through Medicare and Medicaid, 40 percent comes from private insurers, and 20 percent either is paid for by the persons themselves or becomes a financial loss to primary care givers. Increasingly the burden falls on the public sector. The sicker AIDS patients become, the more they rely on public hospitals rather than private physicians for their primary care. Care for AIDS patients now accounts for 28 percent of the costs incurred by public hospitals and 36 percent of their financial loss.[73]

Part of the dilemma is that in order to qualify for Medicaid assistance, AIDS patients must show that they are impoverished (have less than $2,000 in assets). Thus AIDS is not only a disease of the poor but also a disease that impoverishes, as the afflicted "spend down" to that limit in order to be able to "afford" treatment.

Given the provisos noted above, an ethic of stewardship will treat AIDS sufferers with the same care as those who suffer from other debilitating diseases. The goal should not be to impoverish them but to find a

72. Marsha F. Goldsmith, "Costs in Dollars and Lives Continue to Rise," *Journal of the American Medical Association* 266, no. 8 (28 Aug. 1991): 1055.

73. Before they were diagnosed with AIDS, 37 percent of current AIDS patients sampled relied on public hospital outpatient clinics for primary care; after AIDS diagnosis the percentage increased to 79 percent. Goldsmith, p. 1055.

just way to use government funds to supplement both their own resources and those of insurers. Since people with AIDS generally cannot hold a job and accordingly lose private health benefits with their job loss, the question of their dignity becomes a significant caring concern as they struggle to cope both with the disease and with the financial burden that it poses.

The other side of the coin concerns the funds dedicated to AIDS research. The Public Health Service allocated more funds for AIDS research and education ($1.6 billion in 1990) than for any other cause of death, including cancer ($1.5 billion). Yet more Americans die of heart disease every fourteen weeks than have contracted AIDS in the last decade. In 1990 the Center for Disease Control spent on prevention and education $10,000 for each AIDS patient, whereas it spent only $185 for each cancer patient and $3.50 for each sufferer from heart disease. "Total federal research expenditures on AIDS this year [1990] will be more than 100 percent of nationwide patient costs; in the case of cancer, the corresponding ratio of research-and-development spending to patient cost is about 4.5 percent, in the case of heart disease about 2.9 percent, and in the case of Alzheimer's disease, less than 1 percent."[74]

Part of the result of this is that money for research on other, more prevalent diseases, such as cancer, is being scaled back, as money and researchers are being wooed into the fight against AIDS. "AIDS research has now weakened cancer research to the point where the National Cancer Institute's ability to fund promising new proposals is lower than at any time in the past two decades."[75]

Justice involves calculation — calculation of equalities and inequalities, and calculation of how the inequalities are to be dealt with so as to maximally provide for the interests of all and to care for those who manifest more need or, where relevant, more merit. Hence, some kind of calculation is necessary to determine how to allocate resources both among diseases and among the legitimate competitors for public funds. How then does one determine how much of the public's limited resources are to be spent on particular diseases?

On the one hand, the financial comparison given above in terms of amount spent per present sufferer from AIDS and other diseases is somewhat unfortunate, since there is a significant difference between AIDS

74. Michael Fumento, "Are We Spending Too Much on AIDS?" *Commentary* 90, no. 4 (Oct. 1990): 51.

75. Fumento, p. 52.

and, for example, cancer. It is true that AIDS is a disease that one can, at least to some extent, prevent oneself from getting, and in that sense perhaps it merits less attention than diseases that one cannot prevent oneself from getting. Changes in life-styles that involve sexual promiscuity or drug use and the practice of abstinence or "safer sex" can protect (to some extent) against the disease. However, given the contemporary climate of sexual liberation and prevalence of drug usage, AIDS has the potential for being an epidemic, as indeed it already is in some African countries. Hence, the question of a just and caring allocation of research resources must take into account not only the numbers of present sufferers from the disease but also the numbers of those who could contract AIDS and the heavy financial burden this would pose.

On the other hand, an ethic of caring must urge those who make public policy to resist the kinds of pressures that often make the government lurch between programs and throw fiscal caution to the wind. AIDS is a case in point, for it is one of the few diseases for which Congress, under the influence of strong, vocal pressure groups, appropriated specific funds for its research and treatment. There is no doubt that AIDS sufferers need governmental assistance for research. Yet they do not stand alone; victims of multiple sclerosis, cancer, heart disease, arthritis, Alzheimer's, and numerous other diseases that lack vocal, powerful lobbies can make similar claims. In this sense a depoliticization of AIDS and other disease research is called for.

There is no precise calculation that we can use to determine how to distribute resources justly. This is no less true for balancing demands within the health care system than it is for adjudicating between health care needs and the other pressing needs of society. A stewardship ethic cannot provide a formula. What it can do is to make us aware of the diversity of the needs to which justice demands that we attend and to call attention to our obligations to care for others over whom God has made us stewards. It can also demand that life-styles and practices be altered for the good of society, as has occurred with cigarette smoking. It can also point out that diminished quality of life for the Alzheimer's patient, for example, is no less dehumanizing, debilitating, and potentially lethal than for the AIDS sufferer.

In place of precise calculation we must use both reason and compassion. This applies to all of those who are suffering and struggling to maintain their existence, whether their suffering results from disease, abuse, malnutrition, or war. Over the centuries God has called human beings to be stewards who are providers of health care, givers of comfort and food, and peacemakers. The threat of AIDS makes our obligations to care for others, and the risks this presents, no less demanding than they have ever been.

Testing and Obligations to Inform

A final concern has to do with the conflict between preserving confidentiality and the dissemination of critically important information. Caring for others applies not only to those who suffer from AIDS but also to those whom they endanger through sexual relations, sharing used needles, medical practice, or in some countries the donation of blood. To care for others means that these other persons need to be informed about the HIV status of those with whom they have relations (sexual or otherwise) that could endanger them. This then raises two questions: (1) Should there be general or selective screening of persons for HIV? (2) Should information about a positive result on an HIV antibody test be shared with people other than those tested?

Several problems confront those who advocate either a general screening of the population or selective screening of particular groups of persons, such as health care providers. One problem concerns the usefulness of the screening itself. A negative report on the HIV antibody test affirms that possibly at the time of the test the person tested did not have the virus. The "possibly" comes from the fact that the most common test screens for the presence of HIV antibodies, but these antibodies do not form until about four weeks after infection.[76] It also provides no guarantee that the person has not contracted the virus since the test. Hence, screening at best can indicate that the testee has HIV; it cannot provide a guarantee that the person is free from the virus. Hence it has a limited usefulness — though it is nonetheless important for that.

A second problem concerns how the screening is conducted. If it is *voluntary,* then the autonomy of those who are screened is protected. However, this provides little comfort regarding the HIV status of the general populace or of particular segments of it (for example, health care providers), since one does not know the status of those who did not volunteer to be tested. On the other hand, if the screening is *mandatory,* how will it be enforced, and how will the autonomy and privacy of those who are tested be protected? Enforcing mandatory screening within a limited population (health care

76. Steinbock, p. 217.

providers in hospitals) would be feasible (though not necessarily warranted) through making the test a requirement for continued licensure to practice. Enforcing it for the general populace, however, would seem to be impossible without grossly infringing on basic freedoms. The cost and general harms would greatly exceed any possible benefit that could be derived. The one bright spot is that once a cure for HIV is found, the resistance to mandatory testing should dissipate.

A third problem is perhaps the most difficult of all: What use will be made of the results of HIV antibody tests, whether these are obtained voluntarily or involuntarily? If the testee agrees that the results may be shared with his or her sexual partners or, in the case of health care providers, with his or her patients, no problem exists. However, in one study as many as 25 percent of the respondents indicated that if they tested positive for HIV, they would not want it reported to their sexual partners.[77]

But do not those who have access to the test results — health care providers — have obligations to the sexual partners of those who are tested to warn them of the possible deadly danger that they face?[78] The argument that those who are in charge of the test results should keep the information confidential rests on the concern for respecting the autonomy of the testee and on the fear that if confidentiality is breached, people will not voluntarily undergo HIV testing. The latter is a genuine fear; not protecting confidentiality certainly has this significant risk. However, the former — concerns for autonomy — cannot be made an absolute but must be considered in light of our other obligations. As we argued in Chapter Seven, stewardship responsibilities for others provide the moral grounds for the obligation to care for those who might be harmed, in this case by transmission of HIV. It is true that the consequences of sharing this information might be harmful. For example, it might break up a marriage if one person was unaware of the risky behavior indulged in by their spouse. Though this might be regretted, the failure of spouses who engage in risky behavior to be honest with their spouse is itself immoral. Had there been honesty between the spouses in the first place, the fear of revelation would not be problematic.

Furthermore, the danger from HIV to the unsuspecting partner greatly outweighs the privacy and confidentiality concerns of the testee,

---

77. Steinbock, p. 220.

78. The legal precedent for this is found in the Tarasoff decision in California. California Supreme Court; 1 July 1976. 131 California Reporter 14.

since the result can be deadly. In such cases the danger is real, for the evidence shows that the more one is exposed to HIV, the more likely one will be infected by it. For example, it is indeed tragic when one person engages deceptively in AIDS-risky behavior and then infects his or her spouse, causing death. It is tragic not only for the infected spouse but for the entire family. The fact that one spouse has not cared for the other in such cases provides no grounds for absolving the health care provider of the responsibility of informing the endangered spouse.

The case of health care providers who are infected by HIV is more ambiguous. Here the risk to others — potential clients — is not as clear. Instances in which patients have contracted the disease from their health care providers have been rare, the noted case of Kimberly Bergalis being an important (and deadly) exception. Here the concern to protect the confidentiality, reputation, and possible contributions of the health care provider must be weighed against the obligation that health care providers have to fully inform their patients about matters affecting their health. Health care providers have taken an oath to cause no needless harm to their patients, and this oath should weigh heavily in their own decision to inform their patients about the possible risks of being treated by infected persons. Where a health care provider fails in his or her duty to inform patients fully, and where there is a reasonable risk, others have the obligation to inform as stewards caring for other persons.

❋   ❋   ❋

In our individualist culture, the argument is frequently advanced that since sexual relations are engaged in privately, they should, for the most part, be exempt from moral and legal scrutiny. Scrutiny is warranted, according to this argument, only when sex is used to destroy, dominate, or denigrate another. This privacy argument would be persuasive if, in fact, sexuality were predominantly private and did not also concern matters of individual character and how one treats others. Yet the contrary is true. In sexuality, what is most private, from our genes to our bedroom behavior, can have very public consequences — from how we look and feel, how we develop physically and psychologically from children to adults, how we behave in the presence of others, and the diseases from which we suffer, to the children we conceive, the maintenance or destruction of our family, and the trauma we cause for others. It also affects how we treat and consider ourselves. Consequently, it is an appropriate topic for ethical reflection.

341

At the same time, our emphasis must turn from primarily a microscopic concern with particular sexual acts to a concern for human beings as sexual beings: for our identity and character, for how we view ourselves and how we treat each other, and for the variety of relations that derive from our sexuality. Thus biological concerns meld into psychological concerns for human mental wholeness, into behavioral concerns about growing up and relating to others, into sociological concerns for the family, into political concerns for protecting persons who enter into sexual and marital relations — as well as for the persons who result from those relations — and into spiritual concerns for well-being and healing *(shalom)*. Each of these involves persons relating to and caring for others in the context of being good stewards.

CHAPTER ELEVEN

Epilogue

T HE ARRIVAL of the colonists' space vehicles on Mars went more smoothly than anyone had predicted. All four vehicles landed in their designated areas within a half mile of each other. After the initial euphoria and celebration, the colonists began to erect the geodesic domes that would be their first homes in the new world. They established permanent communication links between the four sites and home base on Earth and commenced research on what would be needed to survive in their new environment.

The mission botanist established an experimental station where various oxygen-producing microorganisms were tested for their ability to survive high ultraviolet exposure with minimal water. Previous exploratory missions had revealed water trapped deep in the Martian rocks that could be tapped and used for irrigation. After numerous unsuccessful experiments, she found that by administering regular but brief and widely spaced sprays of water, a particular microorganism whose genes she had manipulated could survive and produce large amounts of oxygen. She also engineered it to reproduce rapidly, so that it soon covered the rocky soil in the test area.

On the basis of her initial experiments, she argued before the New Organization Administrative Hierarchy (NOAH) that they should authorize a large-scale program to distribute the organisms at various sites on Mars. The water for the spray could be stored in small tanks and piped to where the organisms would grow. These tanks could hold a two-year supply before they would need replacing. Some of the organisms would be grown in enclosed areas so that their emitted oxygen could be trapped

343

and combined with hydrogen to recycle the water used to irrigate them. She felt that with time and more experimentation the dispersed systems of microorganism farms could become self-sufficient, needing only the addition of more piping and sprays to cover the extensive growth of the microorganisms. She also argued that eventually the organisms would develop a tolerance to the dry climate of Mars and would spread without the need for water other than what they put into their immediate environment. Over a long span of time these could create the atmosphere that would make settlement of Mars by oxygen-breathing animals a reality.

The committee was enthusiastic about her proposal, envisioning a planet that could be colonized more easily because it had a protective and nurturing atmosphere. One member, however, questioned what would be the consequences of releasing these organisms on Mars. What effects would they have on the soil and atmosphere of the planet? Would there come a point when the colonists might want to control the spread of the organisms, and did they have the means to do so? Did the colonists have any stewardship responsibilities to the planet itself and to future generations of inhabitants?

NOAH decided that the matter of impact on the planet was moot since no life forms had been discovered; hence the released organisms would not threaten any native species. Their obligations to future generations were taken to be the main argument for proceeding expeditiously with the microorganism dispersal, since the generation of oxygen was a top priority for successful colonization. The vote, when taken, was seven to one in favor of the immediate release of the microorganisms around the colony base and the establishment of other sites when the long-distance probes were sent out.

The first major problem the colonists faced concerned the Assisted Reproduction Containers (ARCs). One of the power cells responsible for keeping the embryos frozen failed and was not repairable. NOAH had to make a decision about which embryos to destroy and which to maintain. Two members argued that they should destroy some of each kind of embryo. They would have fewer embryos to work with, but at least they would preserve all the species brought on the expedition. Two others expressed concern that saving only a few of each species would drastically reduce the genetic variability of the animals with which Mars would be colonized and would threaten the success of the project. Inbreeding of those that remained would destroy the quality of the animals. They argued that they should preserve only select species capable of forming a stable ecological unit while destroying the rest. Two others argued that since the colonists had the greatest responsibility to the frozen human embryos, these should have first priority in the remaining space in the ARCs. The planners had given the colonists charge over these embryos; hence they were responsible for their safety. The colonists' primary responsibility was to preserve their own kind. Others in NOAH rejected this as too clearly an example of speciesism. Why, they argued, should they give the human species preference over other species? After a long and heated debate, NOAH could not come to any consensus. While they debated the matter, the technicians in charge of the ARCs took matters into their own hands and preserved some from every species, in the process destroying about half of the frozen human embryos.

Through their constant companionship and mutual interests, two of the colonists working in the geodesic dome construction unit fell in love and were living together in the housing unit. Though initially this was winked at by NOAH, it became a problem for the community when the couple decided that they wanted to have their own children. This was a matter for consideration by NOAH, since prior to the mission the group had unanimously determined that only the frozen human embryos would be brought to term. These had been carefully selected for their genetic fit, but NOAH could not exercise such quality control over normally generated children. They offered the couple the opportunity to be the first eligible surrogates when reproduction would be allowed, but the couple refused the offer. The couple reaffirmed that they wanted to have children who would be their own genetic offspring. One NOAH member argued that the colonists' primary responsibility was to the embryos that were already conceived. Another emphasized that the couple had voluntarily given up the freedom to reproduce when they had agreed to participate in the Mars mission and that it would be immoral to break their agreement at this point, even in the name of love. A third suggested in their favor that it would be preferable to have children generated out of and nurtured in a loving atmosphere. Children so raised might be better suited to the community on Mars than those selected because of their genetic composition. But others rejected his opinion. The committee voted to separate the couple into different work units.

One of the most difficult decisions faced by NOAH occurred as the result of a viral invasion. NOAH never ascertained its origin, but its symptoms were unmistakable. It attacked the brain, gradually destroying the amygdala in which it lodged. The outbreak began with two persons working in the genetics research division but soon spread to two other colonists who had contact with them. The first symptom was an uncontrolled rage that developed at the smallest occasion. A statement or incident that seemed innocuous to everyone else precipitated verbal outbursts and even physical attack. Other members of the colony tried to avoid the two persons whenever possible, and even when they were in their company they kept their distance, quickly moving to the other side of the room. Some wondered aloud whether there had been some serious mistake in choosing these two for the mission, especially since they came from the same ethnic group. Surely, they argued, they could control their tempers and their language in mixed company.

At first the colonists attributed the strange behavior to the results of stress, since the genetics research group was under pressure to adapt organisms as quickly as possible to the Martian environment. But after NOAH relieved them of their position, the incidents of anger not only persisted but increased, so much so that they placed the two individuals in isolation, except for regular visits by a psychotherapist. When two more of the colonists began to display similar symptoms, one of the mission physicians decided to take several brain biopsies. After numerous tests and cultures, the virus was isolated and identified. The question remained, however, what to do with those afflicted. Should NOAH isolate them permanently from the rest of the group so as not to spread the virus? Should NOAH screen other colonists to see whether the virus had already spread? The colonists agreed upon both. As a result of the screening, they identified one other colonist as having the virus and isolated her with the other four.

The colonists found it hard to determine to what extent the afflicted persons could control their outbursts. At first there was considerable animosity toward them. Since they were highly trained members of the mission, selected for the psychological traits of being cool-headed, sociable, and of outstanding moral character, much could be expected of them. Significant character flaws must somehow have escaped the careful psychological screening they underwent. This initial response was followed by a stage in which the community completely exonerated those afflicted on the ground that the viral infection over which they had no control caused their outbursts. The disease, however, seemed to be progressive, and the behavior of the five gradually deteriorated, with more frequent and violent fits, so that NOAH had to isolate them in padded rooms to protect both themselves and others. NOAH made research to find a cure

346

for the virus a top priority, while it established safety procedures to insure that the virus did not spread, either from those afflicted or from the experimental cultures.

On one of the exploratory missions the solar-powered Mars Rover tipped over on a steep hill, killing two of the colonists. Since NOAH had previously decided that they should replace any members who were killed and since the Embryo Development Unit (EDU) was not yet operational, NOAH chose two women to be surrogate mothers for two of the embryos. The embryos selected were of opposite genders to foster gender balance in the new generation. Two women were selected at random and impregnated. However, one of the women, Genera, changed her mind after experiencing two months of nausea during her pregnancy. Her physician felt that the nausea was not severe enough to endanger her health, though she could make no prognosis about how long the nausea and discomfort would last. Genera questioned both the propriety and the safety of carrying an embryo to term and wished to terminate the pregnancy.

Her case was taken to NOAH for review. One member argued that Genera had accepted responsibility for the embryo and hence had an obligation to carry it to term, since otherwise it would die. She also had an obligation to the mission, whose numbers were diminishing. Others, however, argued that her primary responsibility was to herself and that this should be her first consideration. Prior agreements were not necessarily binding.

NOAH was about to take a vote when the radio operator rushed in with a message from one of the exploratory teams. They had found some primitive life forms near the polar cap while setting up the microorganism unit in that area. They needed an immediate decision on whether to proceed, since this was their final stop and they were rapidly running out of oxygen. They feared that the released microorganisms would destroy the native species, though it seemed likely that the native species was much more primitive and incapable of creating an oxygen atmosphere, at least in the foreseeable future. They were also unsure whether it would be safe to bring a sample of the native species back to the colony. They were awaiting further instructions.

✳   ✳   ✳

The Martian colonists got off to what might be called euphemistically a rocky start. We cannot explore the entire adventure. Undoubtedly some future historian will chronicle it.

What we can say is that the colonists found that, although they had escaped Earth, they had not escaped having to deal with the moral dimensions of their human existence and their power over nature. Questions about right and wrong actions, about what things are good and ought to be produced or bad and ought not to be produced, and ultimately about what sorts of persons they ought to be, were still central.

Some colonists continued the moral quest from the perspective of determining what it means to be God's good stewards in this new and harsh environment. They puzzled over how they would carry out their obligation to *fill* this new planet, both quantitatively and qualitatively. What kinds of beings — human beings, microorganisms, plants, and animals — should be designed and for what purposes? To what extent, if at all, should they modify human beings to survive on this planet? Are there some features of humans they should not modify? How should they change and not change the planet? Should they preserve the organisms they found, and if so, how? Should they welcome other colonists?

How should they *rule* this planet over which they are now assuming dominion? What kinds of relations should hold between them and that over which they rule, whether indigenous to the new planet, brought with them from Earth, or their own creations? What should be the true extent of their power, and how will they bring God and his concerns into this colonization?

How should they *care* for what they create? Mars and all that they transported on their spaceship ark are entrusted to them. How should they conserve and invest that trust to profit the created things, the other stewards, and ultimately the Landlord of all? Will their mission inaugurate a new understanding of stewardship, or will the history of selfishness they share with Earth be repeated?

——— *For Further Reading* ———

I N THIS BOOK we have not attempted to cover all areas of biology, nor have we surveyed extensively the range of views about ethics in the areas we have covered. For the benefit of those who might wish to read further, the following books provide additional treatment and at times differing perspectives. We acknowledge comments and suggestions that have come from a number of colleagues, although the final selection has been ours.

General Ethics

Boulton, Wayne G.; Kennedy, Thomas D.; and Verhey, Allen, eds. *From Christ to the World: Introductory Readings in Christian Ethics.* Grand Rapids: Eerdmans, 1995.
Pojman, Louis P. *Ethics: Discovering Right and Wrong.* Belmont, CA: Wadsworth, 1990.

Biomedical Ethics

Books

Beauchamp, Tom L., and Childress, James F. *Principles of Biomedical Ethics,* 4th ed. New York: Oxford University Press, 1994.

Bouma, Hessel, III, et al. *Christian Faith, Health, and Medical Practice.* Grand Rapids: Eerdmans, 1989.

Kilner, John F. *Life on the Line: Ethics, Aging, Ending Patients' Lives, and Allocating Vital Resources.* Grand Rapids: Eerdmans, 1992.

Lammers, Stephen E., and Verhey, Allen, eds. *On Moral Medicine: Theological Perspectives in Medical Ethics.* Grand Rapids: Eerdmans, 1987.

Munson, Ronald, ed. *Intervention and Reflection: Basic Issues in Medical Ethics,* 4th ed. Belmont, CA: Wadsworth, 1992.

O'Rourke, Kevin, and Boyle, Philip, eds. *Medical Ethics: Sources of Catholic Teaching,* 2nd ed. Washington, DC: Georgetown University Press, 1993.

Sommerville, Ann. *Medical Ethics Today: Its Practice and Philosophy.* London: British Medical Association, 1993.

Veatch, Robert M., ed. *Medical Ethics.* Boston: Jones & Bartlett, 1989.

Other

Bioethicsline. This is an online database of bibliographic references concerning ethical and public policy issues in health care and biomedical research, supported by the National Library of Medicine. Access can be gained by calling the National Library of Medicine (800-638-8480). Assistance with searches can be obtained from the Kennedy Institute of Ethics (800-MED-ETHX).

Science as a Human Endeavor

Kuhn, Thomas S. *The Structure of Scientific Revolutions,* 2nd ed. Chicago: University of Chicago Press, 1970.

Ratzsch, Del. *Philosophy of Science: The Natural Sciences in Christian Perspective.* Downers Grove, IL: InterVarsity Press, 1986.

Toulmin, Stephen. *Human Understanding.* Princeton: Princeton University Press, 1972.

The Environment

Attfield, Robin. *The Ethics of Environmental Concern*, 2nd ed. Athens, GA: University of Georgia Press, 1991.

Berry, R. J., ed. *Environmental Dilemmas: Ethics and Decisions*. London: Chapman & Hall, 1992.

Oelschlager, Max. *Caring for Creation: An Ecumenical Approach to the Environmental Crisis*. New Haven: Yale University Press, 1994.

Rolston, Holmes, III. *Environmental Ethics: Duties to and Values in the Natural World*. Philadelphia: Temple University Press, 1988.

Wilkinson, Loren, ed. *Earthkeeping in the '90s: Stewardship of Creation*. Grand Rapids: Eerdmans, 1991.

Assisted Reproduction

Alpern, Kenneth D., ed. *The Ethics of Reproductive Technology*. New York: Oxford University Press, 1992.

Chadwick, Ruth F. *Ethics, Reproductive and Genetic Control*. London: Croom Helm, 1987.

Coughlan, Michael J. *The Vatican, the Law and the Human Embryo*. Iowa City: University of Iowa Press, 1990.

Hull, Richard T., ed. *Ethical Issues in the New Reproductive Technologies*. Belmont, CA: Wadsworth, 1990.

Jones, Gareth. *Manufacturing Humans: The Challenge of New Reproductive Technologies*. Leicester: IVP, 1987.

The Human Genome

Bartels, Dianne M.; LeRoy, Bonnie S.; and Caplan, Arthur L., eds. *Prescribing Our Future: Ethical Challenges in Genetic Counseling*. New York: Aldine de Gruyter, 1993.

Nelson, J. Robert. *On the New Frontiers of Genetics and Religion*. Grand Rapids: Eerdmans, 1994.

Suzuki, David, and Knudtson, Peter. *Genethics: The Clash Between the New Genetics and Human Values*. Cambridge, MA: Harvard University Press, 1989.

Science and Technology

Barbour, Ian. *Ethics in an Age of Technology*. San Francisco: HarperSan-
  Francisco, 1993. (This volume and the next provide an excellent
  overview of these two related fields.)
————. *Religion in an Age of Science*. San Francisco: HarperSanFrancisco,
  1990.
Jonas, Hans. *The Imperative of Responsibility: In Search of an Ethics for the
  Technological Age*. Chicago: University of Chicago Press, 1984.
Monsma, Steven V., ed. *Responsible Technology*. Grand Rapids: Eerdmans,
  1986.
Orlans, F. Barbara. *In the Name of Science: Issues in Responsible Animal
  Experimentation*. New York: Oxford University Press, 1993.

Brains, Genes, and Moral Responsibility

Eccles, Sir John C., and Robinson, Daniel N. *The Wonder of Being Human*.
  New York: Free Press, 1984.
Jones, Gareth. *Our Fragile Brains: A Christian Perspective on Brain Research*.
  Downers Grove, IL: InterVarsity Press, 1981.
Mackay, Donald M. *Behind the Eye*. Oxford: Basil Blackwell, 1991.
Searle, John R. *The Rediscovery of the Mind*. Cambridge, MA: MIT Press,
  1992.

Human Sexuality

Curran, Charles E., and McCormick, Richard A., eds. *Dialogue About
  Catholic Sexual Teaching*. Readings in Moral Theology, vol. 8. New
  York: Paulist Press, 1993.
Nelson, James B., and Longfellow, Sandra P., eds. *Sexuality and the Sacred:
  Sources for Theological Reflection*. Louisville: Westminster John Knox,
  1994.

Index